D1811188

9780080096278

ABSORPTION SPECTRA
AND CHEMICAL BONDING
IN COMPLEXES

ADIWES INTERNATIONAL SERIES

IN CHEMISTRY

This book is in the

ADDISON-WESLEY SERIES IN ADVANCED CHEMISTRY

Sidney Golden, Consulting Editor

ABSORPTION SPECTRA AND CHEMICAL BONDING IN COMPLEXES

C. K. Jørgensen

Cyanamid European Research Institute
Cologny, Geneva, Switzerland

PERGAMON PRESS

OXFORD · LONDON · PARIS · FRANKFURT

1962

ADDISON-WESLEY PUBLISHING COMPANY, INC.
READING, MASSACHUSETTS

Copyright © 1962

PERGAMON PRESS LTD.

U.S.A. Edition distributed by
Addison-Wesley Publishing Company, Inc.
Reading, Massachusetts, U.S.A.

———————

Library of Congress Card Number 61–12437

Set in Times New Roman 10 on 12pt. and Printed in Great Britain by
PAGE BROS. (NORWICH) LTD.

CONTENTS

PREFACE

OUR knowledge of the electron clouds of gaseous atoms and ions is based on the energy levels, studied in atomic spectroscopy. It has not been generally realized among chemists that the study of energy levels, i.e. the absorption spectra, is equally fruitful in helping our understanding of the chemical bonding, not only in transition group complexes, but in every type of molecule. Actually, these spectra have yielded very much useful information in the last few years, and the development has only been delayed because physicists are generally more interested in nuclei than in bound electrons, and because chemists think that spectroscopy is a matter for physicists and group theorists.

The main purpose of this book is to summarize the results of this recent evolution and attempts to win over some of the mental inertia which has caused most chemists to think of chemical bonds in terms of a valency-bond and hybridization description. Actually, this approximation does not agree at all with the knowledge gained from absorption spectra, while molecular orbital theory is shown to be adequate aid to the classification of the energy levels. It will probably be argued that the treatment presented is too complicated and too mathematical. The present writer originally intended to gather the detailed arguments into separate appendixes. However, he soon realized that nearly all the text would have to be put into these appendixes, which would not be a great advantage. Hence, the reader should not feel himself obliged to read the chapters consecutively; less difficulties would probably arise by reading the different chapters several times.

I am very grateful for many valuable discussions with Dr. Claus E. Schäffer, Kemisk Laboratorium A, Danmarks tekniske Højskole, Copenhagen, and to my other colleagues there and the Director, Professor Jannik Bjerrum. Further, I am much indebted for most illuminating visits to Dr. Per-Olov Löwdin, Kvantkemiska Gruppen, Uppsala, Dr. Leslie E. Orgel, Department of Theoretical Chemistry, Cambridge, and Professor Marcel Delépine, Collège de France, Paris.

LIST OF SYMBOLS

(Page numbers indicate the positions where the symbol is first defined.)

A   angular function, p. 25
A   proportionality constant for interelectronic repulsion, p. 154 (eqn. 146)
A   activation energy, p. 260 (eqn. 222)
A Racah parameter, p. 40 (eqn. 35)
A_p homogeneous polynomium in x, y, z, p. 25 (eqn. 3)
a   coefficient to Δ, p. 74
a   number of γ_5-electrons
a_0 Bohr radius, p. 22
a_n activity, p. 248
B Racah parameter, p. 40 (eqn. 35)
b   coefficient to B, p. 74
b   number of γ_3-electrons
C Racah parameter, p. 40 (eqn. 35)
C_M and C_L total concentrations of M and L, p. 252 (eqn. 202)
$C_{\infty v}$ linear symmetry without centre of inversion, p. 48
c   coefficient to C, p. 74
c   velocity of light *in vacuo*
c   molar concentration, p. 89 (eqn. 98)
D   term with $L = 2$
D   optical density, p. 89 (eqn. 98)
D   proportionality constant for spin-pairing energy, p. 154 (eqn. 146)
D   dielectric constant, p. 267
D_2 orthorhombic symmetry, p. 68
D_{4h} tetragonal symmetry, p. 68
$D_{\infty h}$ linear symmetry with centre of inversion, p. 51
d   orbital with $l = 2$
E   energy
E^3 Racah parameter, p. 40 (eqn. 36)
$(E_1 - E_2)$ expression equivalent to Δ
E_p polynomium, corresponding to secular determinant, p. 37
e   electric charge of proton
e degeneracy number, pp. 24, 30
eV   electron volt, p. 85
F   term with $L = 3$
$F^2, F^4, \ldots, F^k$ Slater–Condon–Shortley parameters of interelectronic repulsion, p. 39 (eqn. 32)
f   orbital with $l = 3$
f   spectrochemical function of ligands, p. 113 (eqn. 123)
f_n activity coefficient, p. 248
G   term with $L = 4$
G   free energy, p. 246
ΔG change of free energy at a reaction
G_l exchange integral according to Slater, p. 185
g   two-electron operator, p. 38 (eqn. 27)
g   spectrochemical function of central ion, p. 113 (eqn. 123)

g   *gerade* = even, as subscript, p. 59

g gyromagnetic factor, p. 106

H   term with $L = 5$

H   enthalpy (heat content), p. 246

ΔH   change of enthalpy at a reaction

h   Planck's constant

h   nephelauxetic function of ligands, p. 138 (eqn. 137)

H_h change of enthalpy of hydrogen ion formation (eqn. 186, p. 237)

I   term with $L = 6$

I   ionization energy

I and I_0 light intensities, p. 89 (eqn. 98)

J total angular momentum quantum number of system, p. 30

J(aa) or J(ab)   Coulomb integral, p. 39 (eqn. 31)

j total angular momentum quantum number of orbital

K   kayser = cm^{-1}, wave number unit

°K   degree Kelvin, unit for absolute temperature

K   absorption spectrum, caused by X-ray excitation of 1s-electrons, p. 189

K(ab)   exchange integral, p. 39 (eqn. 31)

K_n stepwise formation constant, p. 245 (eqn. 189)

k   various constants

k   Boltzmann constant (when multiplied by T)

k   nephelauxetic function of central ion, p. 138 (eqn. 137)

kK   kilokayser = $1000\ cm^{-1}$

L   ligand in general formulae, p. 245

L orbital angular momentum quantum number of system, p. 31

l   layer thickness, p. 89 (eqn. 98)

l orbital angular momentum quantum number of orbital, p. 22 (eqn. 1)

M   central ion in general formulae, p. 245

M   molar concentration, before a chemical formula

M_S magnetic quantum number of system, p. 70

m   electronic mass

m   effective mass, p. 95 (eqn. 106)

m_s magnetic quantum number of electron, p. 38

N   maximum co-ordination number, p. 245

N_c characteristic co-ordination number, p. 245

n principal quantum number of orbital, p. 22

$\bar{n}$ formation number, p. 253 (eqn. 205)

O_h cubic-octahedral symmetry, p. 51

P   term with $L = 1$

P   perturbation, p. 32

P   pressure

P oscillator strength, p. 92 (eqn. 102)

p   orbital with $l = 1$

pK   acid strength constant p. 247 (eqn. 199)

q   various numbers, e.g. of electrons in a shell

R   radial function, p. 25

R   gas constant (when multiplied by T)

r   distance from nucleus

r_{ion} ionic radius

r_L internuclear ligand–central ion distance, p. 55

ry   Rydberg constant, p. 22

S   term with $L = 0$

S   entropy, p. 246

ΔS change of entropy at a reaction
S_{ab} overlap integral between a and b, pp. 60, 64
S spin quantum number of a system, p. 30
ΔS change of spin
s   orbital with $l = 0$
T   absolute temperature
T_d tetrahedral symmetry, p. 128
U   potential, p. 98 (eqn. 109)
U   electrochemical potential, p. 246 (eqn. 197)
U_0 standard oxidation potential, p. 246 (eqn. 197)
$U(r)$ central field, p. 27 (eqn. 4)
u   *ungerade* = odd, as subscript, p. 59
V   volt
V_0 ligand field of spherical symmetry, p. 55
V_{oct} ligand field of octahedral symmetry, p. 55
v seniority number p. 73
W expression for average value of r^{-1}, p. 39 (eqn. 33)
x   various quantities, pp. 35, 82
x_c characteristic amplitude vibration, p. 95 (eqn. 107)
x_l Rydberg correction, p. 23
x_A electronegativity of atom A, p. 309
Z atomic number
Z_* effective charge, compared to hydrogenic ions, p. 28
Z_{eff} effective charge, compared to data of atomic spectroscopy, p. 142
$Z_0 - 1$ ionic charge in spectroscopic arguments, p. 23
z   ionic charge as a chemical property, p. 248
Å   Ångstrom unit of wavelength $= 10^{-8}$ cm, p. 85
Brackets on chemical symbols, [Ne], [A], [Kr], [Xe], and [Em], indicate closed-shell cores in electron configurations, and otherwise in the last chapters, molar concentrations of a given species for use in mass-law expressions.
a Racah–Trees correction, p. 46
a Madelung constant, p. 235 (eqn. 185)
a correlation factor, p. 211 (eqn. 174)
β nephelauxetic ratio, p. 77, 134
β_n total formation constant, p. 245 (eqn. 191)
Γ_n group-theoretical quantum number (analogous to L) for systems, p. 51
γ_n group-theoretical quantum number (analogous to l) for orbitals, p. 51
Γ_J (analogous to J) for systems, p. 51
γ_j (analogous to j) for orbitals, p. 51
Γ_{tn} and γ_{tn} for tetragonal symmetry, p. 68 (eqn. 65).
Δ orbital energy difference in octahedral complexes, p. 54, 132
Δ term with $\Lambda = 2$
δ half width, p. 92 (eqn. 100)
$\delta(-)$ half width towards smaller wave numbers
$\delta(+)$ half width towards larger wave numbers
δ orbital with $\lambda = 2$
δ_n disproportion constant, p. 256 (equ. 212)
ϵ molar extinction coefficient, p. 89 (eqn. 98)
ϵ_0 molar extinction coefficient of maximum
ζ multiplet splitting factor
ζ_{nl} Landé multiplet splitting factor, characteristic for the partly filled n, l-shell, p. 48 (eqn. 41)
η scaling parameter, p. 29

Θ	quantity related to M.O. deviation from $l = 2$, p. 79 (eqn. 84)
ϑ	„ , p. 79
ϑ	Weiss correction, p. 182 (eqn. 157)
Λ	orbital angular momentum quantum number in linear symmetry, for system, p. 57 (eqn. 52)
λ	„ , for orbital, p. 57
λ	wave length, p. 85
μ	magnetic moment in Bohr units, p. 182 (eqn. 157)
μ	ionic strength, p. 248
$m\mu$	wave length unit ($=10^{-7}$ cm)
ν	frequency, p. 85
Π	term with $\Lambda = 1$
π	orbital with $\lambda = 1$
ρ	ligand field stabilization parameter, p. 98, 125
Σ	summation sign
Σ	term with $\Lambda = 0$
σ	orbital with $\lambda = 0$
σ	wave number, p. 85
σ_c	characteristic wave number of a vibration, p. 95 (eqn. 106)
τ	variables for wave function, p. 38
φ	an angle, p. 36 (eqn. 23)
χ_m	magnetic (molar) susceptibility, p. 182 (eqn. 157)
χ_{corr}	diamagnetic correction, p. 182 (eqn. 157)
Ψ	wave function of system, p. 20
ψ	wave function of orbital, p. 25
Ω	quantum number (analogous to J) of system in linear symmetry, p. 57
ω	quantum number (analogous to j) of orbital in linear symmetry, pp. 57, 159

CHAPTER 1

HISTORICAL INTRODUCTION

IN THE nineteenth century, chemists realized that Dalton's atomic theory gave a fair description of the composition and reactions of the chemical compounds, assuming their molecules to consist of a small number of different atoms. It was clear that some of the bonds between the atoms were much stronger than others. Thus, it became convenient to talk about sub-entities, "radicals", in the molecules; and before the determination of atomic numbers, it was hard to prove that the entities, assumed to be atoms, were not themselves radicals, composed of several atoms. For many years, uranium dioxide UO_2 was considered to be an element, while chlorine Cl_2 was thought to be a higher oxide of the hypothetical element, murium, forming hydrochloric acid HCl (since all acids were thought to contain oxygen).

Even before Arrhenius' theory of electrolytic dissociation, inorganic qualitative analysis described properties of the individual metals and acid radicals in salts. Thus, NaOH, NH_3, Na_2S, H_2S, Na_2CO_3, etc., give characteristic precipitates of different colours, which may dissolve in an excess of the reagent. While K_2SO_4 or $BaCl_2$ are relatively simple compounds, containing only one metal and one acid radical each, it was discovered that some compounds having a constant composition (thus proving that they are not simple mixtures) contain several metals or acid radicals. Thus, the alums of the type $K_2SO_4 \cdot Al_2(SO_4)_3 \cdot 24H_2O$ or the Tutton salts of the type $K_2SO_4 \cdot NiSO_4 \cdot 6H_2O$ give in solution all the reactions of their respective metals and of sulphate. They were called "molecular compounds" or "double salts". However, other compounds exist, which must have a type of chemical bonding different from that in the simple salts. The formulation $4KCN \cdot Fe(CN)_2$ does not indicate that the ferrocyanides show none of the reactions of cyanide, are not poisonous, and react very slowly with acids (where KCN immediately forms the volatile acid HCN). Further, a large number of ferrocyanides can be isolated, all containing the group $Fe(CN)_6$ as if it were a tetravalent acid radical. It was decided to call such salts

1

"complex" compounds. Actually, the sulphate group SO_4 has distinctly the same character, i.e. it is extremely difficult to remove oxygen atoms from it.

It is not easy to give a consistent definition of a *complex* without including nearly every type of molecule. In view of the phenomenon of ionic dissociation, we may say that a complex ion or molecule (the neutral "inner-salts") contains one or more *central atoms*, bound rather strongly to a well-defined number of *ligand* groups. The central atom is for most practical applications of the word "complex" a metal atom; but there is no logical distinction between this case and the sulphur atom in the sulphate ion. The ligands may either be single atoms (of the type, which in ionic salts would be Cl^-, F^-, O^{2-}, etc.) or polyatomic anions (SO_4^{2-}, CO_3^{2-}, NO_3^-) or neutral molecules (H_2O, NH_3, CO, pyridine C_5H_5N, etc.). It is nearly always one definite atom of the polyatomic ligands which is directly attached to the central atom, thus oxygen of the oxo-anions mentioned and of water, while it is nitrogen in ammonia or pyridine. The number of such atoms directly attached to the central atom is called its *co-ordination number*. A given ligand may occupy more than one such co-ordination place. Thus, the ligands oxalate $OOCCOO^{2-}$, glycinate $NH_2CH_2COO^-$, and ethylenediamine $NH_2CH_2CH_2NH_2$ are all *bidentate* (bound by an oxygen or nitrogen atom in each end), diethylenetriamine $NH_2CH_2CH_2NHCH_2CH_2NH_2$ is *tridentate* (bound by the three N), nitrogentriacetate $N(CH_2COO)_3^{3-}$ is *quadridentate* (one N and three O), and ethylenediaminetetra-acetate $(OOCCH_2)_2NCH_2CH_2N(CH_2COO)_2^{4-}$ may be *sexadentate* (two N and all four O) or *quinquedentate* (two N and three O). These multidentate ligands are said to form *chelate* complexes. However, a given complex may very well contain both multidentate and uni-dentate ligands, e.g. one ammonia and two ethylenediamine molecules and one chloride anion.

The description of chemical bonding has for more than a century oscillated between two extreme points of view, corresponding to the division of the old valency concept into the concepts of *oxidation number* and *bond number*. Inorganic chemists have nearly always conceived valency as having a sign. The atoms in the salts NaCl or KBr are not sufficiently described merely as univalent; the electrical charges are definitely $+1$ for the sodium and potassium ions and -1 for the chloride and bromide ions, and they can freely be substituted to the other salts NaBr and KCl. Admittedly, the molecules Cl_2,

ClBr, Br_2 and the metallic alloy NaK exist, but they have properties very different from those of the ionic salts. In modern times oxidation number is defined in the following way:

(1) The sum of oxidation numbers of the atoms in a molecule or an ion equals its electrical charge in units of the proton charge. The oxidation number of a monatomic entity always equals its charge.

(2) The oxidation number of hydrogen is $+1$ (except in H_2 and metallic hydrides, where it is 0, and in salt-like hydrides, where it is -1).

(3) The oxidation number of oxygen is -2 (except in O_2, where it is 0, in some rare superoxo compounds, where it is $-\frac{1}{2}$, and in H_2O_2 and other peroxo compounds, where it is -1, and in OF_2, where it is $+2$).

(4) The oxidation number of fluorine is -1 (except in F_2 where it is 0).

These rules do not specify the oxidation numbers in every discrete compound (such as N_4S_4 or $Pb(CH_3)_4$), but, generally, reasonable analogies can be established with well-known cases. It must be remembered that the oxidation numbers are useful concepts in the classification of chemical compounds, but they are not measurable quantities (except in the case of monatomic entities). A very electrostatically minded chemist may perhaps think of the carbon atom in CO_2 or the silicon atom in SiO_2 as bearing a hypothetical charge of $+4$, because the oxidation number is $+4$ according to rule (3) above. However, there is no doubt that the actual electron distribution inside the molecule CO_2 and the crystalline SiO_2 does not correspond to such a charge distribution, but rather approaches electroneutrality, having about the same number of electrons in the neighbourhood of each nucleus, as indicated by its atomic number. On the other hand, as will be shown later, the concept of oxidation number has deteriorated much less than the argument about electrical charges might indicate, and we shall use the Roman numerals in parenthesis after the symbol or name of each element when giving the oxidation number. We will commit the anachronism of using a minus sign for negative values, Cl($-$I), and the Arabic zero, Pd(0).

Criticism of the concept of oxidation numbers generally comes from organic chemists. The compounds CH_4, CH_3Cl, CH_2Cl_2, $CHCl_3$ and CCl_4 do not have essentially different properties, but the oxidation number of the carbon atoms rises in stages, -4, -2, 0, $+2$ and

+4. On the other hand, the co-ordination number of carbon in this series is always 4. This is not invariably true in carbon chemistry, since ethylene H_2CCH_2 and acetylene HCCH also exist. But if the bonds to H, Cl, etc., are considered single, and the bonds to O in formaldehyde H_2CO, etc., are considered double, we may support a constant bond number of four in carbon, if the first series of compounds has single carbon–carbon bonds, while the ethylenic type of compounds have double and the acetylenic type triple carbon–carbon bonds. Carbon is the element which has the most constant bond number, and it is not at all possible to classify the facts of inorganic chemistry with the aid of bond numbers in the same easy way as in organic chemistry.

We might use this as an argument for dividing chemical bonds into two types: *electrovalent* and *covalent* bonds, the former being most adequately described by oxidation numbers and the latter by bond numbers. Thus, acetic acid CH_3COOH forms ionic salts, containing the acetate ion CH_3COO^-. By chlorine substitution, the acids $CH_2ClCOOH$, $CHCl_2COOH$ and CCl_3COOH are formed, all resembling the original acetic acid. The corresponding anions CH_2ClCOO^-, etc., also closely resemble the acetate ion. The hydrogen atoms bound to the first carbon atom are obviously bound covalently, and cannot be exchanged with metals of a positive oxidation state, forming for instance CH_2NaCOO^-. On the other hand, the metal ions substituting the hydrogen atom bound to the carboxyl group are obviously bound ionically. We cannot substitute them with Cl, forming CH_3COOCl, while the acetyl chloride CH_3COCl does not seem to be a salt of CH_3CO^+.

In many cases, the division of compounds into those bound with electrovalent or covalent bonds seems very successful. Among the simple chlorides, it is surprising to see a clear division into compounds melting and boiling below 400°C (HCl, CCl_4, NCl_3, Cl_2O, ClO_2, $AlCl_3$, $SiCl_4$, PCl_3, PCl_5, Cl_2, $TiCl_4$, $SnCl_4$, $SbCl_3$, $SbCl_5$, $HgCl_2$, etc.) and salts melting and boiling well above 400°C (LiCl, NaCl, $MgCl_2$, $CrCl_3$, $NiCl_2$, $CdCl_2$, $LaCl_3$, $PbCl_2$, $BiCl_3$, $ThCl_4$, UCl_4, etc.). The compounds in the latter group dissolve in water forming metal ions and chloride ions, while the compounds in the former group in some cases do not react with water and in some they react vigorously, forming hydrolysis products from which the chloride cannot be regenerated by distilling water from the solution. Pure HCl, especially, has quite different properties (the liquid state does not conduct electricity; the absolutely anhydrous solution in toluene does not precipitate with $AgClO_4$; etc.) from those of the aqueous solution of hydrochloric acid, containing H_3O^+ and Cl^-.

One of the most interesting features of complexes is that they definitely show that the division into electrovalent and covalent compounds cannot be a sharp one, but that a continuous series of intermediate cases occurs. This result implies very important consequences for our understanding of chemical bonding, and we shall see later how molecular orbital theory is the best instrument for describing the bonding in any type of compound.

The bond number description of complexes was developed by Blomstrand about 1870 and S. M. Jørgensen about 1880. By analogy to the CH_2-chains in normal hydrocarbons such as n-butane $CH_3CH_2CH_2CH_3$, it was assumed that the yellow cobalt (III) hexammine chloride $Co(NH_3)_6Cl_3$ contains NH_3-chains:

$$Co \begin{array}{l} \diagup NH_3-Cl \\ -NH_3-Cl \\ \diagdown NH_3-NH_3-NH_3-NH_3-Cl \end{array}$$

The three chloride groups are ionically dissociated. It was assumed that chloride groups directly attached to the cobalt atom would not dissociate so readily. Formulae were evolved, and a constant bond number of 5 was assumed for nitrogen (NH_4Cl was thought at that time to have a N—Cl bond). The different ions were given Greek names according to their colours:

Yellow $Co(NH_3)_6^{3+}$, luteo.
Pink $Co(NH_3)_5H_2O^{3+}$, roseo.
Purple $Co(NH_3)_5Cl^{2+}$, purpureo.
Green $Co(NH_3)_4Cl_2^+$, praseo.
Violet $Co(NH_3)_4Cl_2^+$, violeo.

S. M. Jørgensen defended the formulation with NH_3-chains during 20 years of ingenious preparative work, and discussed with A. Werner, whose octahedral model must now be accepted. S. M. Jørgensen prepared the analogous series of chromium (III) ammonia complexes, having colours similar to those given in the table and finally the analogous series of rhodium (III) which are white or yellow. Palmer, in about 1895, prepared the white and pale yellow iridium (III) ammonia complexes, and Grünberg and other Russian chemists the platinum (IV) series.

B

Werner maintained that the six ligands encountered in each complex of these series are situated at the apices of a regular octahedron with the metal ion in the centre. This is analogous to the distribution of the four bonds of a carbon atom in a regular tetrahedron, which was so successful in organic chemistry. The spatial distribution of the six ligands explains the occurrence of *isomers*, e.g. the violet and green $Co(NH_3)_4Cl_2^+$, since the two chloride groups may occupy adjacent positions (the *cis*-isomer) or transposed positions (the *trans*-isomer). Complexes of the type MA_3B_3 (which M. Delépine later prepared as $IrCl_3$ pyridine$_3$) also have two monomeric isomers, 1:2:3- or *cis*- with the three A's on the corners of a small, facial triangle of the octahedron, and 1:2:6- or *trans*- with two of the A's in mutual *trans*-position.

Werner also explained many facts known about platinum (II) complexes with the co-ordination number 4, by assuming the ligands to be arranged in a square around the central atom. Thus, $Pt(NH_3)_2Cl_2$ was shown to have two monomeric isomers with two adjacent (*cis*-) or two diagonal (*trans*-) chloride groups (in addition to salts such as $[Pt(NH_3)_3Cl][Pt(NH_3)Cl_3]$ or $[Pt(NH_3)_4][PtCl_4]$ or $[Pt(NH_3)_3Cl]_2[PtCl_4]$ having the same analytical composition).

All these complexes were characterized by the preparation of solid salts which could be dissolved in aqueous solvents and, at least for some minutes, preserve their characteristic colour, etc., in solution. Some of the ammonia complexes known, such as $Ni(NH_3)_6Cl_2$ or $Cu(NH_3)_4SO_4 \cdot H_2O$, have a high tension of gaseous ammonia in equilibrium with the solid, and the aqueous solution reacts immediately with acids forming NH_4^+ and the ordinary metal aquo ions. On the other hand, $Co(NH_3)_6^{3+}$ does not decompose even by prolonged boiling with strong hydrochloric acid. After 1900, many chemists and physicists became interested in the solution chemistry of complexes. N. Bjerrum remarked that one of the most invariant properties of a given species is its colour, or more precisely, its absorption spectrum. The colour of a transition group complex is dependent to any large extent only on the ligands directly attached to the central ion (the first co-ordination sphere), while solvents, or the formation of solid salts with different anions, have only a very minute influence. This was important for a study of a phenomenon related to Arrhenius' theory of ionic dissociation: the electrical conductivity of most salt solutions increases less with increasing concentration than corresponding to a proportionality. This was once interpreted as incomplete ionic dissociation at high concentration (actually, $HgCl_2$ shows an anomalous low conductivity even at very low concentrations, corresponding to the formation of a very strong complex). However, since the accompanying spectral changes are very small, the conductivity effects are now related to changes of the properties of the solvent and to the formation of ion pairs or second co-ordination sphere complexes. These are very firmly bound between cations and anions of high charge, and in the ultra-violet part of the spectrum of $Co(NH_3)_6^{3+}$, changes due to the association of a sulphate ion or an iodide ion can actually be observed.

The different results of various physicochemical studies applied to this type of problem will be discussed in Chapter 13 of this book. The first connexion between the slowly reacting *robust* complexes discussed above, and the quickly reacting *labile* complexes in equilibria in solution, was established by N. Bjerrum, who investigated the equilibrium constants of the thiocyanate complexes $Cr(NCS)_n$ $(H_2O)_{6-n}^{3-n}$, where all the intermediates $n = 1 \ldots 6$ could be isolated as somewhat robust entities. J. Bjerrum studied many complexes of ammonia and other amines with metal ions, and his book, published in 1941, has inspired very many studies on the formation of complexes in aqueous solution. Special topics are thus G. Schwarzenbach's studies of the multidentate ligands and L. G. Sillén's study of the hydroxo-complexes of metal ions.

One of the interesting results of these chemical studies, connected with the absorption spectra discussed below, is the establishment of the aquo ions as complexes on equal footing with other complexes, e.g. of ammonia. The Tutton salts must now be formulated as $(K^+)_2[Ni(H_2O)_6^{2+}](SO_4^{2-})_2$, and the green aquo ion $Ni(H_2O)_6^{2+}$ must occur in the aqueous solution of ordinary nickel salts and it is strictly comparable to the violet $Ni(NH_3)_6^{2+}$.

Much light has been thrown on the structures of solid compounds by X-ray crystallography, and Werner's idea of an octahedral complex ion has been confirmed, not only for $Co(NH_3)_6^{3+}$, but also for numerous entities such as AlF_6^{3-}, SiF_6^{2-}, $IrBr_6^{2-}$, $PtCl_6^{2-}$, etc. The infinitely extended structures of, for example, the complicated silicates found in mineralogy, can be explained if each silicon atom is surrounded tetrahedrally by four oxygen atoms, which are then bound to other silicon atoms, or to aluminium or other metal atoms. Werner recognized many polynuclear complexes, containing more than one central ion. Thus, the bridged ions $(NH_3)_5CrOCr(NH_3)_5^{4+}$

and $(NH_3)_4Co(OH)_2Co(NH_3)_4{}^{4+}$ are binuclear. Werner demonstrated that the group $(NH_3)_4Co(OH)_2{}^+$, being itself a known monomeric complex, may function as a ligand in the tetranuclear complex $Co[(OH)_2Co(NH_3)_4]_3{}^{6+}$. The latter brown ion is present in the first compound not containing carbon which was resolved in optically active isomers. This is possible because the central cobalt ion is asymmetric in the same way as in the mononuclear complexes, with three oxalate or ethylenediamine groups, where the three bidentate ligands may induce a left-hand or right-hand behaviour towards polarized light.

The distinction between electrovalent and covalent complexes was for many years based on two main criteria: that of robustness of the covalent bonds and that of magnetically anomalous behaviour. In German literature, the two types of complexes were denoted *Anlagerung* and *Durchdringung* complexes. Actually, the study of absorption spectra has demonstrated the impossibility of this distinction. The reaction rates of complexes are undoubtedly connected in a complicated way with the nature of the chemical bonding, but are probably highly dependent on the number of electrons in the partly filled shell of the transition groups, as discussed below. It cannot be concluded that mainly covalent complexes are always robust; thus $HgCl_2$ and $Hg(CN)_2$ are quick reacting.

When the magnetic susceptibility of a material is very close to that of empty space (as in most materials, which are not ferromagnetic), the occurrence of paramagnetism (relative susceptibility slightly above 1) may be related to a value of the total spin S greater than 0 in some of the ions or molecules, while diamagnetism (values just below 1) corresponds to $S = 0$, as found in closed-shell ions with inert-gas configuration. As discussed below, these explanations must be corrected to some extent in actual complexes. While many complexes have almost the paramagnetism expected from the value of S in the ground state of the corresponding gaseous ion (e.g. $Ni(H_2O)_6{}^{2+}$ compared to Ni^{2+}), other complexes have much lower values, and, in particular, cobalt (III), rhodium (III), iridium (III), and platinum (IV) are diamagnetic. The two types of magnetic property are now called *high-spin* and *low-spin* behaviour.

There is no reason to believe that there is an essentially different type of chemical bonding in high-spin compared to that in low-spin complexes. In particular, small numerical changes in the parameters describing the chemical bonding can sometimes shift a given central atom with a given oxidation number from high-spin to low-spin behaviour.

In the section on orbitals in gaseous atoms in spherical symmetry, it will be explained why some electron numbers (2, 10, 18, 36, 54, 86, 118, . . .) correspond to much more stable configurations than others. This was related by Rydberg and by N. Bohr to the chemical inertness of the inert gases. Reviewing the periodic system, we observe two different trends of chemical behaviour as functions of the atomic number Z. Elements with values adjacent to the inert-gas numbers prefer to re-adjust their electron number by forming anions or cations. Thus, F^-, Na^+, Mg^{2+} and Al^{3+} certainly indicate a preference for the neon configuration, and these elements are not known in compounds with other oxidation numbers. Quite a different behaviour is observed in the middle of each interval between two inert gases: the elements near nitrogen ($Z = 7$), phosphorus ($Z = 15$), vanadium to selenium ($Z = 23$ to 34), molybdenum to tellurium ($Z = 42$ to 52), rhenium to polonium ($Z = 75$ to 84), and uranium onwards ($Z = 92$ to at least 95) cannot form ions of inert-gas configuration, owing to the need for an excessive change in the electron number. These elements have a strong tendency to vary the oxidation number (even though a few metals with constant oxidation number, such as zinc and cadmium are distributed in the intervals given), and they have often a strong tendency towards covalent bonding (we have here excepted the group of lanthanides, "rare earths", with $Z = 57$ to 71 forming ionic compounds with the almost constant oxidation number $+3$).

It is unfortunate that so many authors of text-books consider the intervals of eight elements between helium and neon, or between neon and argon, as the only normal positions in the periodic system, regarding the "transition group elements" in all the following periods as anomalous phenomena, which should be hidden from the students. Actually, the usual formulation of Mendelejev's Periodic Table in periods of eight is an attempt to re-adjust the relation $2 \times 8 = 18$. Before the time of Moseley, the mysterious appearance of the triad metals (each three metals being the "side-group" analogy of one inert gas) and the rare earths might very well support Crookes' idea that hundreds of new elements might still be found. After the discovery of Z, it was no longer possible to consider the periodic system as based solely on similarities of chemical properties, but we must consider regularities only as functions of Z. Chemists holding the former point of view could have cited many resemblances, such as Sn–Pt, In–Bi, Ba–Pb, which are at least as conspicuous as many of the "genuine" resemblances, such as N–P, O–S, Sb–Bi, etc. N. Bohr constructed the first periodic table from the point of view of the electron configurations known from atomic spectroscopy. In drawing the dashed lines, indicating sometimes two homologues in the following period, he tried to reconcile the chemical idea of side groups, etc., with the physical facts of spectroscopy (as Rydberg had already done in 1909). Accepting the periodic system as a classification according to Z, we may introduce a sub-classification according to one of two principles:

(1) Analogy between the electron configuration of the ground states (and we may add the other low states) of the corresponding elements (the neutral atoms and their ions in gaseous state).

(2) Analogy between the chemical properties of the corresponding elements.

Of these, (1) is the more objective and consistent, while (2) is often ambiguous and liable to subjective feelings. However, (1) offends the chemist owing to several anomalies. By far the most serious is that helium is irreparably placed in the same group as Be, Mg, Ca, Sr, Ba, Ra and not in the group of the other inert gases Ne, A, Kr, X, Em. Other exceptions of this type are that the chemistry of Pa and U does not resemble that of Pr and Nd. In the writer's opinion, Seaborg's actinide hypothesis is of the type (1).

We may define a transition group as an interval of Z-values, where the oxidation numbers, though often highly variable, have a slowly decreasing average value. This definition of type (2) leads to results different from (1), where it is completely clear that the first transition group includes the elements from Sc to Zn. Even though the Bohr formulation Sc to Zn is certainly preferable to the "side-group to the alkali metals" classification of Cu, it is not entirely satisfactory to the chemists, who may as well consider Sc and Ga as homologues of Al, or even to some extent, both Ti and Ge as the homologues of Si. From the point of view of (2), the typical transition group behaviour begins about V and finishes about Cu, having an average oxidation number slightly above $+3$ in the beginning and about $+2$ at the end. The similar treatment (2) of the second and third transition groups give more restricted intervals, say Mo–Pd and Re–Au, and with higher average oxidation numbers, falling from above $+4$ to about $+3$. Chemically (2), the actinides contain a similar uranide group (as Haïssinsky has proposed) beginning at U and not terminating at a definite element, and with an evolution from above $+5$ to $+3$. The most regular of all transition groups is the lanthanides, having a nearly constant oxidation number $+3$ for fifteen consecutive elements.

However, the chemical arguments (2) do not change the fact, discovered in 1950, that the absorption spectra of complexes with a partly filled shell actually behave according to (1), and that it is a fair approximation to use the names of the orbitals (d and f) known from spherical symmetry to the partly filled shell of complexes of the three ordinary transition groups (3d, 4d, 5d) and the lanthanides (4f) and actinides (5f), respectively.

Following the relation of electron configurations to chemical properties, it became usual to ascribe the covalent bonds to shared electron pairs. In many cases, the classical formula for a molecule could have all the "valency strokes" substituted by each electron pair, and the rest of the electrons distributed on the individual atoms, as "inner shells" and as "lone pairs". In the case of the elements outside the transition groups, the total number of bonding electron pairs and lone pairs associated with a given atom is nearly always four, establishing the "octet rule". It was once supposed that these eight electrons surrounding each atom, were situated at the corners of a cube. This gave a very nice picture of many molecules with single bonds, and Fajans proposed the model of cores (C^{4+}, S^{6+}, etc.) bound together with electron pairs, "quanticules".

However, tremendous difficulties arose in the cases where it was not obvious whether a given bond was simple or double, etc. We may here introduce the concept bond order as the number of electron pairs associated with a given bond. The bond order might be a rational, simple fraction. Thus, it was almost certain that the two individual Kekulé structures are not realized in benzene, but that each carbon–carbon bond has the bond order 1·5, i.e. half-way between a single and a double bond. Similarly the two C—O bonds in the carboxyl group of acetate ions were expected to have the bond order 1·5. An empirical estimate of bond order was later developed from the variation of the interatomic equilibrium distance, which decreases with increasing bond order, producing a tighter bonding and stronger reaction against the repulsion between different cores.

In these cases, the total number of bond orders around an atom was still the original bond number, an integer like 4 for carbon. However, a new question arose: could the bond order be an arbitrary decimal fraction, as the bond numbers are, which vanished for the electrovalent case of bonding? This was equivalent to saying that each atom in a molecule has a charge assuming that each electron pair was shared equally between the two bonded atoms. Thus, the sulphate ion SO_4^{2-} disposes of thirty-two electrons outside the cores S^{6+} and O^{6+}. If the bond order of all four bonds is 1, the charge of the sulphur atom will be $+2$ and of each oxygen -1. On the other hand, the classical picture of two single and two double bonds or the modern picture of a common bond order $1\cdot5$, would lead to neutral sulphur and the charge $-0\cdot5$ on each oxygen atom. It is evident that the truth might be somewhat intermediate, and later the term "resonance between ionic and covalent structures" was used to exhibit the possibility that the charges were perhaps $+0\cdot88$ on sulphur and $-0\cdot72$ on each oxygen. The same situation might also be described in another way by saying that the electron pairs were not symmetrically shared between the sulphur and a given oxygen atom.

Here it might even be asked whether the bond order is actually as high as $+1$ in the series CH_4, NH_3, H_2O, HF. In the model given above, these molecules should be composed of neutral atoms. However, the acid strength increases considerably in this series (equivalent to the way in which the basic strength increases in the anions F^-, OH^-, NH_2^-, CH_3^-), and there is a general feeling that the bonding of the protons grows distinctly more electrovalent and less covalent in the direction of HF. This might then either be ascribed to "ionic character of the single bonds" with unequal sharing of the electron pair, or it might be ascribed to a bond number varying from, say, $3\cdot9$ in CH_4 to $0\cdot6$ in HF. Thus, the simple model of single bonds given above could not be considered as proof that in the series BH_4^-, CH_4, NH_4^+ or in H_3O^+ the central ion was the bearer of the ionic charge, surrounded by four neutral hydrogen atoms. Even in distinctly covalent compounds, it could not be argued that each bond should have a bond order of at least 1; the existence of electron-deficient compounds, of which the most famous is diborane B_2H_6 with only twelve electrons available for seven bonds; or in gaseous state, the definite stability of the ion H_2^+ containing two protons and one electron, was strong evidence against the general validity of electron-pair bonding. Probably very many electron-deficient entities are stable against dissociation in their atomic fragments, but we do not usually observe them, because they rearrange to even more stable molecules, conforming to the electron-pair hypothesis.

The difficulties encountered in the hybridization theory of Pauling will be described in a special chapter; here it will only be argued that the usual tendency of text-books to describe molecules in terms of "resonance" between, for instance, $S^{2+}(O^-)_4$ and $S^{2-}(O)_4$ exaggerates the general importance to chemistry of resonance, of intermixing of wave functions, simply because the original wave functions of S^{2+}, etc., were extremely badly chosen. In molecular orbital theory, it is possible to describe the sulphate ion as having one configuration only, by almost as good an approximation as saying that a neon atom has only one configuration. But the molecular orbitals of the sulphate ions are delocalized in such a way that the charge of each atom (which is very difficult to define unequivocally, since the electron density does not vanish in its minima between the atoms) may very well be $+0·88$ or $-0·72$. It is believed to be very confusing to attempt to ascribe this effect to a resonance between integral values of the atomic charges. Even more subtle are the resonance effects between H^+H^- (and an equal amount of H^-H^+) or between O^+O^- and O^-O^+ in homonuclear molecules.

In the case of the less "polar" compounds of organic chemistry, the translation of the old valence strokes to electron pairs was fairly satisfactory. On the other hand, hardly any octahedral complex or molecule known conforms to this description. Even though very many guesses might be made about the bond order in the central atom–ligand bond in the examples $Ni(H_2O)_6^{2+}$, $Co(NH_3)_6^{3+}$, $Fe(CN)_6^{4-}$, $PtCl_6^{2-}$, SF_6, $RhCl_6^{3-}$, CrF_6^{3-}, it is agreed that it probably is a collection of various decimal fractions. This does not disturb the validity of oxidation numbers, as defined from the approximately valid molecular orbital configurations, and we shall later give the reasons for assigning the definite classifications Ni(II), Co(III), Fe(II), Pt(IV), S(VI), Rh(III), and Cr(III) to these examples.

One further problem arises in the simple electron-pair description. The *bond energy* of two different bonds, both having the definite bond order 1 from the naïve point of view, may be highly different. The dissociation energy of H_2 is five times as large as of Na_2, while no assymmetry (as in HF) can be assigned to the electron pair. This variation is probably connected with the different overlap between the participating orbitals, as we shall show later. But we may fore-cast the existence of single bonds which do not stabilize a given complex very much.

Sidgwick applied the simple electron-pair hypothesis to the special domain of transition group complexes, pointing out that an octa-hedral complex such as $Co(NH_3)_6^{3+}$ might achieve inert-gas con-figuration by formation of six single bonds, consisting of the lone pairs of the ammonia molecules. Since gaseous Co^{3+} with twenty-four electrons is paramagnetic (its partly filled d-shell has the re-sultant $S = 2$ in the ground state), while a krypton configuration with thirty-six electrons consists of filled shells and is diamagnetic, it was concluded that the diamagnetism of $Co(NH_3)_6^{3+}$ was evidence for the formation of six such bonds. With the assumption of equal sharing of the electron pair, this is equivalent to a charge -3 on the cobalt atom. Sidgwick argued that the transition group com-plexes generally were the most stable when a krypton, xenon or emanation configuration was achieved. However, this rule could not be the whole truth, and it was especially conspicuous that very stable transition group complexes might contain an odd number of electrons. Except for compounds like NO, NO_2 and ClO_2, molecules with an odd number of electrons are exceedingly rare outside the transition groups, and they are termed "free radicals" when they occur.

Linus Pauling's papers from 1931 and his book *The Nature of the Chemical Bond* refined Sidgwick's ideas considerably. As will be discussed later from the point of view of group theory, the hybridiza-tions apply atomic orbitals, intrinsically adapted to spherical sym-metry, for forming the molecular orbitals appropriate to the lower symmetry of the complex, e.g. the octahedral one. However, severe restrictions are made on the radial variation of these orbitals in Pauling's theory, and it can be seen in the more general molecular orbital theory that these restrictions are quite unnecessary.

The concept of resonance between ionic and covalent structures was used to imply that the charge of the central atom is intermediate between the positive oxidation number (as it would be in electro-valent bonding) and the generally negative charge obtained by accepting six electrons (half the twelve electrons from the ligands). Later (1948), Pauling recognized that even the "magnetochemical criterion" of high-spin or low-spin behaviour only distinguishes between different proportions of electrovalent and covalent bonding in the complexes. However, many chemists were left with rather confused ideas about resonance, accepting it as a fact similar to the tautomeric behaviour of acetone derivatives, where a proton may move from the ketone form CH_3COCH_3 to the enol form $CH_3C(OH)CH_2$, in just the same way as the transformation of one isomer to another. The idea of a cobalt atom in $Co(NH_3)_6^{3+}$ quickly changing the charge $+3$ to -3 is not too informative about the ground state, and it was only the study of excited states (with which the hybridization theory is definitely in disagreement) which settled the problem of chemical bonding in complexes.

Recently, Gillespie and Nyholm applied Sidgwick's electron-pair ideas to the description of the stereochemical properties of complexes, where a good agreement with experience could be obtained, assuming that the lone pairs and the ligands of a given atom repel each other and demand nearly the same space angle on the surface of the central atom. The translation of this feature into molecular orbital theory will be discussed below.

Actually, even before Pauling's theory, a much more appropriate description of the transition group complexes was initiated by Bethe in 1929. His *crystal field theory* considers the perturbing effect of the electric field, produced by the ligands, on the energy levels of a partly filled shell of the central ion. The energy levels, especially, vary as a function of a perturbation of octahedral symmetry as a function of one parameter only, representing the splitting of the partly filled d-shell into two sub-shells. While this model is entirely electrostatic, Van Vleck pointed out (1935) that molecular orbital theory, assuming partly covalent bonding, would result in the same qualitative behaviour and formation of two sub-shells. The application of these theories to transition group complexes is known as *ligand field theory*.

Originally the ligand field theory was mainly applied by magneto-chemists to explain rather subtle magnetic details and very small splittings in energy of the levels of the ground state. This study, which originally considered the macroscopic magnetic susceptibility only, has now been extended by use of the much more refined *paramagnetic resonance* (electron spin resonance) method. Thus the anistropy of the magnetic behaviour in different directions in a crystal can be investigated, and very dilute solid solutions of para-magnetic ions in diamagnetic crystal lattices can now be studied.

Even though Joos and other authors had remarked empirically that coloured ions were much more common in the transition groups than outside, it had not been possible to identify the excited levels, corresponding to these absorption bands. According to N. Bohr's postulates in 1913, the wave number of the absorbed light is expected to be proportional to the energy difference between the excited state and the ground state. Thus, the usual colourless compounds with absorption bands in the ultra-violet, with higher wave numbers than visible light, have a larger distance to the excited levels than most transition group compounds. Kato (1930) suggested that the excited levels are those of the partly filled shell of the corresponding gaseous ion, only slightly disturbed by the ligands. While this assumption is true for the f-shell of the lanthanides and the actinides, where the distribution of narrow absorption bands is nearly the same in different compounds with the same central ion, Kato's idea can hardly be generally valid for the three transition groups, since the colours vary widely with the ligands.

The first application of the ligand field theory to absorption spectra of transition group complexes, assuming a large influence from the ligands and correspondingly, a considerable energy difference between the sub-shells, was made by Finkelstein and Van Vleck (1940) to the absorption spectra of chromium (III) in alums as $Cr(H_2O)_6^{3+}$ and in the ruby, embedded in Al_2O_3, where each Cr is octahedrally surrounded by six oxygen atoms. These spectra consist of two groups of weak, extremely narrow lines in the red about 14·5 kK (we shall use the wave number unit kK = *kilokayser* = 1000 cm^{-1}) and in the blue about 21 kK, while two broad bands are observed at about 18 and 24 kK.

The narrow bands were identified by Finkelstein and Van Vleck as three spin-forbidden (of which the two first are nearly coincident) transitions from the quartet (i.e. $S = \frac{3}{2}$) ground state to three doublet ($S = \frac{1}{2}$) levels. Finkelstein and Van Vleck remarked that two of the predicted spin-allowed transitions (to excited levels with $S = \frac{3}{2}$) were unfortunately hidden by the two broad bands. Actually, it was later recognized that the broad bands are simply the spin-allowed transitions predicted in the ligand field theory. The reason for the highly differing band width is that the three doublet levels belong to the same sub-shell configuration as the ground state. Therefore, the excited states have the same internuclear equilibrium distances as the ground state. On the other hand, a large number of vibrations are co-excited with the two excited quartet levels, where the electron distribution in the central ion is essentially different, and where the equilibrium distances would correspond to an expanded complex.

The identification of these Cr(III) energy levels remained rather isolated until Ilse and Hartmann (1951) began a systematic study of the first transition group absorption spectra. As a first step these authors considered the titanium (III) hexa-aquo ion $Ti(H_2O)_6^{3+}$, where the corresponding gaseous ion Ti^{3+} has only one 3d-electron outside the closed argon shells. Since a broad band is observed of the purple aquo ion at 20·3 kK, this was a very effective piece of evidence against Kato's theory. As will be discussed below, the configuration [A]$3d$ in gaseous Ti^{3+} corresponds to only two energy levels, separated by 0·38 kK. One possibility might have been that the distance to the configuration [A]$4s$ was five times smaller in $Ti(H_2O)_6^{3+}$ than in Ti^{3+}. However, this is highly improbable for reasons given later, and would also lead to the expectation of broad bands in the visible region of zinc (II) complexes and other d^{10}-systems.

From the identification of the energy difference 20·3 kK in $Ti(H_2O)_6^{3+}$ with the one orbital energy difference between the two d-sub-shells in octahedral symmetry, it was possible to make a unified description of the hexa-aquo complexes of ions containing more than one 3d-electron. Here, we have the problem that the ligand field perturbation energies have the same order of magnitude as the interelectronic repulsion energies (to be discussed below), separating the energy levels of a partly filled shell in spherical symmetry into more multiplet terms, each characterized by quantum numbers S and L. Ilse, Hartmann and Schläfer (1951) originally made the (in most cases rather bad) assumption that only the lowest such multiplet term (according to Hund's rules where the maximum value of S, and with this condition, the highest value of L applied) contributed to the observed energy levels. Furthermore, these authors tried to apply an electrostatic perturbation calculation on a very uncertain choice of $3d$-wave functions for estimating quantitatively the value of the orbital energy difference between the two sub-shells. This neglect of the covalent contributions to the bonding of the complex could not be supported by subsequent studies.

In the years following, several physicists and chemists developed the ligand field theory of first transition group complexes. The most general treatment of the octahedral complexes was given by Tanabe and Sugano (1954). In a series of papers from 1952 to 1955 Orgel compared the electrostatic and covalent descriptions of ligand field

perturbations, as did Owen (1955). McClure and Lacroix had independently obtained similar results at this time, but did not publish
very much. Hartmann's laboratory in Frankfurt-am-Main continued the use of the mathematical apparatus of electrostatic perturbations for obtaining at least qualitatively correct results. Similar
attempts were made by C. J. Ballhausen in Chemistry Department A,
Danmarks Tekniske Højskole, Copenhagen, where the present writer,
since 1954, has measured the absorption spectra of a large number
of complexes and tried to establish the relations with atomic spectroscopy and molecular orbital theory. At the same place, C. E. Schäffer
has intensively studied the numerous chromium (III) and cobalt (III)
complexes.

The second and third transition group complexes have absorption
spectra, the identification of which runs parallel to that of the first
transition group, without showing any intrinsic difference. Except
for a study of molybdenum (III) by Hartmann and Schmidt (1957),
only the present writer seems to have published papers on the octahedral $4d^n$- and $5d^n$-complexes* (1956–1958) where Stevens (1953)
had previously described the magnetic properties.

Two scientific meetings have been of importance for the recent
development of ligand field theory. The Tenth Solvay Council on
Chemistry in Bruxelles invited L. E. Orgel, R. S. Nyholm, J. Bjerrum,
and the present writer for discussion, and review papers were published in *Quelques Problèmes de Chimie Minérale*. The General Discussion of the Faraday Society on Transition Group Ions in Dublin,
September 1958, made it possible for Orgel, J. S. Griffith, Owen,
Pryce, Tanabe, Linnett, Williams, Vlcek, Chatt, Hush, Nyholm,
Van Vleck and the present writer to meet and talk about the recent
developments of the theory of chemical bonding. The proceedings
of this meeting are published in *Discussions of the Faraday Society*
No. 26.

Before considering the actual state of ligand field theory, it will
be useful to consider first the evaluation of atomic spectroscopy.
The principal purpose of the present book is to indicate the close
analogies between the concepts and the difficulties and results encountered in atomic spectroscopy, and molecular orbital theory.
In a certain sense, the theory of chemical bonding is in the same
situation now as was the description of atomic spectra 30 years ago.

---

* However, Moffitt, Goodman, Fred and Weinstock (1959) discussed the
$5d$-hexafluorides.

Chemists often remark that modern evolution reduces chemistry to a minor sub-division of physics, and that it will soon become necessary to make lengthy highly specialized studies in physics before understanding chemistry at all. To some extent, however, it is physics rather which has partly been invaded and conquered by the ideas of chemistry.

In terms of classical physics, as represented by Newton's mechanics and Maxwell's electromagnetic theory, the theory of chemical bonding was entirely unintelligible. The selectivity and the saturation capacities of the valency forces did not correspond to any other phenomenon. That gaseous atoms and ions were shown by spectroscopy, evolved after Bunsen and Kirchhoff (1860), to be very selective emitters and absorbers of light of sharply defined wave numbers was another of the unexplained phenomena. These features of chemistry and atomic spectroscopy succeeded in disposing of most of the ideas of classical physics and resulted in the evolution of quantum mechanics. It is often supposed that the quantum number description of the energy levels in atomic spectroscopy was the result of deductions from the new theory of quantum mechanics. Actually, this discipline did not do much for atomic spectroscopy other than to rationalize all the empirical material already well known; most, if not all, quantum numbers were introduced originally as empirical classifications. It may be expected in a similar way, to be possible for the theory of chemical bonding to obtain valuable results by deduction from the facts. It is not reasonable for the chemist to sit down and wait for the pure theorists to deduce everything *ab initio*; experience shows clearly that the preliminary steps of approximate calculation have an unfortunate tendency of producing nonsensical results from incorrect assumptions, unless these are strictly governed by a realistic knowledge of the experimental results.

CHAPTER 2

ORBITALS IN ATOMIC SPECTROSCOPY

QUANTUM mechanics in its two modern forms, i.e. Heisenberg's matrix mechanics and Schrödinger's wave function formulation (which can be shown to be equivalent), can be applied to electron systems, containing positively charged nuclei. It has not been proved that this application leads to inaccurate results in any specific case, but there are very severe technical difficulties impeding a strict calculation in systems containing more than two electrons. This is not surprising, when we remember that classical mechanics could not give a general expression for the motion of three particles under their mutual gravitational attraction.

The *wave function* Ψ of a system containing q electrons in a stationary state has 4q variables. Of these, 3q are continuous *space co-ordinates*, assuming all real values between $-\infty$ and $+\infty$, while q are *spin co-ordinates*, each assuming only one of two possible values. In the following description, we shall assume that Ψ is a real, not a complex function. This is no essential restriction, when we are not considering magnetic effects, splitting up Kramers doublets (see below). Then, Ψ^2 is the statistical probability of finding the system with its q electrons at the given values of the 4q co-ordinates.

Schrödinger's equation determines Ψ, when the potentials of the acting forces are known. Here, we are in the fortunate situation that the electrostatic potentials, inversely proportional to the distances from the electrical charges, are by far the most important. The electrodynamic and relativistic effects may be treated as small corrections. The unfortunate situation when quantum mechanics is applied to nuclei, and where the potentials acting are not even known, need not trouble us here.

On the other hand, the presence of 4q variables present nearly insurmountable difficulties, when reasonably large values of q are considered. Even if we tried to make a very rough table with only ten values for each continuous variable, the 10^{72} entries in one of the wave functions of a neutral chromium atom exceed the number of atoms in a sphere cut out of the universe with a diameter of a hundred million light-years. The most appropriate way of approximating these awkward wave functions is to assume Ψ to consist of the antisymmetrized product of one-electron wave functions ψ, each with only three space and one spin variable. The antisymmetrization, as described in Condon and Shortley (1953) is a necessary process for preserving the undiscernibility of the electrons. We can never distinguish two individual electrons, and the wave functions Ψ always ascribe the same probability of finding electron no. 1 at A and no. 2 at B as to find electron no. 2 at A and no. 1 at B. To gain a mental picture of the situation, it is certainly more fruitful to imagine a certain probability density of electrons at A and at B than to think about the individual electrons, from which quantum mechanics tries to remove any significant individuality.

It is obvious that when the wave function Ψ is split up into, say, twenty-four electron functions ψ, it is possible to make a catalogue of 24×10^3 or even 24×100^3 entries about it. In spherical symmetry, the most valuable procedure has been to separate ψ into the product of an angular variation and a radical function. In the Hartree–Fock method the best radial function for a given atom may be calculated by use of the Schrödinger equation and a simultaneous comparison with the radial functions of the other electrons, supplying the electrostatic potential together with the nucleus.

For most purposes, the wave function Ψ in its many-dimensional space contains much too much information for its practical use. We may restrict ourselves to regarding the *density matrices* of different order. The *first-order density matrix* (or rather its diagonal elements) represents the integrated electron density in ordinary space with only three variables (and if necessary, the spin variable). This is the electron density observed for example, by refined crystallographic measurements. As seen below, it is not possible to define the energy from the *first-order density matrix* alone, since the average values of the mutual interelectronic distances may be different in two cases having the same electron density distribution in our space. On the other hand, since the forces known to act in electron systems can always be divided into interactions between two electrons at the time, it might be hoped that no order higher than the *second-order density matrix* is necessary for the complete description of the electron system. Here, the probability of different distances between two electrons is given, in addition to the probability of one-electron

c

density in every point of our space. It is obvious that six space variables would be easier to handle than 3q variables. Unfortunately, Löwdin and Shull have shown that many second-order density matrices do not correspond to wave functions of q electrons and would give lower energy than that observed, if this quantity was calculated. It is therefore necessary to find some condition for a second-order density matrix to correspond to a Ψ, before the latter inconvenient concept can be abandoned.

The quantum numbers, belonging to wave functions, originated as a classification of spectroscopical energy levels. The studies of Rydberg in particular, and those of subsequent spectroscopists, of gaseous atoms containing only one or two electrons outside the inert-gas closed shells, led to the classification of series spectra, where a given set of one-electron energy levels could be characterized by a filing number n and a quantum number l which later turned out to be related to the orbital angular momentum. The values of l have special names, viz.

$$l = 0 \quad 1 \quad 2 \quad 3 \quad 4 \quad 5 \quad 6 \quad 7 \quad 8 \ldots \tag{1}$$
$$ \text{s} \quad \text{p} \quad \text{d} \quad \text{f} \quad \text{g} \quad \text{h} \quad \text{i} \quad \text{k} \quad \text{l}$$

where the first four were derived from the empirical names of four different series in the spectrum of neutral sodium, the "sharp", "principal", "diffuse", and "fundamental" series. We use italics for the general symbols of quantum numbers such as l and roman type for the trivial names s, p, d, etc.

The energy levels of a hydrogen atom or other systems with only one electron and a nucleus (He^+, Li^{++}, etc.) can be calculated by quantum mechanics to a very great accuracy. If we neglect small effects, amounting to some 10^{-4} of the wave numbers, the energy of these levels is independent of l, and is $-ry/n^2$. The *Rydberg constant* ry is 109·74 kK in heavy atoms and slightly smaller, 109·68 kK, in the ordinary hydrogen atom. This result was already found by N. Bohr in 1913 from the application of quantum conditions to classical mechanics, and ry was derived from Planck's constant and other universal constants. This semi-classical atomic model is now used by exhibitions and nuclear energy (called atomic energy) institution letter-heads for symbolizing an atom for the public. The one electron in the hydrogen atom might move according to Kepler's law in circular or elliptic orbits, having in the former case the radius n^2a_0, where a_0 is the *Bohr radius* 0·528 Å (1 Å = 1 ångström = 10^{-8} cm) also derived from universal constants. Independent of the ellipticity (introduced for relation with l), the potential energy of such an orbit in the attraction potential of the nucleus is $-2ry/n^2$, while the kinetic energy of the motion in the orbit is $+ry/n^2$.

However, this model is completely unsuited for systems containing more than one electron. It is impossible to describe a helium atom as containing two particles, moving in definite orbits. This is one reason why it is somewhat misleading to compare an atom with a sun surrounded by planets. The latter picture is only correct for the mass ratio between the nucleus and the electrons and to some extent for the very low density in the volume outside the nucleus. On the other hand, the planets are individually different, tangible bodies, while the electron density in an atom is described by the cloudy wave function mentioned above. One further important difference will be discussed below: While the gravity interactions between the planets are very small, compared to the interaction with the sun, the electrostatic interactions between the electrons in an atom with several electrons are by no means negligible, compared to the attraction by the nucleus.

In the neutral alkali metal atoms (Li, Na, K, Rb, Cs) the energy levels were described by Rydberg as $-ry/(n - x_1)^2$ where x_1, the Rydberg correction, is largely dependent on l (being small for l above 2) and almost invariable with n. Thus, we might conclude a series of increasing energies for one electron, writing n before the name of l; and $n > l$:

$$
\begin{aligned}
&1s \\
&2s < 2p \\
&3s < 3p < 3d \\
&4s < 4p < 4d < 4f \\
&5s < 5p < 5d < 5f < 5g
\end{aligned}
\tag{2}
$$

and beginning the series of n with the small values of l. It must be remarked that there is no rule giving the relative energy of two levels with different n. It is often stated in text-books that the first transition group begins where 3d falls below 4s in energy. However, this statement is ambiguous, if the *ionic charge* $Z_0 - 1$ is not given. For neutral atoms, 3d has lower energy than 4s in hydrogen and the first, other light atoms. The energies seem to cross about carbon, and both levels correspond here to highly excited states of the neutral atom. In the interval carbon ($Z = 6$) to calcium ($Z = 20$), 4s has distinctly lower energy than 3d. Around $Z = 22$, it may still be questionable which has the lower energy, but 3d is distinctly creeping below 4s. For $Z = 31$ (gallium) and higher values, 3d belongs in principle to the filled shells to be studied by use of X-ray line spectra, and for high values of Z, the hydrogenic order according to increasing n tends to be re-established.

However, if we consider doubly charged gaseous ions, there is no doubt that 3d has much lower energy than 4s in the whole series Sc^{2+}, Ti^{2+}, . . ., Zn^{2+}, and the domain of reversed energy 4s $<$ 3d around $Z = 15$ is reduced to a near coincidence in Si^{2+} and P^{2+}. If we can actually compare the positive oxidation numbers $+2$, $+3$, and $+4$ of transition group complexes with the corresponding gaseous ions, these must be said to be purely 3d-ions (or 4d or 5d in the next two groups). The one or two s-electrons occurring in the ground state of the neutral atoms in gaseous state (we are not here talking about the metals, which is a subject much more complicated and outside the scope of this book) are removed at first by ionization, and only d-electrons are left in the divalent ions. Similar features are encountered at the beginning of the actinide group, where neutral thorium has about the same energy of 5f, 6d and 7s-electrons. However, for increasing ionic charges, these separate in the order 5f $<$ 6d $<$ 7s, and it is true that for ionic charges above $+2$, the ions have purely 5f-electrons in their ground state.

The great idea of atomic spectroscopy was now to assume *electron configurations* in every atom or ion, even those not giving series spectra. It is assumed that an integral number, at least 0 and at most $4l + 2$, electrons occupy each of the *shells* characterized by n and l, and it is customary to write their number as exponents, as no other typographical place is available. Thus, the electron configuration of the ground state of the neutral sodium atom in $1s^2 2s^2 2p^6 3s$, having one 3s-electron outside the neon configuration. We often write it [Ne]3s, and the excited configuration, corresponding to the yellow lines at 16·96 kK (known from sodium lamps or gas burners, salted with Na) is [Ne]3p.

The maximum number of electrons $4l + 2$ in each shell was determined empirically for obtaining agreement with the periodic system; but actually, quantum mechanics interprets it as a product of two *degeneracy numbers*, 2 belonging to the two possible spin co-ordinates and $2l + 1$ related to the possible directions of the angular functions discussed below. The $2l + 1$ entities of a given shell are called *orbitals*, the independent wave functions for one electron (of each spin direction). We may sometimes prefer to use *spin orbitals*, the wave function for one electron at the most.

As discussed in the next chapter, the electron configuration does certainly not specify the energy as the sum of orbital energies; on the other hand, it has been found by experience to be a very expedient way of classifying the energy levels, and it might be hoped that the actual wave functions Ψ resemble to some extent Slater's antisymmetrized products of orbitals ψ. In spherical symmetry, ψ has the very remarkable property of being separable into the product of an *angular function* A and a *radial function* R, divided by r.

In most text-books, the variables of A are chosen as the angles φ and θ of a polar co-ordinate system, while the variable of R is the distance r from the nucleus. However, since the trigonometric expressions of A in cos φ, etc., are rather complicated and do not give a clear mental picture, and since we shall be especially interested in octahedral complexes later on, we shall write A as a linear combination of homogeneous polynomia A_p of *l*th degree in the Cartesian co-ordinates x, y and z, viz.:

$$A_p = x^a y^b z^c r^{-l}, \text{ where } a + b + c = l \text{ and } r^2 = x^2 + y^2 + z^2 \quad (3)$$

We shall here mention that the *orthogonalization* condition of two wave functions Ψ_1 and Ψ_2 to be independent is that the integral of $\Psi_1 \Psi_2$ over all its variables is zero, while the *normalization* refers to the similar integrals of Ψ_1^2 and Ψ_2^2 being 1.

Table 1 gives the values of A for $l = 0, 1, 2, 3$, i.e. for s-, p-, d- and f-electrons. These were studied by Von der Lage and Bethe (1947) as *cubic harmonics*. The functions A fall into two classes, *even* and *odd parity* according to the even or odd value of *l*. The even functions do not change value by going from the point (x, y, z) to the point $(-x, -y, -z)$ (corresponding to an inversion in the centre of the co-ordinate system at the nucleus) while the odd functions are multiplied by -1 by an inversion. An even and an odd function are necessarily orthogonal, since contributions of opposite signs would always occur by integration. It is evident that only three homogeneous polynomiae A_p exist for $l = 1$. Six A_p's were expected to exist for $l = 2$, viz. x^2, y^2, z^2, xy, xz and yz. While the last three actually occur in Table 1, the first three are not linearly independent. This is caused by the orthogonality condition with the previous even function $(l = 0)$, since $(x^2 + y^2 + z^2)/r^2$ is another way of writing the number one. Therefore, only two linear combinations of x^2, y^2 and z^2 exist as independent wave functions for $l = 2$, and Table 1 illustrates one possibility of choosing these two linear combinations.

TABLE 1. ANGULAR FUNCTIONS IN CARTESIAN CO-ORDINATES

	Octahedral symmetry	Polynomium A (in each case to be divided by r^l)	Normalization factor
$l = 0$ (s-electrons)	γ_1 even	1	1
$l = 1$ (p-electrons)	γ_4 odd	$\begin{cases} x \\ y \\ z \end{cases}$	$\sqrt{3}$ $\sqrt{3}$ $\sqrt{3}$
$l = 2$ (d-electrons)	γ_5 even	$\begin{cases} xy \\ xz \\ yz \end{cases}$	$\sqrt{15}$ $\sqrt{15}$ $\sqrt{15}$
	γ_3 even	$\begin{cases} x^2 - y^2 \\ z^2 - \frac{1}{2}x^2 - \frac{1}{2}y^2 \end{cases}$	$\sqrt{(15)}/2$ $\sqrt{5}$
$l = 3$ (f-electrons)	γ_2 odd	xyz	$\sqrt{105}$
	γ_4 odd	$\begin{cases} z^3 - \frac{3}{5}zr^2 \\ y^3 - \frac{3}{5}yr^2 \\ x^3 - \frac{3}{5}xr^2 \end{cases}$	$5\sqrt{(7)}/2$ $5\sqrt{(7)}/2$ $5\sqrt{(7)}/2$
	γ_5 odd	$\begin{cases} z(x^2 - y^2) \\ y(x^2 - z^2) \\ x(z^2 - y^2) \end{cases}$	$\sqrt{(105)}/2$ $\sqrt{(105)}/2$ $\sqrt{(105)}/2$
$l = 4$ (one of the g-electrons)	γ_1 even	$x^4 + y^4 + z^4 - \frac{3}{5}r^4$	$5\sqrt{(21)}/4$

In a similar way, ten different A_p's were expected for $l = 3$. However, the complicated behaviour in Table 1 results from the need to separate out the three p-functions x, y and z, and only seven genuine f-functions are conserved. Generally, among the $(l + 1)$ $(l + 2)/2$ possible A_p's, $l(l - 1)/2$ polynomiae are separated out for preserving the orthogonality with lower l-values of the same parity, and only $2l + 1$ orbitals A are left, resulting in rather complicated expressions for large values of l.

While the angular functions A are hydrogenic, those of the theory for a one-electron system in spherical symmetry, the radial functions R are *not hydrogenic* for more-electron systems. The derivation of the functions R in the case of hydrogen can be found in many textbooks. We shall here denote by R^2 the electron density per spherical shell, not per volume unit, as is often done in literature. The two definitions differ by the appearance of a factor r in the wave function.

The radial functions R in a more-electron system is governed by Schrödinger's equation, regarding a *central field* potential U(r) of spherical symmetry:

$$\frac{h^2}{8\pi^2 m} \frac{d^2R}{dr^2} + \left(E_{nl} - U(r) - \frac{h^2}{8\pi^2 m} \frac{l(l+1)}{r^2} \right) R = 0 \qquad (4)$$

where E_{nl} is the corresponding orbital energy, m is the mass and $-e$ the charge of the electron, h is Planck's constant, and c is the velocity of light. For negative E_{nl} (bound states), only a discrete set of values of E_{nl} make it possible to choose R according to eqn. (4) in a way that R vanishes for $r \to \infty$.

The orbital energy E_{nl} can be found by integration

$$E_{nl} = e^2 \int_0^\infty -\tfrac{1}{2}R \frac{d^2R}{dr^2} + R^2 U(r) + R^2 \frac{l(l+1)}{2r^2} \, dr \qquad (5)$$

where the atomic units, giving lengths in units of a_0 and energy in units of two Rydberg ($= e^2/a_0$), can be applied.

The potential part of the energy of eqn. (5) is the second term, being a simple integration of the product of the electron density per spherical shell R^2 and the central field potential U(r). The kinetic energy is the first and the third term. The dependence of the first term on the second differential quotient of R is a feature of quantum mechanics and Schrödinger's equation, unparalleled in classical mechanics. The third term in eqn. (5) has a formal resemblance with the potential of a centrifugal force, $l(l+1)/2r^2$, being absent for s-electrons with $l = 0$. Actually, this is the kinetic contribution from the angular dependence A. It explains the order of increasing orbital energies with increasing l in eqn. (2) (when U(r) decreases more quickly than r^{-1} for increasing r). It can be shown from eqn. (4) that only s-electrons have a finite probability density close to the nucleus, while the positive l-values prevent a non-vanishing density, due to the "centrifugal" term of the kinetic energy. Generally, the behaviour of R for small values of r will be as a normalization factor multiplied by r^{l+1}. For $r \to \infty$, R of a stationary state vanishes exponentially as $\exp[-r\sqrt{(-E_{nl})}]$ according to eqn. (4).

Not only for a hydrogen atom, but for any of the central fields found in atoms, the number of zero points of R (excluding those at $r = 0$ and $r \rightarrow \infty$) is $n - l - 1$. Thus, the lowest value of n for a given l(1s, 2p, 3d, 4f, . . .) has only one maximum, and the orthogonality with the other orbitals is secured from the angular dependence alone. The following values of n produce their orthogonality with the first orbital by having different domains with alternatingly positive and negative values of R.

The hydrogenic radial function for the nuclear charge Z_0 in the simple case $n = l + 1$ is, neglecting the normalization factor,

$$R = r^{l+1} \exp(-Z_0 r/na_0) \text{ with } r_{max} = n^2 a_0/Z_0 \qquad (6)$$

with its maximum value occurring at that r_{max} which was the radius of N. Bohr's orbit model of 1913. As *analytical approximations* to the functions, which Hartree and Fock indicated methods of calculating (being *self-consistent radial functions* by having U(r) adjusted to the corresponding electron densities R^2), Slater proposed the hydrogen-like function of eqn. (6) with the effective charge Z_* as a variable parameter, to lie somewhere between the nuclear charge Z and the ionic charge plus one. According to the investigations of Hartree, the radial functions of a given shell in a series of ions with the same charge vary as functions of Z only by *scaling*, i.e. the variations of the functions can be represented by a simple change of the unit of length and a consequent change of the normalization factor. However, these *isomorphous* functions of the same "shape" do not in general correspond to the hydrogenic functions of eqn. (6), but rather to functions

$$R = r^a e^{-br} \qquad (7)$$

with values of "a" between 1 and 2.

The *virial theorem* states that the kinetic energy E_{kin} is related to the potential energy E_{pot} in a potential, proportional to r^n, by the equation $E_{kin} = nE_{pot}/2$. Since $n = 1$ for electrostatic potentials, then for all energies of Hartree–Fock orbitals or of actual systems (the total energy $= E$), we have

$$E_{kin} = -E \text{ and } E_{pot} = 2E \qquad (8)$$

as mentioned above for the classical orbit picture of the hydrogen atom. However, eqn. (8) is only valid for molecules, containing more nuclei, if the corrections for variation of E with the internuclear distances r are introduced:

$$E_{kin} = -E - r \frac{dE}{dr} \quad \text{and} \quad E_{pot} = 2E + r \frac{dE}{dr} \qquad (9)$$

Eqn. (9) reduces to eqn. (8) in the cases $r = \infty$ and $r = r_0$, the equilibrium internuclear distance. Eqn. (5) shows that if two isomorphous radial functions are related to each other by a *scaling parameter* η, having the dimension of a reciprocal radius, then the kinetic energy will be proportional to η^2 and the potential energy (having the opposite sign for stationary states) be proportional to η.

This answers to some extent a question which many students may ask: why has the radial function of a given shell, e.g. the 1s-electron of a hydrogen atom, exactly the extension (e.g. the average radius) observed and not a smaller or a greater extension? If the value of η is fixed to 1 for the wave function actually found, the variation of the energy would be:

$$E = k\eta^2 - 2k\eta \quad \text{and} \quad \frac{dE}{d\eta} = 2k\eta - 2k \qquad (10)$$

Thus, for smaller average radii ($\eta > 1$), the increase in the kinetic energy by packing the electron cloud into a smaller volume more than outweighs the corresponding decrease in potential energy, due to stronger attraction by the central field. On the other hand, for larger average radii ($\eta < 1$), the decrease in kinetic energy is not sufficient for repairing the loss of attraction by the potential.

It is seen that the variation of E with η near to $\eta = 1$ corresponds to the shallow minimum of a parabola; the energy is relatively insensitive to a small change of η, and a change from $\eta = 1$ to 1·1 will certainly not produce a 10 per cent increase of the energy, but only some 1 per cent increase.

CHAPTER 3

THE ENERGY LEVELS OF A CONFIGURATION

THE number of independent wave functions of a given electron configuration is its number of *states*. For a partly filled n, l-shell, containing q electrons, the number of states is

$$\binom{q}{4l+2} = \frac{(4l+2)!}{q!(4l+2-q!)} \tag{11}$$

These states may be assembled in *energy levels*, according to different sorts of coupling schemes. The number of states in a given level is called its *degeneracy number e*. In spherical symmetry, the energy levels are characterized by the *total angular momentum quantum number J*, having $e = 2J + 1$. If we can neglect the electrodynamic and relativistic effects, the energy levels J will be assembled in *multiplet terms*, each characterized by the *spin quantum number S* and the *orbital angular momentum quantum number L*, producing a degeneracy number $e = (2S + 1)(2L + 1)$. If L and S are well-defined "good" quantum numbers, we have *Russell–Saunders coupling*. The main reason for the energy differences between the multiplet terms is the differences in the potential energy of inter-electronic repulsion, evolving from the different distribution of the electrons of the partly filled shell on the angular functions A. In the Slater description of atomic energy levels, as given in Condon and Shortley's *Theory of Atomic Spectra*, the different terms of a given configuration are assumed, as a first approximation, to have the same radial functions of the orbitals corresponding to the same attraction by the central field and the same orbital energies (eqn. 5), but to have energy differences dependent on *interelectronic repulsion parameters* representing the mutual repulsions between the electrons in the partly filled shell. We shall discuss the phenomenon below, in the general terminology of perturbation theory.

It is rather complicated to calculate the number of multiplet terms and their values of L and S, belonging to a given configuration. If the configuration contains more than one partly filled shell, the number of states will be the product of the binomial expressions (11) for each partly filled shell, and the L- and S-values are produced by unrestricted vector sum coupling of the L- and S-values of each shell. However, we shall not treat this problem in general, but rather concentrate our attention on the case of one partly filled shell. The multiplet terms will be designated by names of the L-values analogous to eqn. (1):

$$L = 0 \quad 1 \quad 2 \quad 3 \quad 4 \quad 5 \quad 6 \quad 7 \quad 8 \quad 9 \quad 10 \ldots \qquad (12)$$
$$ \quad S \quad P \quad D \quad F \quad G \quad H \quad I \quad K \quad L \quad M \quad N$$

and the "multiplicity" $2S + 1$ is indicated as a superscript at the left-hand side of the symbol. The single state of a completely filled or empty shell corresponds to one term ^1S having $S = 0$ and $L = 0$. The $4l + 2$ states of one electron in a shell corresponds to one term ^2S, ^2P, . . . with $S = \frac{1}{2}$ and $L = l$. In any type of coupling, the *hole-equivalency principle of Pauli* holds that q "holes" (e.g. $4l + 2 - $ q electrons in a shell) and q electrons produce the same number and species of terms. Thus, both 2 electrons in a partly filled shell (p^2, d^2, . . .) and $4l$ electrons (p^4, d^8, f^{12}, . . .) have their $(2l + 1)(4l + 1)$ states (eqn. 11) distributed on the series of multiplet terms

$$^1S,\ ^3P,\ ^1D,\ ^3F,\ ^1G,\ ^3H,\ ^1I,\ \ldots\ (L \le 2l) \qquad (13)$$

terminating at the highest L-value possible by simple addition of the two equal l-values. According to a proposal by Mulliken, we shall use *capital letters* for the quantum numbers of systems and *small letters* for the quantum numbers of orbitals.

If at least three electrons occur in a partly filled shell, more than one multiplet term may have the same combination of L and S. This is not effective for p^3 (having the terms ^4S, ^2D and ^2P), but d^3 has two terms ^2D in addition to the terms ^4F, ^4P, ^2P, ^2F, ^2G and ^2H. The more complicated d^n- and f^n-systems have a long series of terms. If we use the words *singlet* ($S = 0$), *doublet* ($\frac{1}{2}$), *triplet* (1), *quartet* ($\frac{3}{2}$), *quintet* (2), and *sextet* ($S = \frac{5}{2}$), d^4 has one quintet term ^5D, seven triplet and eight singlet terms, and d^5 has one sextet term ^6S, four quartet terms ^4G, ^4P, ^4D and ^4F, and eleven doublet terms.

According to *Hund's rule*, the lowest term in a given configuration has the maximum value of S, and the highest value of L compatible with this condition. Hund's rule does not consider the order of the other values of L and S, which may be distributed on the energy scale in a rather irregular way.

The orbitals with room for two electrons of opposite spin direction are often symbolized by boxes, containing arrows in upward or downward direction. However, this representation often leads to incorrect results when applied uncritically, e.g. to the number of energy levels. It can be safely used for two types of arguments only: the maximum value of S can always be found as half the number of upward arrows minus half the number of downward arrows, but the distribution of the other S values cannot be simply recognized. Further, the number of states is given correctly by the permutation theory applied to the boxes. However, a fundamental difficulty can be seen in a simple example: one electron in each of two orbitals corresponds obviously to four states. Of these, the two states with parallel spins belong to a triplet state with $S = 1$, but it cannot be concluded that the two states with "compensated" spins both belong to the singlet state with $S = 0$. Actually, they are mixed in such a way that a triplet and a singlet state results.

Most text-books describe *perturbation theory* in a number of steps, leading to some confusion in terminology. These steps are:

(1) "Zero-order" perturbation, where a set of "degenerate" wave functions Ψ_a, Ψ_b, . . . Ψ_n having the same energy before the perturbation are used for forming linear combinations $\Psi_1 = k_a\Psi_a + . . . + k_n\Psi_n$ appropriate for step (2) (securing vanishing non-diagonal elements between two different such linear combinations under the perturbation).

(2) First-order perturbation of the energy, where the *diagonal elements* $E_{11} = \int \Psi_1 P \Psi_1$ of each wave function is calculated, changing the energy from its original value E_1 to $E_1 + E_{11}$ (the latter quantity is often referred to as the diagonal element).

(3) First-order perturbation of the wave function, where the *non-diagonal elements* $E_{12} = \int \Psi_1 P \Psi_2$ between two wave functions is calculated, changing the wave function from the original Ψ_1 to

$$\Psi_1 + \sum_{n>1} \frac{E_{1n}\Psi_n}{E_{nn} + E_n - E_{11} - E_1}$$

multiplied by a normalization factor.

(4) Second-order perturbation of the energy, where the non-diagonal elements E_{1n} change the energy by the last summation term in

$$E = E_n + E_{nn} - \sum_{n \neq m} \frac{E_{nm}^2}{E_m + E_{mm} - E_n - E_{nn}} \qquad (14)$$

In the writer's opinion, it is somewhat artificial to separate out the behaviour of energy and wave functions, and it seems important to discuss two concepts, the steps (1) + (2) called here, simply, *first-order perturbations* and the steps (3) + (4) *second-order perturbations*.

The first-order perturbations are characterized by concerning the same wave function before and after the perturbation P in the integral (we use here an integral sign without variables for indicating an integration over all pertinent variables in their whole domain; other authors use the brackets $<\Psi_1|P|\Psi_1>$ proposed by Dirac). We are here yielding homage to the fact that P might contain operators, affecting the kinetic energy. The total energy of a normalized wave function is $\int \Psi_1 H \Psi_1$ where the Hamiltonian function H partly produces kinetic energy, necessitating in its definition the indication of a wave function before H and one after H; and partly potential energy, which is simply the integration of the product of Ψ_1^2 with a potential. In nearly all cases met in practice, P is only a potential. Thus the diagonal elements of energy are simply the integrals of $P\Psi_1^2$. This is the change of energy expected in classical theory of electricity by putting an extended charge distribution Ψ_1^2 down in a potential P (calculated for electrons, not for positive charges).

On the other hand, the second-order perturbations concern two wave functions Ψ_1 and Ψ_2. If P is a pure potential, the non-diagonal elements will be the integrals of $P\Psi_1\Psi_2$. If P was a constant, this integral would be zero, since $\int \Psi_1 \Psi_2$ of two orthogonal wave functions vanish. Thus, the non-diagonal elements are produced by differing values of P at the places in (the multi-dimensional) space, where $\Psi_1\Psi_2$ is positive and where it is negative. It is one of the principal results of group theory to indicate which non-diagonal elements must necessarily be zero. As shall be discussed later on, this occurs in most cases when the quantum numbers Γ_n assigned to each of the wave functions Ψ_1 and Ψ_2 are different.

Generally, the second-order perturbations change a given wave function Ψ_1 by intermixing it with small parts of other wave functions $\Psi_2, \Psi_3, \ldots$ with which it has non-diagonal elements. The influence on the wave function Ψ_1 is inversely proportional to the distance on the energy scale of the other wave functions, while the influence on the energy is proportional to the squares of the non-diagonal elements. On the other hand, only the energy and not the wave function itself is changed by a first-order perturbation. Thus, there is a close connexion between the chemical concepts of covalent bonding and second-order perturbations; and between electrovalent bonding and first-order perturbations. We shall justify the general use of this notation in the chapters on molecular orbitals and ligand field theory.

The perturbation theory is closely connected with the theory of *secular determinants*. If q different wave functions fulfil the necessary conditions for intermixing (having the same Γ_n, which, for example, may signify the same combination of L and S in spherical symmetry), the energies of the resulting mixed functions will be the q roots E (the *eigenvalues*) of the polynomium of qth degree, written as a symmetrical determinant:

$$\begin{vmatrix} E_{11} - E & E_{12} & E_{13} & \cdots & E_{1q} \\ E_{12} & E_{22} - E & E_{23} & \cdots & E_{2q} \\ E_{13} & E_{23} & E_{33} - E & \cdots & E_{3q} \\ \cdots & & \cdots & & \\ E_{1q} & E_{2q} & E_{3q} & & E_{qq} - E \end{vmatrix} = 0 \quad (15)$$

explaining the terms "diagonal elements" for E_{nn} and "non-diagonal elements" for E_{nm}. The *diagonal sum rule* for such determinants states that the sum of the q eigenvalues always equals the sum of the q diagonal elements.

If one of the diagonal elements, E_{nn}, is separated from all the other diagonal elements by an energy which is considerably greater than the order of magnitude of the non-diagonal elements, it will be a valid approximation that one of the roots E has the approximate value

$$E \cong E_{nn} + \sum_{m \neq n} \frac{E_{mn}^2}{E_{nn} - E_{mm}} \quad (16)$$

representing a sort of additive, repulsive potential, depressing E when E_{mm} is larger than E_{nn}, and increasing E in the opposite case. This is analogous to eqn. (14).

We may consider the simplest case with $q = 2$. The roots E of the determinant

$$\begin{vmatrix} E_{11} - E & E_{12} \\ E_{12} & E_{22} - E \end{vmatrix} = (E_{11} - E)(E_{22} - E) - E_{12}^2 = 0 \quad (17)$$

are

$$E = \frac{E_{11} + E_{22}}{2} \pm \sqrt{\left(\frac{E_{11} - E_{22}}{2}\right)^2 + E_{12}^2} \quad (18)$$

showing that the distance between the two roots is never smaller than $2E_{12}$, twice the non-diagonal element. The validity of eqn. (16) for E_{12} small compared to $(E_{11} - E_{22})$ can be seen by writing the correction $x = E_{11} - E_b = E_a - E_{22}$ (assuming the order $E_b < E_{11} < E_{22} < E_a$) according to eqn. (17):

$$x = \frac{E_{12}^2}{E_{22} - E_{11} + x} \quad (19)$$

It is often paradoxical to the non-theoretical chemist that "resonance", *intermixing* between two wave functions may decrease the resulting energy below even the lowest of the two original energies. This is connected with the energy E_{12}, which, in quantum mechanics, we must ascribe to the mixed product of wave functions $\Psi_1\Psi_2$. If we consider the formation of a linear combination

$$\Psi_b = b\Psi_1 + a\Psi_2 \quad (a^2 + b^2 = 1) \quad (20)$$

the other linear combination must be

$$\Psi_a = a\Psi_1 - b\Psi_2 \quad (21)$$

or the equivalent form, multiplied by -1, for being orthogonal on Ψ_a, and then the energy of the two states can be written

$$\begin{aligned} E_a &= a^2 E_{11} - 2ab E_{12} + b^2 E_{22} \\ E_b &= b^2 E_{11} + 2ab E_{12} + a^2 E_{22} \end{aligned} \quad (22)$$

where E_{11}, E_{12} and E_{22} are the energies assigned to the wave function products Ψ_1^2, $\Psi_1\Psi_2$, and Ψ_2^2. The coefficients a and b can be written as trigonometric substitutions

$$a = \sin \varphi \quad b = \cos \varphi \quad 0 \leq \varphi \leq \frac{\pi}{2} \qquad (23)$$

We may now apply the *variation principle*, choosing for a and b in eqn. (20) the values, resulting in the lowest energy E_b in eqn. (22). The energy minimum can be found by differentiation with respect to $\sin \varphi$ or $\cos \varphi$, but since these are monotonic functions of φ in the considered interval, it is just as useful to differentiate with respect to φ:

$$\frac{dE_b}{d\varphi} = 2 \cos \varphi \sin \varphi \, (E_{22} - E_{11}) + 2(\cos {}^2\varphi - \sin {}^2\varphi)E_{12} = 0 \quad (24)$$

giving the result

$$\frac{E_{11} - E_{22}}{E_{12}} = \frac{1 - 2 \sin {}^2\varphi}{\cos \varphi \sin \varphi} = -2 \cot 2\varphi \qquad (25)$$

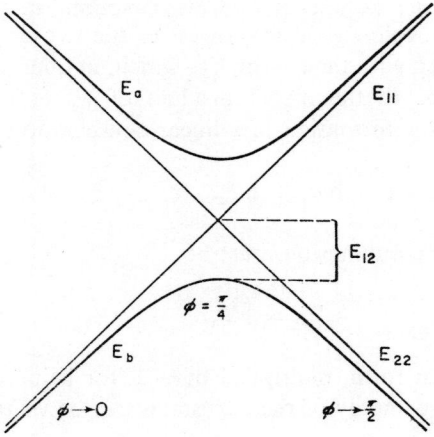

FIG. 1. The hyperbolic behaviour of two interacting levels. The diagonal elements are represented by the straight lines E_{11} and E_{22}, the non-diagonal element E_{12} is half the distance between the actual energy levels, E_a and E_b, produced by the intermixing, for the case $E_{11} = E_{22}$. The angle φ is given in eqn. (23).

Fig. 1 gives E_a and E_b as function of $(E_{11} - E_{22})$. It is a hyperbola with the straight lines E_{11} and E_{22} as asymptotes and the distance $2E_{12}$ between the two levels for $E_{11} - E_{22} = 0$, agreeing with eqn. (18). The stabilization of the lowest state by intermixing is given by the value of x from eqn. (19)

$$x = E_{11} - E_b = \frac{E_{12}^2}{E_{22} - E_b} = -\frac{a}{b} E_{12} = -E_{12} \operatorname{tg} \varphi \qquad (26)$$

Thus, the result of the intermixing of a small part, $a\Psi_2$, of a new wave function into Ψ_1 is exactly the opposite of that expected, when E_{12} is neglected. Instead of increasing the energy by the amount $a^2(E_{22} - E_{11})$ according to eqn. (22), the energy is actually decreased by approximately $-a^2(E_{22} - E_{11})$, since $2abE_{12}$ turns out to be about equal to $-2a^2(E_{22} - E_{11})$ for $a \ll b$.

The physical significance of all these arguments is not changed, if E_{12} is everywhere assumed to be multiplied by -1. This is connected with the fact that each of the q wave functions producing the secular determinant, eqn. (15) has an arbitrary sign. Thus, among the $q(q - 1)/2$ different non-diagonal elements, the $q - 1$ first (e.g. in the first row, corresponding to the signs of the products $\Psi_1\Psi_n$) have arbitrary signs. There exists only two different polynomiae E_p, corresponding to a certain third degree secular determinant with arbitrary choice of its three signs of the non-diagonal elements; and only eight different E_p's for $q = 4$. We shall not discuss further the complications of "choice of phases" for wave functions.

Until now, we have considered the diagonal or non-diagonal elements of a certain operator P in $\int \Psi_1 P \Psi_1$ and $\int \Psi_1 P \Psi_2$. We have not specified the number of electrons affected by P, which is highly important for the problem of how to express these elements in terms of P acting on orbitals ψ rather than on the total wave functions Ψ. Actually, since the potentials encountered, at least in electron systems, can be expressed as the sum of interactions between two particles, it is only necessary to consider *one-electron operators* and *two-electron operators*. The one-electron operators (determining, for example, orbital energies according to eqn. (5) as diagonal elements) take over all the results of the discussion above, substituting ψ for Ψ in all expressions. The only two-electron operator of consequence for our treatment is the electrostatic interaction between two electrons,

D

being a potential of the form $g = e^2/r_{12}$. The non-diagonal elements of this operator affect at most four orbitals and are expressed

$$(abgcd) = (ac:bd) = \int\int\psi_a(\tau_1)\psi_b(\tau_2)g\psi_c(\tau_1)\psi_d(\tau_2)d\tau_1 d\tau_2 \quad (27)$$

The first notation was used by Condon and Shortley for conforming to the traditional order of wave functions before and after the operator g. For the present purpose, it is preferable to separate the wave functions according to the order of the two integrations (with the sets of variables τ_1 and τ_2) and to write (ac:bd), since eqn. (27) simply expresses the electrostatic interaction between the two extended charge distributions $\psi_a\psi_c$ and $\psi_b\psi_d$.

We can find the elements of g, acting on two different electron configurations with an integral number of electrons (at most two, with different signs of the spin component $m_s = +\frac{1}{2}$ or $m_s = -\frac{1}{2}$) in each distinct orbital ψ_n. If the two configurations are different in more than two of the electrons, the non-diagonal element is zero. If the two electrons a^k and a^t in one configuration are substituted by the two electrons b^k and b^t in the other configuration, then the non-diagonal element is

$$\pm [(a^kb^k: a^tb^t) - (a^kb^t: a^tb^k)] \quad (28)$$

One of these two terms often vanishes since it is a necessary condition for (ac:bd) $\neq 0$ that a has the same value of m_s as c, *and* that b has the same value of m_s as d.

If the two configurations are different in only one electron, being a^k in the first and b^k in the second case, the non-diagonal element is a summation over the contributions from all the other electrons, a^t, which are the same in both configurations:

$$\pm \sum_t [(a^kb^k:a^ta^t) - (a^ka^t:b^ka^t)] \quad (29)$$

The signs in eqn. (28) and (29) are related to the choice of phases of the wave functions.

The diagonal element of a given configuration consists of a double summation

$$\sum_{k>t} [(a^ka^k:a^ta^t) - (a^ka^t:a^ka^t)] \quad (30)$$

Here, only two electrons are involved at a time, and it is convenient to define two types of integrals

$$\text{Coulomb integral } J(a, b) = (aa:bb)$$
$$\text{exchange integral } K(a, b) = (ab:ab) \tag{31}$$

where the first is the classical interaction between the two extended charge distributions ψ_a^2 and ψ_b^2 while the exchange integral is the less familiar interaction of $\psi_a\psi_b$ with itself.

It is not always possible to express the energies of the multiplet terms as multiples of J- and K-integrals alone, since the partly filled shells with l of at least 1 actually correspond to degenerate energies for several electron distributions on the individual orbitals, and therefore, non-diagonal elements of type eqn. (28) must necessarily be considered.

However, this is of no importance to the n,l-shells in spherical symmetry, since Condon and Shortley indicate methods of relating all their interelectronic repulsion energies to the parameters F^k, defined from the radial functions R

$$F^k = e^2 \int\limits_0^\infty \left[\int\limits_0^{r_2} \frac{r_1^k}{r_2^{k+1}} R^2 dr_1 + \int\limits_{r_2}^\infty \frac{r_2^k}{r_1^{k+1}} R^2 dr_1 \right] R^2 dr_2 \tag{32}$$

The integrations of the angular distributions A produce certain rational multiples of F^k for a given value of L and S. All the multiplet terms of a partly filled shell l^q contain a common contribution $F^0 q(q - 1)/2$. The following integrals, F^2, F^4, F^6, . . . (at most F^{2l}) determine the energy differences between the multiplet terms of the same configuration, assuming identical radial functions.

The present writer has previously discussed how F^k is a slowly decreasing function of k for any reasonable shape of the radial function R, and how the numerical values of F^k are related to the integral

$$W = e^2 \int\limits_0^\infty \frac{R^2}{r} dr \tag{33}$$

being a measure of the average value of r^{-1} of the partly filled shell. For a set of isomorphous wave functions, related to each other by pure scaling (see eqn. 10), the values of W and F^k are all proportional to this average value, and are thus a measure of the spatial extension of the partly filled shell.

Only one parameter, F^2, is necessary for the description of the energy differences between the multiplet terms of p^n-systems, while two parameters, F^2 and F^4, are necessary for d^n-systems, and three for f^n-systems. Linear combinations of these parameters are often constructed for simplifying the multiples involved in the energy expressions. Thus for d^n-systems, parameters with subscript k are defined

$$F^2 = 49F_2 \quad \text{and} \quad F^4 = 441F_4 \tag{34}$$

while more precise quantities have been recommended by Racah

$$A = F^0 - \frac{1}{9}F^4, \, B = \frac{1}{49}F^2 - \frac{5}{441}F^4, \, C = \frac{5}{63}F^4 \tag{35}$$

for obtaining the advantage that the energy differences between terms with the maximum value of S is a multiple of B alone, viz. $15B$ between ^3F and ^3P of d^2 and d^8 and between ^4F and ^4P of d^3 and d^7. Similar expressions have been constructed for f^n-systems, e.g. the parameter E^3 analogous to B is defined

$$E^3 = \frac{1}{135}F^2 + \frac{2}{1089}F^4 - \frac{175}{42471}F^6 \tag{36}$$

Table 2 gives some of the term distances, relative to the lowest term, of the d^n-systems. We may use this theory in two senses, a *strong sense*, where F^k is calculated according to eqn. (32) from the

TABLE 2. EXCITATION ENERGIES OF SOME MULTIPLET TERMS IN SPHERICAL SYMMETRY, EXPRESSED IN RACAH'S PARAMETERS OF INTERELECTRONIC REPULSION. $(C \sim 4B)$.

	d^2 and d^8	d^3 and d^7		d^5	
^3F	0	^4F	0	^6S	0
^1D	$5B + 2C$	^2G	$4B + 3C$	^4G	$10B + 5C$
^3P	$15B$	^4P	$15B$	^4P	$7B + 7C$
^1G	$12B + 2C$	^2P	$9B + 3C$	^4D	$17B + 5C$
^1S	$22B + 7C$	^2H	$9B + 3C$	^4F	$22B + 7C$

Hartree–Fock self-consistent (HFSC) radial functions R and compared with the experimental values (as calculated for the levels compiled in Charlotte Moore's tables of atomic energy levels), and a *weak sense*, where agreement with experiment is obtained by a reasonable choice of F^k as numerical parameters without any reference to the wave function. This choice is not completely free, since the values of F^k must be slowly decreasing with k. Previously, the levels with lower S than the ground state have been incorrectly identified in the lanthanides, due to the assumption of a very small value of F^6.

TABLE 3. THE DISTANCE BETWEEN ^3F AND ^3P (ACTUALLY BETWEEN THE LEVELS $^3F_3$ AND $^3P_1$, WHICH ARE NOT DISTURBED BY ELECTRODYNAMIC EFFECTS WITHIN d^n FOR ANY VALUE OF ζ_{nl}) IN SOME GASEOUS IONS. THE ENERGY UNIT IS $K = cm^{-1}$

Sc^+	$3d^2$	7218	Y^+	$4d^2$	5690
Ti^{2+}	$3d^2$	10,420	Zr^{2+}	$4d^2$	7644
V^{3+}	$3d^2$	12,920	Nb^{3+}	$4d^2$	9037
Cr^{4+}	$3d^2$	15,171	Mo^{4+}	$4d^2$	10,277
Mn^{5+}	$3d^2$	17,311	Zr	$4d^2 5s^2$	3806
Fe^{6+}	$3d^2$	19,361	Rh^+	$4d^8$	8114
Ti	$3d^2 4s^2$	8322	Pd^{2+}	$4d^8$	10,241
Co^+	$3d^8$	12,454	La^+	$5d^2$	4602
Ni^{2+}	$3d^8$	15,617	Hf	$5d^2 6s^2$	4216
Ni	$3d^8 4s^2$	14,402	Pt	$5d^8 6s^2$	8450
Cu^+	$3d^8 4s^2$	17,074	Au^+	$5d^8 6s^2$	6925 ?
			Ac^+	$6d^2$	4066
			Th^{2+}	$6d^2$	3820

Slater's theory in the weak sense is very successful for the gaseous d^n-ions (and the f^n-systems in chemical compounds). Table 3 shows the variation of the energy difference in d^2 and d^8, which should be $15B$, as function of ionic charge, the presence of s-electrons, and the number of the transition group (3d, 4d, 5d, 6d). Actually, the variation of the interelectronic repulsion parameters is very regular for a given electron configuration as function of the ionic charge $Z_0 - 1$, being proportional to $Z_0 + Z_c$, where Z_c is a small constant =

$$
\begin{array}{llll}
2s^2 2p^2 & 1\cdot 0 & 3s^2 3p^2 & 1\cdot 6 \\
2s^2 2p^3 & 1\cdot 5 & 3s^2 3p^3 & 2\cdot 3 \\
2s^2 2p^4 & 2\cdot 1 & 3s^2 3p^4 & 3\cdot 0
\end{array}
\tag{37}
$$

On the other hand, recent investigations have shown that Slater's theory is not correct in the strong sense, but that Hartree-functions invariably give 10–40 per cent larger values of F^k than the values derived from observation. Thus, the electron configurations cannot be considered as entirely pure with the same radial function for different terms of the same configuration. Dr. Klaus Appel, Kvant-kemiska Gruppen, Uppsala, has kindly calculated the values of W and F^k for the Hartree-function 3d of ^6S, Mn^{2+}. The following ratios are calculated:

$$\left.\begin{array}{l} W = 2\cdot3922 \text{ Rydberg } (\bar{r} = 1\cdot082\ a_0,\ \overline{1/r^{-1}} = 0\cdot8361\ a_0,\ \overline{r\cdot r^{-1}} = \\ \quad 1\cdot294).\ \text{(for hydrogenic radial functions, } \overline{r\cdot r^{-1}} = \\ \quad \dfrac{3}{2}\left(1 - \dfrac{l(l+1)}{3n^2}\right) = 1\cdot167 \text{ for 3d).} \\[2mm] F^0 = 1\cdot6996 \\ F^2 = 0\cdot7952 \text{ (observed } \sim 0\cdot612). \\ F^4 = 0\cdot4966 \text{ (observed } \sim 0\cdot423). \\ \text{HFSC} \qquad W/F^2 = 3\cdot01,\ F^2/F^4 = 1\cdot601 \text{ (obs. } \sim1\cdot45). \\ \text{Hydrogenic 2p } W/F^2 = 2\cdot84. \\ \text{Hydrogenic 3d } W/F^2 = 2\cdot45,\ F^2/F^4 = 1\cdot53. \\ \text{Hydrogenic 4f } W/F^2 = 2\cdot22,\ F^2/F^4 = 1\cdot50,\ F^4/F^6 = 1\cdot34. \end{array}\right\} \quad (38)$$

Much of this material supports the idea of eqn. (7) that HFSC radial function to some extent has the same shape as hydrogenic functions with a very small l.

The deviation from the strong sense theory may partly be explained by slightly expanded radial functions in the more excited terms of a given configuration. Dr. Per-Olov Löwdin, Uppsala, has pointed out the effect of the virial theorem (eqn. 8, p. 28) on eqn. (10), if the interelectronic repulsion energy is added* as $k_1\eta$ to E, and suitable units chosen:

* Eqn. (39) is often illustrated in text-books by the particular example of a two-electron system (such as helium) having two hydrogenic 1s-electrons with the effective charge Z_* (eqn. 6), where the orbital energy of each electron is $Z_*^2 - 2Z_*Z$ and the interelectronic repulsion energy $\frac{5}{8}Z_*$ in Rydberg units. By minimization of this expression with respect to Z_*, it is found $Z_* = Z - \frac{5}{16}$ and $E = -2Z^2 + \frac{5}{4}Z - \frac{25}{128}$. Thus, for $Z = 2$, the energy calculated is $-5\cdot696$ ry, in fair agreement with the observed value of $-5\cdot807$. It is interesting to compare the values of the ionization energy He $\rightarrow$ He$^+$ (observed value $1\cdot807$ ry) for the direct subtraction of the two cases ($Z_* = 2$ for He$^+$, $Z_* = 1\cdot6875$ for He), 434 ry/256 $= 1\cdot6953$ ry, with the calculation of $-E_{1s} + J(1s, 1s)$ retaining $Z_* = 1\cdot6875$, giving the value 459 ry/256 $= 1\cdot793$ ry. It is an interesting case of Koopman's theorem mentioned p. 243 that the ionization energies calculated without consideration of the change of the orbitals of the remaining electrons by ionization nevertheless are fairly accurate.

$$E = \eta^2 - 2\eta + k_1\eta \text{ has } \frac{dE}{d\eta} = 2\eta - 2 + k_1 = 0$$

$$\text{at } \eta = 1 - \frac{k_1}{2} \tag{39}$$

Thus, the energy difference between two terms, characterized by k_1 and k_2, is not $k_2 - k_1$ as expected from Slater's first-order perturbation theory, but

$$k_2 - k_1 - \frac{k_2^2}{4} + \frac{k_1^2}{4} \tag{40}$$

However, this effect is probably too small for explaining all the deviation observed.

This redistribution of kinetic and potential energy (eqn. 39) expresses the part of the *electronic correlation effect* which can be described by assuming different, but isomorphous, radial functions (and identical for each electron in the partly filled shell) for different terms of the same configuration. It does not express the part corresponding to a change of shape of the radial function (it will probably be small), nor the part corresponding to different radial functions for the electrons in the same shell (the *radial* correlation), nor finally the part corresponding to other angular functions than those defined by the usually assumed value of *l* (the *angular* correlation). An interesting problem arises as to whether the actual radial distribution of the partly filled shell of a given gaseous ion is not much different from the HFSC behaviour. The latter is defined to give the lowest energy possible of a single, pure configuration; but we saw in eqn. (10) that the orbital energy is fairly insensitive to a moderate variation of the average radius. It might be that the correlation effects in the inner shells decrease their average radii so much that the most loosely-bound, partly filled shell has a weaker central field and consequently is expanded. Hence, the deviations between semi-empirical and HFSC-calculated values of the interelectronic repulsion parameters (eqn. 38) might be partly explained by a deviation of the partly filled shell (defined, for example, from the density of uncompensated electron spin) in the actual ion from the HFSC-radial function.

Actually, the radial and angular correlation effects invalidate to some extent the whole concept of electron configurations, since a given configuration inevitably has some non-vanishing non-diagonal elements of the two-electron operator g with other configurations, and therefore, the actual wave functions must be intermixings of many configurations.

The example most studied is the two-electron system helium. Many calculations since the time of Hylleraas have definitely shown that even the best (HFSC) radial function in a configuration $1s^2$ gives an energy 98·7 per cent of the observed value. If radial correlation with all other configurations such as $2s^2$, $3s^2$, . . . having the same l, is introduced, about half of the remaining energy is secured, giving 99·5 per cent of the observed value. Finally, angular correlation may be allowed for by intermixing with $2p^2$, $3p^2$, . . ., $3d^2$, . . . etc., reproducing the observed energy with a very high accuracy. The quickest convergence of this configuration interaction is obtained, if the intermixed configurations have orbitals with well-defined l, but radial functions not corresponding to any stationary state, and having large K-integrals (eqns. 28 and 31) with the orbital 1s. These "virtual" orbitals belong partly to the continuum of states, having higher energy of a given atom than necessary for the ionization of at least one electron to infinity. Virtual states may sometimes be rather well-defined in atomic spectroscopy; the first example discovered was the configuration 3d4d of neutral Ca, containing more energy than the ionized 4s of Ca^+. We shall not discuss the auto-ionization processes and the behaviour of such virtual states here, but we may remember that very stable objects may be in a virtual state (e.g. the α-radioactive uranium nucleus is unstable against part-dissociation, but it has a half-life of milliards of years).

It might seem a rather good performance of the HFSC-functions that ~99 per cent of the energy of the ground state of He can be explained by a single configuration $1s^2$. However, the rest of it, the *correlation energy* (defined as the difference between the lowest obtainable HFSC-energy with one configuration and the observed value, corrected for relativistic effects) is not small compared to the usual chemical bonding energies in molecules. If it changes by the formation of molecules from atoms, it will most severely influence the bond energies. The correlation energy is remarkably constant, 9·2 to 9·9 kK, in the isoelectronic series He, Li^+, Be^{2+}, B^{3+}, C^{4+}. This may qualitatively be explained as a second-order perturbation effect, including a summation of quantities, where the non-diagonal elements are proportional to Z and the distances to the perturbing levels slightly below and slightly above the lower limit of the continuum are proportional to Z^2. According to eqn. (16), this will produce an energy decrease proportional to Z^2/Z^2, i.e. mainly invariant with Z.

Since the correlation energy in the hydrogen molecule is now known to have the same magnitude as in He, we may still hope that the invariance with electron number also holds in the case of chemical bonding. Dr. Anders Fröman, Uppsala, has calculated the correlation energy in gaseous F^- and Al^{3+} to be 87 kK, by coincidence nearly the same value as found in the isoelectronic molecule HF. It is surprising for a chemist to know that a neon atom is stabilized with 250 kcal/mole more than would have been possible, if it had the lowest energy compatible with a pure inert-gas configuration $1s^2 2s^2 2p^6$.

However, we shall proceed in this book to show that all observations are in agreement with the assumption of relatively pure molecular orbital configurations in complexes corresponding to the case of relatively pure electron configurations in gaseous atoms and ions. We must here distinguish the absolute validity of configurations as descriptions of the wave functions Ψ and their usefulness as *classification*. Group theory never appropriates the former point of view, only the "capital letter" quantum numbers of a system, such as L and S in spherical symmetry with Russell–Saunders' coupling, are strictly defined; the distribution of electrons on orbitals may or may not be a good approximation, but it is always an approximation. Since the number of states does not change by intermixing, it is obvious that a classification according to configurations is useful, if all or most of the wave functions are linear combinations, "superpositions" of configurations, among which one has by far the largest weight.

Since the configurations are intermixed, due to non-diagonal elements of interelectronic repulsion, the possibility of the classification is dependent on whether the energy differences between the molecular orbitals are reasonably large, compared to the parameters of interelectronic repulsion. In some cases, the configurations may be badly defined due to *accidental degeneracy of orbitals*, e.g. in neutral (but not highly ionized) atoms of the first transition group, the orbitals 3d and 4s have nearly the same energy, and therefore terms with same L and S from the configurations $3d^n 4s^2$, $3d^{n+1} 4s$

and $3d^{n+2}$ are highly intermixed. But excepting these somewhat extraordinary cases, the validity of configurations will be highly dependent on whether we have chosen the orbitals in the most appropriate way. There is a fundamental difference between the non-diagonal elements of eqn. (28) and eqn. (29). The former always subsist, even with the best choice of the orbitals, while the latter actually behave almost as one-electron operator quantities, ("mixing" the orbitals a^k and b^k), being negligible if the orbitals are self-consistent. In the next chapter we will discuss how the concept of the effective microsymmetry is important to the question of whether the energy differences between a set of nearly degenerate molecular orbitals are sufficiently larger than the parameters of interelectronic repulsion to secure a sound basis for the classification according to configurations.

Perhaps it will be possible to find some approximate rules for the effects of correlation, the second-order perturbations of configurations. As seen above, there seems to be a general effect (eqn. 40) decreasing all term distances, and the more the higher the term is excited. In addition to this effect, the angular correlation, mixing low L-values with configurations, containing orbitals with lower l, seems to decrease the energy of low L-values more than of high L-values. The two effects are not easy to separate empirically, since the excited terms of a configuration, having large first-order perturbation energies in the Slater model, are simply those with low values of L and S. Racah and Trees have proposed a positive correction $\alpha L(L+1)$ to the term energies and have obtained considerably better agreement in the "weak sense" with experiments by assuming values of α about 0.1 kK. However, the correction may also be negative, proportional to $1/(2L+1)$ or a similar quantity.

Even though these corrections make it impossible to draw direct conclusions from the "weak-sense" values of F^k to the geometrical extension of a partly filled shell (the average value of r^{-1}), it seems justified to assume the correction has nearly the same relative effects in all atoms and ions, and especially, that the evolution of F^k in an isoelectronic series is a fair measure of increasing scaling parameter of the radial function.

It is often assumed that the correlation energy is mainly caused by the electrons having opposite spin directions in the configuration. The argument is that the anti-symmetrization of the wave function, acting on the space co-ordinates in the case of parallel spins, creates a "Fermi-hole" around each electron, preventing other electrons with the same spin direction from coming close to it. Consequently, the interelectronic repulsion is less effective than it would be in the case of a nearly constant density of other electrons close to our electron, as expressed by the subtraction of the K-integral (eqn. 31) in the energy of the electron pair a^+b^+. It is now argued that a similar "Coulomb-hole" is created around each electron, also preventing electrons with parallel spin from coming too close. This situation destroys the validity of pure configurations. We might express this idea in a phenomenological way by subtracting new parameters, $-K^*(ab)$ in the case of a^+b^- and a^-b^+ and $-K^*(aa)$ in the case of a^+a^-.

However, it is dubious whether the two kinds of hole are comparable, since the electrostatic interaction e^2/r_{12} is actually not a "short range force". In particular, the electrostatic energy from a nearly constant density of other electrons does not present any singularity in $r_{12} = 0$, but is even proportional to $r_{12}dr$ per spherical shell with the thickness dr around it. Hence, it is not certain that a^+b^+ has a much lower correlation energy than a^+b^-.

In the $3d^n$-configurations of gaseous ions, the distances between the terms with maximum value of S are just as decreased, compared to pure HFSC-configurations, as the distances to lower values of S. Since the former distances would not be affected by parameters of the type $K^*(ab)$, they do not express the whole truth. Watson (1959) found (by using a somewhat different set of experimental values of F^2 than in eqn. (38)) that the ratio between F^2 for HFSC and the observed value is near 1·35 for the ionic charge $+1$, 1·24 for $+2$, 1·20 for $+3$ and 1·17 for $+4$. The values for neutral atoms vary largely from 2·2 (Sc) to 1·6 (Co) but refer to highly excited configurations (without 4s-electrons) in these atoms. While the calculated values of F^2 vary between 0·43 ry (Sc^+) and 1·05 ry (Ni^{4+}) for the ions, the experimental values are all smaller by nearly the same amount, 0·18 ry. This corresponds to a certain "retardation" in the ionic charge, the HFSC of a given ion resembling the experimental value of a higher charge.

Löwdin and Shull demonstrated that most correlation in the He-like systems could be explained by the intermixing of $1s^2$ with small amounts of $(\alpha s)^2$ and $(\alpha p)^2$, αs and αp representing the "virtual" orbitals from the continuum, discussed above. The configurations $1s^2 2s^2 2p^n$ are expected to intermix with $1s^2 2p^{n+2}$ and $1s^2 2s^2 2p^{n-2}(\alpha s)^2$, among other configurations. In both cases, it can be shown that the terms ^{1}S will be depressed 3K* and ^{2}P 2K*, relative to the other terms, where K* is proportional to the squares of K(2s, 2p) and K(2p, αs). This produces a deviation from the Slater ratios in eqn. (77) However, the decreased value of these interval ratios in the iso-electronic series C, N$^+$, . . . is connected with the interaction with $1s^22p^{n+2}$ rather than with "virtual" configurations, since the ratio is remarkably constant for increasing ionic charge. In ions like O^{4+} and F^{5+}, the highest term ^{1}S of $1s^22p^2$ is strongly increased in energy, caused by interaction with the ground state ^{1}S of $1s^22s^2$. In these ions, K(2s, 2p) can be estimated as half the distance between ^{3}P and ^{1}P of $1s^22s2p$ and turns out to vary as a linear function of the ionic charge (in kK for Be 10·3, B$^+$ 18·0, C^{2+} 25·0, . . ., Na^{7+} 58·2, . . .) while the orbital energy difference between 2s and 2p also is a linear function of the ionic charge. Hence, the deviations from the Slater ratios eqn. (77) are due to an accidental effect of a not too distant configuration and are not connected intimately with the main problem of correlation energy.

Until now, we have not mentioned the effects of electrodynamic (magnetic) and relativistic corrections on the energy levels of atomic spectroscopy. These effects can be ascribed in Slater's model to one-electron quantities, the *Landé multiplet splitting factors* ζ_{nl}, defined from the central field U(r) and the radial function R:

$$\zeta_{nl} = 5 \cdot 8\text{K} \cdot \int\limits_0^\infty \frac{R^2}{r} \frac{dU(r)}{dr} \, dr \qquad (41)$$

The occurrence of the gradient of the central field in the integrals makes it much more difficult to control the "strong sense" validity of eqn. (41) than the similar eqn. (32) for F^k, the latter being dependent only on purely geometrical quantities related to R and not directly on U(r) (even though R of course is determined in principle by eqn. (4) from U(r)). Many physicists have utilized Sommerfeld's formula for ζ_{nl}, assuming hydrogenic radial functions. This is a rather dubious procedure, especially since the normalization factor of r^{l+1} near to the nucleus is highly dependent on the rest of the wave function. Since the gradient of U(r) is so large near to the nucleus, many authors have assumed that the average value of r^{-3} can be estimated from the value of ζ_{nl}. This is of importance for the interpretation of hyperfine structure, introduced in atomic spectro-

scopy and in paramagnetic resonance experiments from the interaction of the electric and magnetic moments of the nucleus with the
electron cloud, especially of partly filled shells. However, these
numerical interpretations have not been proved as yet by comparison
with actual HFSC-functions and eqn. (41).

On the other hand, ζ_{nl} has a large success as a "weak sense"
quantity, describing the splitting of L,S-multiplet terms into their
levels characterized by J. In Russell–Saunders coupling, the first-
order effect on a given term can be calculated as a rational multiple
of ζ_{nl} of the partly filled shell or shells. Notably, the terms with
maximum value of S of one partly filled shell l^q have a distance
between the level J and the level $J + 1$, which is $\zeta_{nl}/2S$ (J has the
lowest energy for q $< 2l + 1$ and the highest energy "inverted
multiplet terms", for q $> 2l + 1$). Since the J-levels of given multiplet terms have

$$J = L + S, \quad L + S - 1, \quad L + S - 2, \ldots, \quad |L - S| \quad (42)$$

the total first-order width of a multiplet term with maximum
S (and $L > S$) is $(L + \tfrac{1}{2})\zeta_{nl}$. Magnetochemists are often interested
in the quantity $\pm\zeta_{nl}/2S$ denoted by A. So far the J-levels can still
be assigned significantly to a given multiplet term, and they are
indicated as a right-hand subscript. Thus, the excited configuration
[Ne]3p of sodium corresponds to the two energy levels $^2P_{1/2}$ and
$^2P_{3/2}$ producing the well-known doublet structure of the yellow
spectral lines, while similarly, the excited configuration [78]6s6p of
mercury produces the four levels $^3P_0$, $^3P_1$, $^3P_2$ and $^1P_1$.

When ζ_{nl} is large, we have conspicuous effects of *intermediate
coupling*, i.e. the intermixing of different values of S (and L) for the
same value of J. Thus, it is hardly significant in mercury (with
$\zeta_{6p} = 4\cdot26$ kK) to assign one of the two multiplet terms ^3P and ^1P to
the two levels with $J = 1$. The intermediate coupling is described by
secular determinants (eqn. 15) having as diagonal elements the
energies of the multiplet terms (including a rational multiple of
ζ_{nl}) and as non-diagonal elements (between energy levels with the
same J) complicated multiples of ζ_{nl}.

It is known empirically that ζ_{nl} is approximately proportional to
$Z^2Z_0^2$, increasing strongly with the atomic number Z and the ionic
charge plus 1, Z_0. The effects of intermediate coupling are important for the absorption spectra of $4f^n$ (lanthanides) and $5f^n$ (actinides).
This is not only because the energy levels are shifting, but also

because transitions between states of different S are forbidden. The latter rule is weakened to the same extent as intermediate coupling mixes the values of S, and the intensity ratio between spin-forbidden and spin-allowed bands is therefore proportional to ζ_{nl}^2.

In the same way as the parameters of interelectronic repulsion can be used for defining an effective charge Z_* (cf. eqn. 37), it is possible to write ζ_{nl} as a quadratic function $k_{nl}Z_*^2$, where the proportionality constant k_{nl} varies surprisingly little within a given p- or d-shell. In this case, Z_* is a linear function of the number of electrons q in the partly filled shell, being $Z_* = Z_0 + 0.5 + 0.7q$ for $4p^q$ and $Z_* = Z_0 - 0.8 + 0.8q$ for $4d^q$. The factors are $k_{4p} \sim 0.1$ kK and $k_{4d} \sim 0.029$ kK, respectively. Z_0 is the ionic charge plus 1. The fundamental reason why ζ_{nl} is proportional to Z_*^2 is not at all as well understood as why B is proportional to Z_*.

Actually, the intermediate coupling is less important in the complexes of the three transition groups than might have been expected. In the gaseous ions with charge $+2$ or $+3$, ζ_{3d} in the first transition group varies between 0.1 and 0.9 kK, while ζ_{4d} varies from 0.3 to 2.3 kK and ζ_{5d} of the third transition group increases up to some 6 kK. However, in complexes, these values are decreased, compared to the gaseous ions, and the relative intensities of spin-forbidden bands indicate values about 0.8 kK for $4d^n$- and 2 kK for $5d^n$-complexes. Recently, Tanabe and Kamimura, and Griffith, and Liehr and Ballhausen, have calculated some determinants of intermediate coupling in octahedral complexes.

Moffitt, Goodman, Fred and Weinstock discussed recently (1959) the ligand field spectra of the gaseous molecules ReF_6, OsF_6, IrF_6 and PtF_6, and found in all cases the narrow lines compatible with internal transitions in the 5d-shell, having $\zeta_{5d} = 3.4$ kK. The corresponding values can be approximately extrapolated in the gaseous ions to increase from 4 kK in Re^{6+} to some 10 kK for Pt^{6+}. The decrease of ζ_{5d} is less accentuated than that of the interelectronic repulsion parameters to be discussed in the section on the nephelauxetic series (the ratio being 0.47 for OsF_6, 0.38 for IrF_6 and 0.30 for PtF_6). Some values are compiled in Table 9, p. 84.

As will be discussed in Chapter 10, the effects of large ζ_{np} are even more conspicuous in the series Tl(I), Pb(II), Bi(III), having transitions which in the gaseous ions would be $6s^2 \rightarrow 6s6p$, and in the electron transfer spectra of halide complexes, where in the gaseous atoms ζ_{np} is strongly increasing with Z (eqn. 151).

CHAPTER 4

MOLECULAR ORBITALS AND
MICROSYMMETRY

GROUP theory tells us the conditions for an energy level to be invariant under changes of the co-ordinate system (rotations, etc.) which do not change the physical properties of the electron system considered. Thus, the $(2L + 1)$-fold degeneracy of multiplet terms and $(2l + 1)$-fold degeneracy of orbitals is closely connected with the assumption of spherical symmetry, surrounding a gaseous atom or ion. In spherical symmetry, the co-ordinate axes can have any direction without introducing any physically significant differences.

In a molecule, containing more than one nucleus, the symmetry is degraded from the extremely high spherical one, to a lower symmetry. We shall here mainly consider three relatively high symmetries, the *cubic-octahedral* O_h, the *linear symmetry with centre of inversion* $D_{\infty h}$, and the *linear symmetry without centre of inversion* $C_{\infty v}$. These three cases can be exemplified by a regularly octahedral complex ML_6, by a homonuclear, diatomic molecule A_2, and a heteronuclear, diatomic molecule AB.

Unfortunately, it is not possible to describe here the theory of character tables and irreducible representations which is admirably presented in several text-books on group theory. We shall content ourselves to mention that the quantum number L in the Russell–Saunders coupling in spherical symmetry is substituted by Γ_n in the lower symmetries, and that J is substituted by Γ_J, while S preserves its denotation. The orbitals, characterized by n and l in spherical symmetry, are substituted by *molecular orbitals* (M.O.) having a file number n and a definite value of γ_n, having the same possibilities as Γ_n.

Bethe discussed (1929) the behaviour of terms L and orbitals l in O_h and many other, lower symmetries. Any given entity L or l corresponds to a set of one or more values of Γ_n or γ_n. On the other hand, a given wave function Ψ in the low symmetry, having a definite Γ_n, does not necessarily correspond to a definite value of L, belonging to the spherical symmetry, which is no longer present in the molecular complex. A given orbital ψ does not either necessarily have a well-defined l_1 even though it has a definite γ_n. Moreover, the *M.O. description of covalent bonding* is that the orbitals of different atoms, having the same γ_n, are intermixed. If only the *bonding* orbitals of lower energy formed are filled, while the *anti-bonding* orbitals are partly or completely empty, a contribution to stabilization of the molecule is obtained.

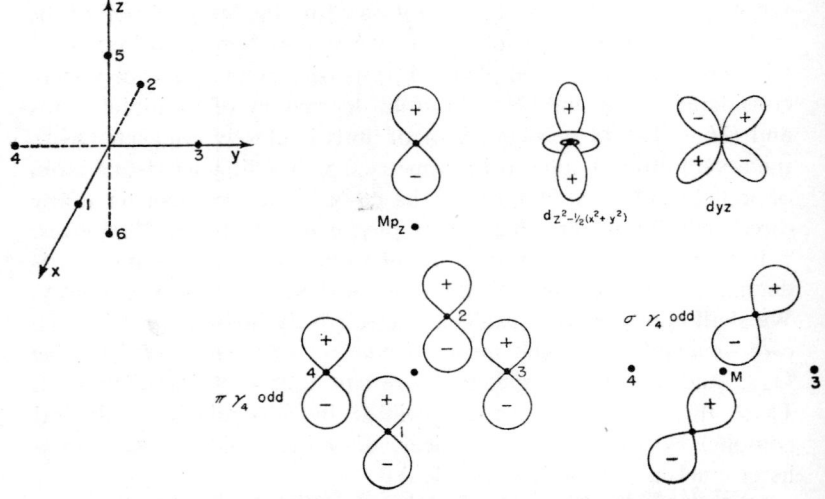

FIG. 2. The co-ordinate system to be used for octahedral complexes. One p and two d orbitals of the central atom M are indicated, and also two of the linear combinations of the ligands' orbitals from Table 4. The $+$ and $-$ indicate the signs of the one-electron wave functions.

The values of Γ_n possible in the cubic-octahedral symmetry are ten. The *parity* is even or odd and combined with one of five values, for which Mulliken has proposed other names, according to their degeneracy number e:

$$
\begin{array}{lccccc}
\text{Bethe:} & \Gamma_1 & \Gamma_2 & \Gamma_3 & \Gamma_4 & \Gamma_5 \\
\text{Mulliken:} & A_1 & A_2 & E & T_1 & T_2 \\
e = & 1 & 1 & 2 & 3 & 3
\end{array}
\qquad (43)
$$

and the orbitals have the analogous names γ_n (or a_1, a_2, . . .). Some English authors use also the five letters α, β, γ, δ, ϵ, talking, for example, about $d\gamma$ and $d\epsilon$ electrons for γ_3 and γ_5. The products of different L-values are:

S $\quad \Gamma_1$ $\qquad$ G $\quad \Gamma_1 + \Gamma_3 + \Gamma_4 + \Gamma_5$

P $\quad \Gamma_4$ $\qquad$ H $\quad \Gamma_3 + 2\Gamma_4 + \Gamma_5$ $\qquad$ (44)

D $\quad \Gamma_3 + \Gamma_5$ $\qquad$ I $\quad \Gamma_1 + \Gamma_2 + \Gamma_3 + \Gamma_4 + 2\Gamma_5$

F $\quad \Gamma_2 + \Gamma_4 + \Gamma_5$ $\qquad$ . . .

combined with even or odd parity (as also in spherical symmetry, where, for instance, ^{4}S from p^3 has odd parity, since $3l = 3$ is odd). The orbitals are even for even values of l and odd for odd values, having the same distribution of γ_n-values as Γ_n in eqn. (44). Thus, the d-shell in spherical symmetry is divided into two *sub-shells*, even γ_5 and even γ_3. We may talk about *sub-shell configurations*, assigning an integral number a between 0 and $2e = 6$ to the first sub-shell and a number b between 0 and 4 to the latter sub-shell, writing $\gamma_5{}^a\gamma_3{}^b$.

Even though we do not propose to reproduce the arguments for the structure of eqn. (44), many of the reasons for it can be seen directly from the Table 1. It is obvious that the three angular functions x, y and z, corresponding to three p-orbitals, are equivalent in O$_h$ where the naming of the three Cartesian axes has no physical influence. If we consider a lower symmetry, where one of the three axes is distinct from the two others, we obtain a splitting of the p-shell into more sub-shells. It is also obvious that the three d-orbitals xy, xz and yz of Table 1 have the same energy in a complex, where the names of the Cartesian axes can be cyclically permutated. They form the sub-shell even γ_5. It is not quite as obvious that the two last d-orbitals are quivalent in O$_h$. But if we consider their charge distributions, i.e. the squares of the wave functions in Table 1, these are (with the radical function R defined p. 25)

$$\left. \begin{aligned} \psi^2(\gamma_3\alpha) &= \tfrac{15}{4}(x^4 + y^4 - 2x^2y^2)r^{-4}R^2 \\ \psi^2(\gamma_3\beta) &= \tfrac{5}{4}(4z^4 + x^4 + y^4 - 4x^2z^2 - 4y^2z^2 + 2x^2y^2)r^{-4}R^2 \end{aligned} \right\} \quad (45)$$

If these squares are multiplied by a potential V$_{oct}$, independent of the naming of the three axes, both orbitals α and β will be subject to the same perturbation, giving the diagonal element $\tfrac{15}{2}[E(x^4) - E(x^2y^2)]$ in both cases. $E(x^4)$ is the perturbation energy on an arbitrary of the fourth-power terms in eqn. (45), and $E(x^2y^2)$ the perturbation energy on every one of the mixed square-products. The perturbation of V$_{oct}$ acting on one of the γ_5-orbitals in Table 1 gives an energy $15E(x^2y^2)$. Thus, we may define the *orbital energy difference* Δ between the γ_3-sub-shell and the γ_5-sub-shell in the

E

electrostatic model (i.e. considering only first-order perturbations) by the equation

$$\Delta = E(\gamma_3) - E(\gamma_5) = \tfrac{5}{2}[E(x^4) - 3E(x^2y^2)] \qquad (46)$$

Van Vleck, Schlapp and Penney denoted Δ by 10Dq, referring to a certain model of perturbing charges. Ilse and Hartmann later introduced the symbol $(E_1 - E_2)$ for this energy difference. We shall here use the more convenient letter Δ. If the ligands in an octahedral complex are approximated by negative point charges, or by dipoles, turning the negative end towards the central ion, it can be seen that Δ of eqn. (46) is positive. The charge distribution, corresponding to the sum of the three squared wave functions of γ_5 is (Table 1):

$$\Sigma\psi^2(\gamma_5) = 15(x^2y^2 + x^2z^2 + y^2z^2)r^{-4}R^2 \qquad (47)$$

concentrated in the eight directions between the six ligands and thus having a low potential energy in the ligand field, while the two γ_3 of eqn. (45) gives a charge distribution

$$\Sigma\psi^2(\gamma_3) = 5(x^4 + y^4 + z^4 - x^2y^2 - x^2z^2 - y^2z^2)r^{-4}R^2 \qquad (48)$$

directed against the six ligands. It may be remarked that the sum of the two sub-shells eqn. (47) and (48), corresponding to the complete d-shell of spherical symmetry, is

$$5(x^2 + y^2 + z^2)^2r^{-4}R^2 \qquad (49)$$

having complete spherical symmetry without any angular preferences. This is also true, for example, for the three p-orbitals, giving together a charge distribution $3(x^2 + y^2 + z^2)r^{-2}R^2$. It is erroneous to believe (as chemists sometimes do) that the p-shell has to some extent a tendency to be directed along the three axes in a Cartesian co-ordinate system, though the three individual orbitals themselves can be made to point in these three directions.

In the electrostatic model, it is convenient to divide the ligand field V into components of decreasing symmetry. Especially, an octahedral field may be divided into two components, V_0 of spherical symmetry, and V_{oct} that part of "genuine" octahedral symmetry, which cannot be incorporated in V_0. If V is expanded in a series near to the central ion in powers of r^n/r_L^{n+1}, it can be shown that

$$V_0 = 6q \cdot \frac{1}{r_L}$$

$$V_{oct} = \tfrac{35}{4}q(x^4 + y^4 + z^4 - \tfrac{3}{5}r^4)r_L^{-5} + \ldots \tag{50}$$

from six charges q at the distance r_L from the central nucleus. With this definition, the *baricentre rule* (or rule of centres of gravity) is preserved for the perturbations from V_{oct} so that the complete d-shell, consisting of six γ_5- and four γ_3-electrons, is only affected by V_0 but not by V_{oct}. Thus, the deviations from the mean energy of the d-shell in V_0 (weighted by the degeneracy numbers e of the orbitals) are $+\tfrac{3}{5}\Delta$ for γ_3 and $-\tfrac{2}{5}\Delta$ for γ_5. The first term of the series for V_{oct} in eqn. (50) gives a contribution (in Rydberg units) to

$$\Delta = \tfrac{5}{3}q \cdot \frac{\overline{r^4}}{r_L^5} \tag{51}$$

where $\overline{r^4}$ is the mean value of the fourth power of the distance from the central nucleus to the electron in the partly filled shell. However, the series in eqn. (50) only converges quickly, if $\overline{r^2}/r_L^2$ is a small quantity. Actually, the quantitative features of the electrostatic model can be shown to be entirely incorrect; the deviations from O_h occurring in complexes of tetragonal symmetry (*trans*-$Co(NH_3)_4Cl_2^+$) can only be explained by the electrostatic model, if $\overline{r^2}/r_L^3$ (giving a large contribution to V_{tetr}) was smaller than $\overline{r^4}/r_L^5$; and the small splittings of levels in lanthanides imply sometimes that $\overline{r_2}/r_L^3$ is much smaller than $\overline{r^6}/r_L^7$. This is one of the many reasons why it is certain that partly covalent bonding (second-order perturbations) are important in the complexes. The strongest arguments for this will be given in the section on the nephelauxetic series, discussing the variation of the interelectronic repulsion parameters in complexes, compared to the gaseous ion.

The characteristic feature of the electrostatic model, as compared with the more general M.O. theory is that in the former the wave functions of the partly filled shell are only new linear combinations of the wave functions for the gaseous ion, while by allowing second-order perturbations, we are altering the basis of wave functions to be used. We may express this in terms of a *Hilbert space*, generally of infinite dimension, where the wave functions are represented by vectors. The cosine of the angle between two vectors Ψ_1 and Ψ_2 is the *overlap integral*, $\int \Psi_1 \Psi_2$, being zero for orthogonal wave functions. Since the cosine of a right angle is zero, this is one meaning of the word "orthogonal". The eqn. (20) and (21), used previously for the description of the intermixing of two wave functions, actually have the same form as those describing a rotation of a planar co-ordinate system in analytical geometry. The angle φ used in eqn. (23) as a mathematical aid is actually the angle in the Hilbert space between the new and the old wave functions. In the same way, the intermixing described by a secular determinant of qth degree is the result of rotating q orthogonal axes in a q-dimensional sub-space of the Hilbert space. Thus, by forming covalent bonds, the d-shell of a transition group ion has its wave functions turned partly "out of plane" with their original sub-space. We may also say that the result of the correlation effects, discussed in some detail above, is to turn the wave function away from the direction of a pure configuration into a new direction, described by very many small deviations from zero of the infinite number of other co-ordinates, each indicating the direction of a pure configuration.

It is often useful to consider an intermediate case between the electrostatic model and the general M.O. theory, i.e. the *expanded radial function* model. All the arguments from the electrostatic model can be re-stated, using another radial function R, than that of the gaseous ion. The major advantage of this treatment is that we get a working hypothesis for the decrease of the interelectronic repulsion parameters, and that we may introduce the idea that the radial function of the d-shell is accommodated to another (less positive) effective charge of the central ion besides that indicated by the oxidation number.

Before discussing the covalent bonding in transition group complexes, it is useful to study the linear symmetries, at first $D_{\infty h}$ with a centre of inversion, as exemplified by a diatomic homonuclear molecule A_2. Since the electronic spectra of diatomic molecules have attracted so much interest (cf. the great treatise by G. Herzberg) a set of symbols for the quantum numbers has been used extensively. The concept analogous to L in spherical and Γ_n in octahedral symmetry is Λ, being a non-negative integer. A given value of L corresponds to all of the values $\Lambda = 0, 1, 2, \ldots, L$. Of these, $\Lambda = 0$

has the degeneracy number $e = 1$, while all the positive values have $e = 2$. The spectroscopic trivial names are:

$$\Lambda = 0 \quad 1 \quad 2 \quad 3 \quad 4 \quad 5 \ldots$$
$$\Sigma \quad \Pi \quad \Delta \quad \Phi \quad \Gamma \quad H \tag{52}$$

The levels Σ exist in two types: Σ^+ and Σ^- according to invariance or a change of sign of the wave function by making the mirror image in a plane, containing the molecular axis. The analogy of J is Ω, having the values $\Lambda - S, \Lambda - S + 1, \ldots, \Lambda + S$. Contrary to J, Ω may thus be negative.

For orbitals, l from spherical and γ_n from octahedral symmetry are substituted by λ. The values of λ have trivial names, analogous to eqn. (52), thus σ-electrons for $\lambda = 0$, π-electrons for $\lambda = 1$, δ-electrons for $\lambda = 2$. A given l-shell is split by a linear ligand field in all its components $\lambda = 0, 1, 2, \ldots, l$. By introducing electrodynamic effects, the σ-sub-shell is not split and has $\omega = \frac{1}{2}$. The other sub-shells have two levels, $\omega = \lambda - \frac{1}{2}$ with the energy $-\lambda/2 \, \zeta_{nl}$ and $\omega = \lambda + \frac{1}{2}$ with $+\lambda/2 \, \zeta_{nl}$.

We may arrange the symmetries in hierarchies, starting with the spherical symmetry as the highest member, and descending towards lower symmetries. Thus, the tetragonal symmetry D_{4h} with the x- and y-axes equivalent, but different from the z-axis, is lower than the cubic-octahedral symmetry O_h, where all three axes are equivalent; and it is higher than the orthorhombic symmetry D_2, where all three axes are different. But we cannot arrange all symmetries in one well-ordered hierarchy; we cannot say unambiguously whether O_h or $D_{\infty h}$ is the highest symmetry. In a certain sense, O_h with the equivalency of the three Cartesian axes is the highest symmetry; but on the other hand, $D_{\infty h}$ has retained the freedom of arbitrarily small or large rotations in one direction; i.e. around the axis of linear symmetry. There is a close connexion between this property and the existence of an infinite number of quantities λ analogous to the finite number of γ_n in most other symmetries.

If we take the z-axis in the orbitals of Table 1 to be the linear ("cylindrical") axis, the one s-orbital is a σ-orbital, as also the p-orbital z. The two others, x and y, are π-orbitals. Among the d-orbitals, $z^2 - \frac{1}{2}x^2 - \frac{1}{2}y^2$ is a σ-orbital, xz and yz both π-orbitals, and finally $x^2 - y^2$ and xy are δ-orbitals. It is seen that within the two latter pairs of equivalent orbitals, xz can be transformed into yz by a 90° rotation around the z-axis, and $x^2 - y^2$ into xy by a 45° rotation. The charge density, i.e. the sum of the squared π-orbitals, and also the sum of the squared δ-orbitals, has rotational symmetry around the linear axis.

TABLE 4. LINEAR COMBINATION OF THE LIGANDS' ORBITALS, PARTICIPATING IN THE COVALENT BONDING OF AN OCTAHEDRAL COMPLEX. THE NUMBERING OF THE SIX LIGANDS IS FROM FIG. 2. N IS THE NORMALIZATION FACTOR, IF OVERLAP INTEGRALS ARE NEGLECTED

		N	Reacting with central ion orbital
σ-orbitals			
even γ_1	$\sigma_1 + \sigma_2 + \sigma_3 + \sigma_4 + \sigma_5 + \sigma_6$	$1/\sqrt{6}$	s: 1
even γ_3 a	$2\sigma_5 + 2\sigma_6 - \sigma_1 - \sigma_2 - \sigma_3 - \sigma_4$	$1/\sqrt{12}$	d: $z^2 - \frac{1}{2}(x^2 + y^2)$
β	$\sigma_1 + \sigma_2 - \sigma_3 - \sigma_4$	$1/2$	d: $x^2 - y^2$
odd γ_4 a	$\sigma_1 - \sigma_2$	$1/\sqrt{2}$	p: x
b	$\sigma_3 - \sigma_4$	$1/\sqrt{2}$	p: y
c	$\sigma_5 - \sigma_6$	$1/\sqrt{2}$	p: z
π-orbitals			
even γ_4 a	$z_3 - z_4 - y_5 + y_6$	$1/2$	
b	$z_1 - z_2 - x_5 + x_6$	$1/2$	
c	$y_1 - y_2 - x_3 + x_4$	$1/2$	
odd γ_4 a	$x_3 + x_4 + x_5 + x_6$	$1/2$	p: x
b	$y_1 + y_2 + y_5 + y_6$	$1/2$	p: y
c	$z_1 + z_2 + z_3 + z_4$	$1/2$	p: z
even γ_5 a	$z_3 - z_4 + y_5 - y_6$	$1/2$	d: yz
b	$z_1 - z_2 + x_5 - x_6$	$1/2$	d: xz
c	$y_1 - y_2 + x_3 - x_4$	$1/2$	d: xy
odd γ_5 a	$x_3 + x_4 - x_5 - x_6$	$1/2$	
b	$y_1 + y_2 - y_5 - y_6$	$1/2$	
c	$z_1 + z_2 - z_3 - z_4$	$1/2$	

If the σ-orbitals consist of the p-shell of each ligand no. 1, 2, . . ., 6, they are

$$
\begin{array}{cccccc}
\sigma_1 & \sigma_2 & \sigma_3 & \sigma_4 & \sigma_5 & \sigma_6 \\
-x_1 & +x_2 & -y_3 & +y_4 & -z_5 & +z_6
\end{array}
$$

λ is related to the orbital angular momentum around the linear axis in the same way as l to the orbital angular momentum in spherical symmetry. Correspondingly, only σ-electrons have a finite probability density on the linear axis, while the orbitals with higher λ are distant from this axis.

If we surround a central ion with six ligand atoms, we would expect σ-bonds, formed by the electrons concentrated in the directions pointing towards the central ion, to be more important than the bonds corresponding to higher values of λ. We may now study how six σ-orbitals, one from each ligand, transform in the octahedral symmetry. If all the possible group-theoretical operations are performed with the six σ-orbitals, it turns out that one linear combination, equally dense in all six bond positions, has the totally symmetric form even γ_1. Three linear combinations are obtained from the linear combination with opposite sign of two σ-bonds in *trans*-position, corresponding to odd γ_4. Finally, two linear combinations are obtained, having the quantum number even γ_3. Fig. 2 gives the axes in such a complex and some of the orbitals, and Table 4 gives the unnormalized linear combinations.

The twelve π-orbitals, two from each ligand, were found by Kimball to transform as the four sets of three equivalent orbitals odd γ_4, odd γ_5, even γ_4 and even γ_5. Flodmark has calculated that the twelve δ-orbitals transform like even γ_2, odd γ_2, even γ_3, odd γ_3, even γ_5 and odd γ_5. (Many authors indicate the parity as *gerade* (even) or *ungerade* (odd) by adding a right-hand subscript g or u to the group-theoretical quantum numbers.)

Seen from the nucleus of one of the ligands, the parity is not well-defined, since this nucleus does not represent a centre of inversion. Therefore, both s- and p-electrons of the ligand can participate in the σ-bonds. On the other hand, s-electrons cannot participate in π-bonds, but only p-, d-, . . . electrons of the ligand. We have constructed Table 4 as if all the σ- and π-bonds were formed only of p-electrons from the ligands.

The interest of relating a classification of orbitals according to λ in the ligands with a classification according to parity and γ_n in the central ion is that only those linear combinations of ligand orbitals which have a certain parity and value of γ_n, can interact with a given central ion orbital having the same parity and γ_n. Thus, we may have the interactions:

$$
\left.
\begin{array}{lll}
\sigma\text{-orbitals even } \gamma_1 & \text{with s-orbitals of the central ion} \\
\text{odd } \gamma_4 & \text{p-orbitals} \\
\text{even } \gamma_3 & \text{d-orbitals} \\
\pi\text{-orbitals odd } \gamma_4 & \text{p-orbitals} \\
\text{even } \gamma_5 & \text{d-orbitals}
\end{array}
\right\} \quad (53)
$$

Now, these interactions have qualitatively the same properties as the second-order perturbations discussed above. One difference is caused by the fact that orbitals of the central ion and of the ligands are not generally orthogonal (if they have the same parity and γ_n) but have an *overlap integral* $S_{12} = \int \psi_1 \psi_2$. The secular determinant eqn. (15) then has the non-diagonal elements $E_{nm} - S_{nm}E$, containing the variable E. As will be discussed below, it is generally assumed that E_{nm} is approximately proportional to S_{mn}, the overlap, so to speak, being the original cause of the interaction of the orbitals.

Van Vleck pointed out (1935) that σ-bonding alone would have the following effect on the partly filled d-shell: that two sets of orbitals, $2\gamma_3$ and $1\gamma_3$ are formed. If the diagonal elements of energy are higher for d-electrons than for the σ-electrons of the ligands, the highest orbital $2\gamma_3$ would be a mixture of much d-electron and a little of the ligand's orbital (the normalization condition of eqn. (20) is now $a^2 + b^2 + 2abS_{ab} = 1$). Slightly below $2\gamma_3$ follows the part γ_5 of the d-shell, which does not participate in the σ-bonding. Lowest is $1\gamma_3$ consisting of a little d-orbital and much ligand-orbital. Thus, the σ-bonding gives a contribution to Δ, lifting the energy of $2\gamma_3$ above that of γ_5 of the d-shell.

Owen and Orgel pointed out in 1955 the additional effects of π-bonding in octahedral transition group complexes. It is necessary to distinguish between (1) π-bonding in the direction from the ligands to the central ion, and (2) π-bonding in the opposite direction. The former case (1) corresponds, as the σ-bonding mentioned, to a lower diagonal element of energy of the ligand orbitals than of the partly filled d-shell. Here, the effect of the intermixing is to increase the energy of the γ_5-sub-shell, being now also a mixture of much d-shell and a little orbital of the ligand, and thus the parameter Δ is decreased. The other case (2) corresponds to the presence of empty π-orbitals in the ligands, having diagonal elements of energy higher than those of the partly filled d-shell. Consequently, the γ_5-sub-shell becomes the lowest, bonding orbital as a linear combination of much d-orbital and a little of the empty π-orbital of the ligands, and it is decreased in energy, producing a higher value of Δ.

We could not *a priori* determine the order of M.O. energies in the octahedral complexes, but the electron transfer spectra to be discussed later on show the following series of energy, neglecting empty orbitals in the ligands:

odd γ_4 $(n + 1)$p $+$ ligand σ $+$ ligand π
even γ_1 $(n + 1)$s $+$ ligand σ
even γ_3 much nd $+$ little ligand σ
even γ_5 much nd $+$ little ligand π
even γ_4 ligand π
odd γ_5 ligand π
even γ_5 much ligand π $+$ little nd
odd γ_4 much ligand π $+$ little $(n + 1)$p $+$ little ligand σ
even γ_3 much ligand σ $+$ little nd
odd γ_4 much ligand σ $+$ $(n + 1)$p $+$ little ligand π
even γ_1 ligand σ $+$ $(n + 1)$s.

$$(54)$$

Of these eleven sets of orbitals (comprising twenty-seven M.O., with room for fifty-four electrons), the seven lowest sets (of eighteen M.O.) are fully occupied in the ground state even in the d^0-complexes. The two following sets are those of the partly filled shell, producing the ligand field absorption bands. The next highest set is filled in complexes of such ions as thallium (I), where the gaseous ion Tl^+ contains two s-electrons in the ground state. In these complexes, transitions can be observed to the highest orbital of eqn. (54).

The orbitals indicated in eqn. (54) are only those representative of the chemical bonding, and which are of importance for the absorption spectrum in the visible and the near ultra-violet. Actually, all the inner shells of the central ion and the ligands can be classified into M.O., having energies lower than those of eqn. (54). Thus, the $42 + 6 \cdot 18 = 150$ electrons of the hexachloro-complex of rhodium (III), $RhCl_6^{-3}$, are distributed on the sub-shells of $1s^2 2s^2 2p^6 3s^2$-$3p^6 3d^{10} 4s^2 4p^6$ according to eqn. (44) (and in the ground state also the filled sub-shell γ_5^6, mainly composed of $4d^6$) and the orbitals, composed of five sets of σ-orbitals and two sets of π-orbitals according to Table 4. Thus, nine filled orbitals even γ_1, numbered from 1 to 9 with increasing energy, occur in the complex, while ten orbitals odd γ_4 are completely occupied. The partly filled shell has the numbers even $7\gamma_3$ and even $4\gamma_5$. In addition to the filled orbitals, the two highest orbitals in eqn. (54) have the numbers even $10\gamma_1$ and odd $11\gamma_4$. The enumeration continues with an infinitely large number of the following $n\gamma_n$.

In literature, M.O. are often divided into "genuine" molecular orbitals, belonging to two or more atoms, and "atomic" orbitals, belonging to a certain atom. This distinction between *delocalized* and *localized* M.O. is only representing some limiting cases, the coefficients of the linear combinations may actually vary continuously from zero to large values not much below 1. Mulliken, who, together with Hund, is the founder of M.O. theory, emphasized as early as 1932 that the enormous advantage offered by M.O. description of the chemical bonding is that the distribution of the orbitals on the individual atoms has the full freedom of variation. This is reflected in the approximation most commonly used, the *linear combination of atomic orbitals* L.C.A.O., where the M.O. is a linear combination of a small number of orbitals from different atoms. However, this system of wave functions (before the intermixing) is overcomplete in the sense that the highly excited orbitals of both the central ion and the ligands cover essentially the same domain in the space, corresponding to very large overlap integrals between some of the ligand and some of the central ion orbitals. Thus, for purely topographical reasons, the orbitals $(n + 1)$s and $(n + 1)$p of the central ions in most transition group complexes are in almost the same region as the orbitals of the ligands. When we indicate as lowest in eqn. (54) a mixture of σ-orbitals from the ligands and the $(n + 1)$s-orbital from the central ion for forming the bonding combination even γ_1, we must admit that the normalization of this orbital is a rather unusual affair, since the overlap integral between the two original orbitals may be large.

For highly excited orbitals of a molecule, as found in the excited states of the *Rydberg bands* in the far ultra-violet, it is often convenient to use the *united atom* description rather than the L.C.A.O. method. The united atom approach may even be illustrative for the ground state of some molecules, such as H_2 considered as a perturbation of a helium atom $1s^2$, or CH_4 as a neon atom, where four protons have been removed from the neon nucleus, forming a regularly tetrahedral arrangement. However, in these cases, we are returning to the approximation of spherical symmetry, neglecting the distance between the nuclei compared to the extension of the electron cloud. Boys and other authors have recommended the treatment of the whole molecule in terms of one-centre orbitals, related to a single point, e.g. a nucleus of a central atom. Thus, the mathematical difficulties involved in the presence of overlap integrals between orbitals on different atoms are removed, but we cannot, in general, expect the resulting wave function to consist mainly of one configuration only; the singularities of nuclei at other points than the centre produce the presence of tightly bound electrons, which can only be described by a strong intermixing of one-centre orbital configurations.

Generally, we must apply the group theory *cum grano salis*. The high symmetries we need for strong arguments are not realized in practical cases. In a terrestrial experiment on atomic spectroscopy, the gaseous atoms or ions are not surrounded by spherical symmetry for a very long distance. In the same way, an octahedral complex nearly always occurs in a solid, where the presence of other ions disturb the high symmetry, or in a solution, where the symmetry is much lower. However, experience shows, according to N. Bjerrum's rule, that at least for the ligand field bands the environment outside the first co-ordination sphere has very little influence. We may therefore treat the complexes according to the *principle of microsymmetry*, utilizing the high symmetry usually found in the immediate environment of a central atom.

There is a certain difference of opinion between the all-or-nothing minded group theoretist, maintaining that all the high symmetry is gone when it has been slightly violated, and on the other hand, the chemical physicist classifying perturbations according to their own effective symmetry. Group theory only indicates to us which matrix elements shall be zero, and rarely the possible size of these matrix elements, which do not necessarily vanish. As will be seen in the discussion of absorption band intensities, or of the splittings of energy levels due to the introduction of small perturbations of lower symmetry, the matrix elements which group theory has completely released control of are not automatically large, but increase slowly in accordance with the magnitude of the low-symmetry influence.

Further, in the discussion of a certain polyatomic molecule, it is reasonable to discuss the behaviour of the orbitals in the neighbourhood of one atom or in the *bond region* between two atoms in terms of a local microsymmetry. Since the bond region has linear symmetry, the orbitals involved in a bond will be classified approximately into one σ-orbital, two π-orbitals, two δ-orbitals, etc. Thus, a double or a triple bond needs the co-operation of one or two π-orbitals in addition to the σ-orbital mainly responsible for a single bond. It may be asked why we cannot have two σ-orbitals involved in the same bond. Since the orthogonality between the two orbitals can no longer be obtained from the angular variation around the linear axis (as for two orbitals with different λ), one of the two σ-orbitals must necessarily have a node, a zero-point between positive and negative values of ψ, on a plane (or a slightly curved surface) between the two nuclei, and experience shows that this phenomenon gives a weaker bonding than one σ-orbital, or even a positive contribution to the energy of the molecule, as will be discussed below. The concept of fractional bond order between 0 and 1 of a given σ-bond does not imply that, for example, 0·7 σ-orbital is involved in the bonding; it means rather that the L.C.A.O. description of the σ-orbital allocates 65 per cent of the electron density to one atom and 35 per cent to the other atom. However, due to the peculiar normalization factors, it is a crude approximation to assume that a bond order x really corresponds to the coefficients $\sqrt{(1 - x/2)}$ and $\sqrt{(x/2)}$ in the L.C.A.O. We have an instinctive feeling that it is slightly nonsensical to talk about a 5d- or a 6p-orbital of the central ion in a complex such as $Co(NH_3)_6^{3+}$, while the 3d-shell has a rather clear meaning. In the same way, the 3p- or the 4s-shells of the nitrogen atoms have a rather more imaginary existence than the 1s-shell. The quantitative expression of these feelings is the concept of microsymmetry, indicating only small deviations from spherical to octahedral symmetry for orbitals near to the cobalt nucleus, while the nitrogen nucleus is surrounded by a region of trigonal, nearly spherical microsymmetry. The mutual interaction between the two microsymmetries on their boundary, the bond region, appears as the influence of the energies and the wave functions (overlap integrals, etc.) of the 3d-, 4s-, 4p-shells of the cobalt atom on the six σ-orbitals of the nitrogen atom (the lone pairs), being arranged in the three sets: even γ_1, even γ_3 and odd γ_4. On the other hand, the splitting of the 3d-shell into the two sub-shells: even γ_5 and even γ_3 is mainly caused by the influence of the σ-electrons on the latter sub-shell.

The secular determinant for two orbitals ψ, having the overlap integral S_{12} is

$$\begin{vmatrix} E_{11} - E & E_{12} - S_{12}E \\ E_{12} - S_{12}E & E_{22} - E \end{vmatrix} = 0 \qquad (55)$$

having the diagonal sum of the two roots $E_1 + E_2 =$

$$(E_{11} + E_{22} - 2S_{12}E_{12})/(1 - S_{12}^2) \qquad (56)$$

which is not necessarily constant (cf. eqn. (18) for the case $S_{12} = 0$). Mulliken proposed a formula for the non-diagonal element

$$E_{12} = kS_{12}(E_{11} + E_{22})/2 \qquad (57)$$

with k an empirical constant, which has been assigned values such as 1·6 or 2·0. Wolfsberg and Helmholz applied eqn. (57) for the description of the tetrahedral complexes CrO_4^{2-} and MnO_4^-, using overlap integrals S_{mn} derived from analytical approximations to HFSC radial functions. Recently, several authors have applied to octahedral complexes the *Wolfsberg–Helmholz method*, using the less realistic hydrogenic radial functions, for which overlap integrals with hydrogenic functions on other atoms are tabulated in literature. However, as discussed in the chapter on electron transfer spectra, it is necessary to collect empirical information for estimating the diagonal elements of orbital energies E_{nn}. The diagonal sum of eqn. (56) when the assumption eqn. (57) is inserted, is

$$(E_{11} + E_{22}) \frac{1 - kS_{12}^2}{1 - S_{12}^2} \qquad (58)$$

being constant for $k = 1$. However, with $k > 1$, the diagonal sum is increased, since $E_{11} + E_{22}$ is generally a negative energy. This corresponds to the empirical fact that closed shells, forming at the same time the *bonding* and *anti-bonding* linear combinations, repel each other, and increase their energy strongly with increasing square of their overlap integral. In analogy to eqns. (20)–(21) the bonding orbital has the wave function

$$\psi_b = a\psi_1 + b\psi_2 \quad (a^2 + b^2 + 2abS_{12} = 1) \qquad (59)$$

and the anti-bonding orbital

$$\psi_a = c\psi_1 - d\psi_2 \quad (c^2 + d^2 - 2cdS_{12} = 1) \qquad (60)$$

with the orthogonality condition

$$ac - bd + (bc - ad)S_{12} = 0 \qquad (61)$$

The total electron density of the bonding and anti-bonding orbital, when both are occupied, is

$$\psi_a^2 + \psi_b^2 = (\psi_1^2 + \psi_2^2 - 2S_{12}\psi_1\psi_2)/(1 - S_{12}^2) \qquad (62)$$

having a small density decrease in the *overlap region* $\psi_1\psi_2$, while $\psi_b{}^2$ alone has a strongly increased density in the overlap region, and $\psi_a{}^2$ has a considerable density minimum in the overlap region, and especially has a node there (at $c\psi_1 = d\psi_2$).

For orthogonalizing the two wave functions, it is not necessary to change both wave functions. Thus, ψ_1 is orthogonal on

$$\psi_c = \frac{1}{\sqrt{(1 - S_{12}^2)}} \psi_2 - \frac{S_{12}}{\sqrt{(1 - S_{12}^2)}} \psi_1 \tag{63}$$

corresponding to the transformation of an oblique co-ordinate system in the Hilbert space to a rectangular co-ordinate system. However, the minimization of the energy may necessitate the rotation of the new co-ordinate system in the Hilbert space, corresponding to an intermixing of ψ_1 and ψ_c according to the second-order perturbation theory for orthogonal wave functions.

CHAPTER 5

THE INTERELECTRONIC REPULSION IN M.O. CONFIGURATIONS

THE theory needed for the effect of the two-electron operators, representing the interelectronic repulsion, is essentially the same in molecules as that given in eqns. (28)–(30). The condition for reasonably pure M.O. configurations is the same as in spherical symmetry, that the orbital energy differences shall be large, compared to the non-diagonal elements of the two-electron operator. In an actual molecule, this may sometimes be impossible to achieve, due to an accidental degeneracy between the energies of some orbitals; but in general, the problem can be solved in a fairly satisfactory way by choosing nearly self-consistent M.O., i.e. those orbitals which are not changed to linear combinations of other orbitals by the interelectronic repulsion, acting as a virtually one-electron operator in the *core field* (which does not have spherical symmetry as has the central field in atomic spectroscopy). As a matter of classification, experience shows that we may actually use the M.O. configurations for a consistent description of the energy levels in the same way as shell configurations are used for gaseous atoms and ions.

An important special case of secular determinants, containing the non-diagonal elements from eqn. (28) and diagonal elements from eqn. (30) is the *Heisenberg determinant*, expressing the interaction between two degenerate configurations a^+b^- and a^-b^+ (the signs indicate the signs of the spin component m_s). Since the non-diagonal element is K(ab), the two energies are with the notation of eqn. (31)

$$\text{singlet term } (S = 0) \quad \text{J(ab)} + \text{K(ab)}$$
$$\text{triplet term } (S = 1) \quad \text{J(ab)} - \text{K(ab)} \tag{64}$$

The triplet energy can also be found as the diagonal element of interelectronic repulsion of the configuration a^+b^+, having $M_S = 1$. We shall also define Heisenberg determinants of third degree, where the diagonal elements all equal J and the non-diagonal elements $\pm K$. If an odd number of the three non-diagonal elements is $+K$, the eigenvalues will be $J + 2K$, $J - K$ and $J - K$, while if an even number has the positive sign of K, the eigenvalues will be $J - 2K$, $J + K$ and $J + K$. Generally, a determinant with q identical diagonal elements J and all the $q(q - 1)/2$ non-diagonal elements $= K$ has the eigenvalue $J + (q - 1)K$ and $(q - 1)$-fold the eigenvalue $J - K$.

In octahedral complexes, it is not possible to evaluate the energies of interelectronic repulsion without making a distinction between the equivalent orbitals in a given set γ_3, γ_4 or γ_5. This can be done by reference to a virtual, lower symmetry, such as the *tetragonal* D_{4h} or the *orthorhombic* D_2. In both cases, Bethe and Mulliken have proposed different symbols for γ_n:

$$
\begin{array}{llll}
\text{tetragonal} & & \text{orthorhombic} & \\
\gamma_{t1} & a_1 & \gamma_{r1} & a_1 \\
\gamma_{t2} & a_2 & \gamma_{r2} & b_1 \\
\gamma_{t3} & b_1 & \gamma_{r3} & b_2 \\
\gamma_{t4} & b_2 & \gamma_{r4} & b_3 \\
\gamma_{t5} & e & &
\end{array}
\right\} \quad (65)
$$

All these orbitals have $e = 1$ except the degenerate γ_{t5}-orbital with $e = 2$. The cubic quantum numbers γ_n can be related to γ_{tn} or γ_{rn}, even though sometimes the choice between two quantum numbers of the lower symmetry is arbitrary. We may trace the evolution of the partly filled shell (of the three ordinary transition groups) in the lower symmetries:

$$
\begin{array}{llll}
\text{octahedral} & \text{tetragonal} & \text{orthorhombic} & l = 2 \\
\gamma_3 \left\{ \begin{array}{l} \gamma_{t3} \\ \gamma_{t1} \end{array} \right. & \begin{array}{l} \gamma_{r1} \\ \gamma_{r1} \end{array} & \begin{array}{l} x^2 - y^2 \\ z^2 - \tfrac{1}{2}(x^2 + y^2) \end{array} \\
\gamma_5 \left\{ \begin{array}{l} \gamma_{t4} \\ \gamma_{t5} \end{array} \right. & \begin{array}{l} \gamma_{r2} \\ \left\{ \begin{array}{l} \gamma_{r3} \\ \gamma_{r4} \end{array} \right. \end{array} & \begin{array}{l} xy \\ xz \\ yz \end{array}
\end{array}
\right\} \quad (66)
$$

At the right-hand side of eqn. (66) is indicated the orbitals of Table 1 occurring for the special case of $l = 2$. We shall now use the tetragonal numbers n in the classification of the J- and K-integrals. For a partly filled γ_5-sub-shell, only three parameters occur: $J(4, 4)$, $J(4, 5)$ and $K(4, 5)$. The first is the J-integral between one of the three equivalent γ_5 electrons and itself; the latter are the J- and

K-integrals between two different electrons in the set. Even though l is not necessarily 2, we can always choose the three orbitals in a way that these relations hold.

TABLE 5. THE MULTIPLICATION TABLE FOR Γ_n IN CUBIC-OCTAHEDRAL SYMMETRY O_h (AND TETRAHEDRAL SYMMETRY T_d)

	Γ_1	Γ_2	Γ_3	Γ_4	Γ_5
Γ_1	Γ_1	Γ_2	Γ_3	Γ_4	Γ_5
Γ_2	Γ_2	Γ_1	Γ_3	Γ_5	Γ_4
Γ_3	Γ_3	Γ_3	$\Gamma_1 + \Gamma_2 + \Gamma_3$	$\Gamma_4 + \Gamma_5$	$\Gamma_4 + \Gamma_5$
Γ_4	Γ_4	Γ_5	$\Gamma_4 + \Gamma_5$	$\Gamma_1 + \Gamma_3 + \Gamma_4 + \Gamma_5$	$\Gamma_2 + \Gamma_3 + \Gamma_4 + \Gamma_5$
Γ_5	Γ_5	Γ_4	$\Gamma_4 + \Gamma_5$	$\Gamma_2 + \Gamma_3 + \Gamma_4 + \Gamma_5$	$\Gamma_1 + \Gamma_3 + \Gamma_4 + \Gamma_5$

In the same way as $L_1 \times L_2$ can be made to form a vector sum L_3 having different values ($= L_1 + L_2, L_1 + L_2 - 1, L_1 + L_2 - 2,$ . . ., $|L_1 - L_2|$) in spherical symmetry, we can form the products of two values of Γ_n in the other symmetries. Tables 5 and 6 indicate the products Γ_n in octahedral, tetragonal, and rhombic symmetries. In the cases $\gamma_a \gamma_b \gamma_c$ . . . with one electron in each sub-shell a, b, . . ., the values of Γ_n of the resultant terms are obtained by simple multiplication (which always is associative and commutative) of $\Gamma_a \cdot \Gamma_b \cdot \Gamma_c$ . . . according to Tables 5 or 6. The accompanying

TABLE 6. THE MULTIPLICATION TABLE FOR Γ_{tn} IN TETRAGONAL SYMMETRY D_{4h} AND FOR Γ_{rn} IN ORTHORHOMBIC SYMMETRY D_2. THE SAME VALUES OF n (THE PART Γ_t OR Γ_r IS NOT INDICATED IN THE TABLE) APPLY TO BOTH CASES, EXCEPT $n = 5$ PARTICULAR TO THE TETRAGONAL SYMMETRY

	1	2	3	4	5
1	1	2	3	4	5
2	2	1	4	3	5
3	3	4	1	2	5
4	4	3	2	1	5
5	5	5	5	5	$1 + 2 + 3 + 4$

values of S are the possible vector sums of the spin components $m_s = \pm\frac{1}{2}$ of the individual electrons. As is also the case in atomic spectroscopy, more complications arise, when several equivalent

F

electrons are present in the same sub-shell. Not all the possible combinations of Γ_n and S occur. However, the terms $^{2S+1}\Gamma_n$ can be derived for any sub-shell configuration $\gamma_5{}^a\gamma_3{}^b$ when the results are known for $\gamma_5{}^2(^3\Gamma_4,\ ^1\Gamma_1,\ ^1\Gamma_3,\ ^1\Gamma_5)$, $\gamma_5{}^3(^4\Gamma_2,\ ^2\Gamma_3,\ ^2\Gamma_4,\ ^2\Gamma_5)$ and $\gamma_3{}^2(^3\Gamma_2,\ ^1\Gamma_1,\ ^1\Gamma_3)$, since according to Pauli's hole-equivalence rule, the same number and species of terms occur for $\gamma_5{}^a$ and $\gamma_5{}^{6-a}$ and for $\gamma_3{}^b$ and $\gamma_3{}^{4-b}$. Finally, the mixed sub-shell configurations $\gamma_5{}^a\gamma_3{}^b$ can be treated by direct multiplication.

In terms of J- and K-integrals, the energies of interelectronic repulsion are for $\gamma_5{}^2$:

$$
\left.
\begin{array}{ll}
^3\Gamma_4 & J(4, 5) - K(4, 5) \\
^1\Gamma_5 & J(4, 5) + K(4, 5) \\
^1\Gamma_3 & J(4, 4) - K(4, 5) \\
^1\Gamma_1 & J(4, 4) + 2K(4, 5)
\end{array}
\right\} \qquad (67)
$$

and for $\gamma_5{}^3$:

$$
\left.
\begin{array}{ll}
^4\Gamma_2 & 3J(4, 5) - 3K(4, 5) \\
^2\Gamma_3 & 3J(4, 5) \\
^2\Gamma_4 & J(4, 4) + 2J(4, 5) - 2K(4, 5) \\
^2\Gamma_5 & J(4, 4) + 2J(4, 5)
\end{array}
\right\} \qquad (68)
$$

The results in eqn. (67) and eqn. (68) can most easily be obtained from the classification according to the orthorhombic symmetry of the three equivalent orbitals γ_{rn} with n = 2, 3, 4. In both cases, we may get the maximum value of S from the determinant with $M_S = 1$ (2^+3^+) and $M_S = \frac{3}{2}$ $(2^+3^+4^+)$, respectively, and eqn. (30).

For $M_S = 0$, we have one Heisenberg determinant of third degree (Γ_{r1} from the three configurations 2^+2^-, 3^+3^- and 4^+4^-) and three Heisenberg determinants of second degree (e.g. 2^+3^- and 2^-3^+). The three latter Heisenberg determinants reproduce the energies of $^3\Gamma_4$ and $^1\Gamma_5$ as eigenvalues, while the first determinant determines $^1\Gamma_1$ and $^1\Gamma_3$.

Analogously, for $M_S = \frac{1}{2}$ of $\gamma_5{}^3$, we have one Heisenberg determinant of third degree (Γ_{r1} from the three configurations $2^+3^+4^-$, $2^+3^-4^+$ and $2^-3^+4^+$) and three Heisenberg determinants of second degree (e.g. Γ_{r2} from $2^+3^+3^-$ and $2^+4^+4^-$). Here, the ground state $^4\Gamma_2$ and one other term, $^2\Gamma_3$, belong to the determinant of third degree, while the two other doublet terms belong to the three equivalent determinants of second degree.

Pauli's hole-equivalency rule is also valid for the relative energy differences of the terms, being the same for γ_5^4 and γ_5^2. However, the zero point of energy is shifted and, for instance, $^1\Gamma_1$ of the filled sub-shell γ_5^6 has the rather complicated energy expression

$$3J(4, 4) + 12J(4, 5) - 6K(4, 5) \tag{69}$$

The interelectronic repulsion energy of the configuration γ_3^2 can be found from a classification according to tetragonal symmetry, γ_{t1} and γ_{t3}. For $M_S = 1$, we have only one configuration, 1^+3^+, giving the energy $J(1, 3) - K(1, 3)$ of the triplet term $^3\Gamma_2$. For $M_S = 0$, we have two Heisenberg determinants of second degree, having the capital quantum numbers Γ_{t1} and Γ_{t3}, respectively. The former, consisting of the configurations 1^+1^- and 3^+3^-, has the two equal diagonal elements $J(1, 1)$ and the non-diagonal element $K(1, 3)$, producing the eigenvalues $J(1, 1) \pm K(1, 3)$ according to eqn. (64). The latter determinant, comprising the configurations 1^+3^- and 1^-3^+, has the eigenvalues $J(1, 3) \pm K(1, 3)$. It is obvious that the lowest of these levels is the $M_S = 0$ component of the triplet, found above from 1^+3^+. Since $J(1, 1)$ presumably is larger than $J(1, 3)$ (because an orbital is closer to itself than to another orbital), the singlet $^1\Gamma_3$ must be distributed between the two different Heisenberg determinants, its component $^1\Gamma_{t1}$ having the energy $J(1, 1) - K(1, 3)$ and $^1\Gamma_{t3}$ $J(1, 3) + K(1, 3)$. Since these two components necessarily must have the same energy, because we actually have octahedral and not only tetragonal symmetry, we have proven the theorem

$$J(aa) = J(ab) + 2K(ab) \tag{70}$$

for a and b being the two equivalent orbitals of the set γ_3. We found hence for γ_3^2:

$$\left.\begin{array}{lll} ^3\Gamma_2 & J(1, 3) - & K(1, 3) \\ ^1\Gamma_3 & J(1, 3) + & K(1, 3) \\ ^1\Gamma_1 & J(1, 3) + & 3K(1, 3) \end{array}\right\} \tag{71}$$

For a mixed sub-shell configuration such as $\gamma_5\gamma_3$, we find from the determinants for Γ_{t2} and Γ_{t4}, having $M_S = 1$ and 0:

$$\left.\begin{array}{lll} ^3\Gamma_5 & J(1, 4) - K(1, 4) \\ ^3\Gamma_4 & J(3, 4) - K(3, 4) \\ ^1\Gamma_5 & J(1, 4) + K(1, 4) \\ ^1\Gamma_4 & J(3, 4) + K(3, 4) \end{array}\right\} \tag{72}$$

while for the rather important case $\gamma_5{}^5\gamma_3$, it can be found in a similar way

$$
\left.
\begin{array}{ll}
{}^3\Gamma_4 & 8J(4, 5) + 2J(4, 4) - 4K(4, 5) + 3J(1, 4) + 2J(3, 4) \\
& \quad - \tfrac{3}{2}K(1, 4) - \tfrac{3}{2}K(3, 4) \\
{}^1\Gamma_4 & 8J(4, 5) + 2J(4, 4) - 4K(4, 5) + 3J(1, 4) + 2J(3, 4) \\
& \quad - \tfrac{3}{2}K(1, 4) + \tfrac{1}{2}K(3, 4) \\
{}^3\Gamma_5 & 8J(4, 5) + 2J(4, 4) - 4K(4, 5) + 2J(1, 4) + 3J(3, 4) \\
& \quad - \tfrac{3}{2}K(1, 4) - \tfrac{3}{2}K(3, 4) \\
{}^1\Gamma_5 & 8J(4, 5) + 2J(4, 4) - 4K(4, 5) + 2J(1, 4) + 3J(3, 4) \\
& \quad + \tfrac{1}{2}K(1, 4) - \tfrac{3}{2}K(3, 4)
\end{array}
\right\} \quad (73)
$$

by using the following theorem, which can be proved from group theory:

$$
\begin{aligned}
J(1, 5) &= \tfrac{1}{4} J(1, 4) + \tfrac{3}{4} J(3, 4) & J(3, 5) &= \tfrac{3}{4} J(1, 4) + \tfrac{1}{4} J(3, 4) \\
K(1, 5) &= \tfrac{1}{4}K(1, 4) + \tfrac{3}{4}K(3, 4) & K(3, 5) &= \tfrac{3}{4}K(1, 4) + \tfrac{1}{4}K(3, 4)
\end{aligned} \quad (74)
$$

Thus, the interaction of a filled $\gamma_5{}^6$-sub-shell containing four γ_{t5} and two γ_{t4} is the same with a γ_{t1}- or a γ_{t3}-electron.

Also the interelectronic repulsion energy in more complicated sub-shell configurations can be treated by this method. Thus, the energy of the lowest term of $\gamma_5{}^3\gamma_3{}^2$ is

$$
\begin{aligned}
{}^6\Gamma_1 \quad & J(1, 3) - K(1, 3) + 3J(1, 4) - 3K(1, 4) \\
& + 3J(3, 4) - 3K(3, 4) + 3J(4, 5) - 3K(4, 5) \quad (75)
\end{aligned}
$$

If we consider the excitation energies of the quartet terms of $\gamma_5{}^3\gamma_3{}^2$, measured relative to ${}^6\Gamma_1$, it is useful to introduce the notation of *parent terms* in the individual sub-shells.

TABLE 7. THE INTERELECTRONIC REPULSION ENERGIES FOR THE SUB-SHELL CON-FIGURATION $\gamma_5{}^3\gamma_3{}^2$, RELATIVE TO ${}^6\Gamma_1$ AS ZERO POINT. THE DIAGONAL ELEMENTS ARE GIVEN FOR THE TWO TERMS ${}^4\Gamma_3a$ AND ${}^4\Gamma_3\beta$, THEY INTERACT WITH A NON-DIAGONAL ELEMENT $(K(1, 4) - K(3, 4))\sqrt{(3)}/2$

| | Parent terms | | Excitation energy | | | | | |
	$\gamma_5{}^3$ · $\gamma_3{}^2$		$K(1, 3)$	$K(1, 4)$	$K(3, 4)$	$J(4, 4)$	$J(4, 5)$	$K(4, 5)$
${}^4\Gamma_1$	${}^4\Gamma_2$	${}^3\Gamma_2$	0	$\tfrac{5}{2}$	$\tfrac{5}{2}$	0	0	0
${}^4\Gamma_2$	${}^4\Gamma_2$	${}^1\Gamma_1$	4	$\tfrac{3}{2}$	$\tfrac{3}{2}$	0	0	0
${}^4\Gamma_3a$	${}^4\Gamma_2$	${}^1\Gamma_3$	2	$\tfrac{3}{2}$	$\tfrac{3}{2}$	0	0	0
${}^4\Gamma_3\beta$	${}^2\Gamma_3$	${}^3\Gamma_2$	0	1	1	0	0	3
${}^4\Gamma_4$	${}^2\Gamma_5$	${}^3\Gamma_2$	0	1	1	1	−1	3
${}^4\Gamma_5$	${}^2\Gamma_4$	${}^3\Gamma_2$	0	1	1	1	−1	1

Table 7 shows that the excitation energy can be divided into two parts: on the one hand the internal excitation in each sub-shell, as represented by the coefficient to $K(1, 3)$ according to eqn. (71) and the coefficients to $J(4, 4)$, $J(4, 5)$ and $K(4, 5)$ according to eqn. (68). On the other hand is the interaction between the two sub-shells, which is $3J(1, 4) - 3K(1, 4) + 3J(3, 4) - 3K(3, 4)$ in the ground-state $^6\Gamma_1$. For the quartet terms in Table 7, this quantity is increased by an amount $S'(K(1, 4) + K(3, 4))$ where $S' = \frac{3}{2}$ for the quartet–singlet and $S' = 1$ for the doublet–triplet parent terms. In the case of the two terms $^4\Gamma_3$, the energy is not a linear combination of J- and K-integrals, but a complicated square-root expression of the type of eqn. (18). Similar behaviour occurs in atomic spectro-scopy, when two terms of a given configuration present the same combination of L and S.

The physical significance of the excitation energies within each configuration can be expressed as *spin-pairing energy* and energy due to decreased *seniority number*. The spin-pairing energy by going from S to $S - 1$ may simply be $2S$ times a certain K-integral (cf. $^1\Gamma_5$ of $\gamma_5{}^2$, $^2\Gamma_3$ of $\gamma_5{}^3$ and $^1\Gamma_3$ of $\gamma_3{}^2$) or it may be differences between J-integrals, when singly-occupied orbitals are transformed into doubly-occupied orbitals. The seniority number, which also is important in atomic spectroscopy, indicates the lowest number v of electrons in a given orbital, which already has produced one of the terms with the same $^{2S+1}\Gamma_n$ as that discussed. Thus, all the terms given in eqns. (67), (68) and (71) have v equal to the number of electrons, except $^1\Gamma_1$ of $\gamma_5{}^2$ and $\gamma_3{}^2$ (having $v = 0$) and $^2\Gamma_5$ of $\gamma_5{}^3$ ($v = 1$). If more than one $^{2S+1}\Gamma_n$ occurs, the seniority numbers are intermixed by the interelectronic repulsion, but generally, the low values of v correspond to highly increased energy.

If the electrostatic model or the expanded radial function model are assumed, the wave functions of the orbitals can be separated into the product of a radial function and hydrogenic angular functions, corresponding to a well-defined l. If $l = 2$ of the sub-shells γ_5 and γ_3, the nine parameters used above can be expressed as linear com-binations of Racah's three parameters for d-electrons (eqn. 35):

$$
\left.
\begin{array}{ll}
J(1, 3) = A - 4B + C & K(1, 3) = 4B + C \\
J(1, 4) = A - 4B + C & K(1, 4) = 4B + C \\
J(3, 4) = A + 4B + C & K(3, 4) = C \\
J(4, 4) = A + 4B + 3C & K(4, 5) = 3B + C \\
J(4, 5) = A - 2B + C &
\end{array}
\right\} \quad (75)
$$

Table 8 gives several of the excitation energies within each d^n-system, calculated in terms of $a\Delta + bB + cC$. These diagonal elements were, in several cases, known previously, but Tanabe and

TABLE 8. THE DIAGONAL ELEMENTS OF ENERGY $a\Delta + bB + cC$ OF SUB-SHELL CONFIGURATIONS IN OCTAHEDRAL d^n-SYSTEMS. THE ATOMIC GROUND-STATES ARE CHOSEN AS ZERO-POINTS.

d^n	config	term	a	b	c
d^2	γ_5^2	$^3\Gamma_4$	−0·8	3	0
		$^1\Gamma_3$	−0·8	9	2
		$^1\Gamma_5$	−0·8	9	2
		$^1\Gamma_1$	−0·8	18	5
	$\gamma_5\gamma_3$	$^3\Gamma_5$	0·2	0	0
		$^3\Gamma_4$	0·2	12	0
		$^1\Gamma_5$	0·2	8	2
		$^1\Gamma_4$	0·2	12	2
	γ_3^2	$^3\Gamma_2$	1·2	0	0
		$^1\Gamma_3$	1·2	8	2
		$^1\Gamma_1$	1·2	16	4
d^3	γ_5^3	$^4\Gamma_2$	−1·2	0	0
		$^2\Gamma_3$	−1·2	9	3
		$^2\Gamma_4$	−1·2	9	3
		$^2\Gamma_5$	−1·2	15	5
	$\gamma_5^2\gamma_3$	$^4\Gamma_5$	−0·2	0	0
		$^4\Gamma_4$	−0·2	12	0
		...			
	$\gamma_5\gamma_3^2$	$^4\Gamma_4$	0·8	3	0
		...			
d^4	γ_5^4	$^3\Gamma_4$	−1·6	6	5
		$^1\Gamma_3$	−1·6	12	7
		$^1\Gamma_5$	−1·6	12	7
		$^1\Gamma_1$	−1·6	21	10
	$\gamma_5^3\gamma_3$	$^5\Gamma_3$	−0·6	0	0
		...			
	$\gamma_5^2\gamma_3^2$	$^5\Gamma_5$	0·4	0	0
		...			
d^5	γ_5^5	$^2\Gamma_5$	−2·0	15	10
	$\gamma_5^4\gamma_3$	$^4\Gamma_4$	−1·0	10	6
		$^4\Gamma_5$	−1·0	18	6
		$^2\Gamma_2$	−1·0	12	9
		...			
	$\gamma_5^3\gamma_3^2$	$^6\Gamma_1$	0·0	0	0
		$^4\Gamma_1$	0·0	10	5
		$^4\Gamma_3$	0·0	10	5

d^n	config	term	a	b	c
		$^4\Gamma_5$	0·0	13	5
		$^4\Gamma_3$	0·0	17	5
		$^4\Gamma_4$	0·0	19	7
		$^4\Gamma_2$	0·0	22	7
		...			
	$\gamma_5^2\gamma_3^3$	$^4\Gamma_4$	1·0	10	6
		$^4\Gamma_5$	1·0	18	6
		...			
d^6	γ_5^6	$^1\Gamma_1$	−2·4	5	8
	$\gamma_5^5\gamma_3$	$^3\Gamma_4$	−1·4	5	5
		$^3\Gamma_5$	−1·4	13	5
		$^1\Gamma_4$	−1·4	5	7
		$^1\Gamma_5$	−1·4	21	7
	$\gamma_5^4\gamma_3^2$	$^5\Gamma_5$	−0·4	0	0
		...			
	$\gamma_5^3\gamma_3^3$	$^5\Gamma_3$	0·6	0	0
		...			
d^7	$\gamma_5^6\gamma_3$	$^2\Gamma_3$	−1·8	7	4
	$\gamma_5^5\gamma_3^2$	$^4\Gamma_4$	−0·8	3	0
		...			
	$\gamma_5^4\gamma_3^3$	$^4\Gamma_5$	0·2	0	0
		$^4\Gamma_4$	0·2	12	0
	$\gamma_5^3\gamma_3^4$	$^4\Gamma_2$	1·2	0	0
		...			
d^8	$\gamma_5^6\gamma_3^2$	$^3\Gamma_2$	−1·2	0	0
		$^1\Gamma_3$	−1·2	8	2
		$^1\Gamma_1$	−1·2	16	4
		...			
	$\gamma_5^5\gamma_3^3$	$^3\Gamma_5$	−0·2	0	0
		$^3\Gamma_4$	−0·2	12	0
		$^1\Gamma_5$	−0·2	8	2
		$^1\Gamma_4$	−0·2	12	2
		...			
	$\gamma_5^4\gamma_3^4$	$^3\Gamma_4$	0·8	3	0
		$^1\Gamma_3$	0·8	9	2
		$^1\Gamma_5$	0·8	9	2
		$^1\Gamma_1$	0·8	18	5

Sugano completed their calculation (and gave also all the non-diagonal elements) in 1954. The values may be compared with the terms in gaseous ions, given in Table 2.

The condition of low-spin behaviour of octahedral d^n-complexes can be seen from Table 8 to be a question of competition between the orbital energy difference Δ between the two sub-shells γ_3 and γ_5 on the one hand and the interelectronic repulsion energy parameters on the other. Assuming pure sub-shell configurations, low-spin magnetism occurs for

$$
\begin{aligned}
d^4: &\quad \Delta > 6B + 5C \\
d^5: &\quad 2\Delta > 15B + 10C \\
d^6: &\quad 2\Delta > 5B + 8C \\
d^7: &\quad \Delta > 4B + 4C
\end{aligned}
\qquad (76)
$$

The possibilities $S = 0$ for d^4, $S = \frac{3}{2}$ for d^5 and $S = 1$ for d^6 are not energetically possible for the ground state of a regularly octahedral complex, according to Table 8. However, it will later be discussed how deviations from octahedral symmetry, e.g. caused by the Jahn–Teller effect, may disturb this simple picture. The best known exception is the planar, diamagnetic d^8-complexes, where the tetragonal distortion from octahedral symmetry is sufficiently strong to decrease the energy of the component $^1\Gamma_{t_1}$ from $^1\Gamma_3$ below that of $^3\Gamma_2$. A closer analysis shows that the necessary promotion energy for going from $^3\Gamma_2$ to this component (having nearly the pure configuration $\gamma_{t_5}^4\gamma_{t_4}^2\gamma_{t_1}^2$) is $3K(1, 3)$.

The occurence of low-spin behaviour is no proof of a particular covalent bonding; the energy difference Δ may have various causes—first-order electrostatic perturbation energy, second-order perturbations from covalent bonding with the ligands, etc.—but they cannot easily be experimentally separated, and group theory tells us that no influence whatsoever may separate further the two equivalent γ_3-orbitals or the three equivalent γ_5-orbitals in a regularly octahedral complex. The fact that low-spin behaviour is so much more common in the second and third transition groups, compared to the first, is explained from the absorption spectra, indicating that Δ is large in the $4d^n$- and $5d^n$-complexes and that the parameters of interelectronic repulsion are small. There is no reason to believe in an inherent difference in the chemical bonding in the two cases.

Since nine parameters from the general M.O. theory can be expressed as linear combinations of only two, B and C, for pure d-electrons (eqn. 75), several relations are valid for this special case.

Among the more remarkable, special relations are two: that $J(aa) = A + 4B + 3C$ for each of the five d-orbitals, chosen by "zero-order" perturbation in octahedral symmetry (this is not valid for the orbitals $\pm m_l = \lambda$ chosen by Condon and Shortley for linear symmetry) and that the theorem eqn. (70) also is valid for γ_5-electrons, when $l = 2$. Slater's rules for p^2 and p^3 represent actually the same regularity, when $l = 1$ is considered as a set of γ_4-electrons, for which all the arguments about rhombic sub-groups given above are also valid. Thus, the terms are divided into three groups, the middle one consisting of two degenerate terms, and the two energy intervals having the size 2K and 3K.

$$
\begin{array}{l}
p^2 \qquad\quad {}^3P \\
\gamma_5{}^2 \equiv \gamma_4{}^2 \quad {}^3\Gamma_4
\end{array}
\left.\begin{array}{}\ \\ \ \end{array}\right\}
\begin{array}{c}\text{interval}\\ 2K\end{array}
\left\{\begin{array}{l}{}^1D\\ {}^1\Gamma_3, {}^1\Gamma_5\end{array}\right\}
\begin{array}{c}\text{interval}\\ 3K\end{array}
\left\{\begin{array}{l}{}^1S\\ {}^1\Gamma_1\end{array}\right.
$$

$$
\begin{array}{l}
p^3 \qquad {}^4S \\
\gamma_4{}^3 \qquad {}^4\Gamma_1 \\
\gamma_5{}^3 \qquad {}^4\Gamma_2
\end{array}
\left.\begin{array}{}\ \\ \ \\ \ \end{array}\right\}
\begin{array}{c}\text{interval}\\ 3K\end{array}
\left\{\begin{array}{l}{}^2D\\ {}^2\Gamma_3, {}^2\Gamma_5\\ {}^2\Gamma_3, {}^2\Gamma_4\end{array}\right\}
\begin{array}{c}\text{interval}\\ 2K\end{array}
\left\{\begin{array}{l}{}^2P\\ {}^2\Gamma_4\\ {}^2\Gamma_5\end{array}\right\}
\qquad (77)
$$

The validity of eqn. (70) for p-electrons and γ_5(d)-electrons is connected with the fact that $\sqrt{2}\,(x + y)/2$ is a new p-orbital and $\sqrt{2}\,(xz + yz)/2$ is a new d-orbital, while it will not in general be valid that among the three equivalent orbitals a, b, c, the linear combination $d = \sqrt{2}\,(a + b)/2$ has the same physical shape as the original orbitals a, b, c. It will be valid

$$
J(dd) = \tfrac{1}{4}\{J(aa) + J(bb) + 2J(ab) + 4K(ab)\} \qquad (78)
$$

with $J(aa) = J(bb)$, since the charge distribution ab generally has another γ_n than the totally symmetric γ_1 of a^2 and b^2 and therefore, $(ab:a^2)$ and $(ab:b^2)$ (eqn. 27) vanish identically. If d has the same physical shape as a and b (as is the case for the p- and d-orbitals mentioned, which were rotated $45°$ around the z-axis), it is evident that $J(dd) = J(aa)$, and then eqn. (78) is equivalent to eqn. (70).

The ligands certainly influence the energy levels of a transition group complex, and this can be approximately described by saying that the interelectronic repulsion parameters are all multiplied by a factor β below 1 compared to those of the d-shell of the corresponding gaseous ion. This effect was first discussed by Tanabe and Sugano and is here called the *nephelauxetic* effect.* As discussed in the chapter on the nephelauxetic series, this effect is, in general, M.O. theory related not only to *central-field* covalency, but also to *symmetry-restricted* covalency, corresponding to the delocalization of the γ_3- and γ_5-orbitals by their intermixing with the orbitals of the ligands and their corresponding deviation from $l = 2$. However, it is a remarkably good working hypothesis to assume the expanded radial function model, considering only the central-field covalency. Here, the spherically symmetrical part, V_0, of the ligand field is added to the central field for determining the radial function (still having $l = 2$) according to eqn. (4). We may think of V_0 merely in an electrostatic manner as the largest component of the ligand field (cf. eqn. 50). However, only the variation of V_0 away from a constant (as it is internally in a sphere, surrounded by negative charge) is of importance for the radial function and for the interelectronic repulsion parameters. We may also think of V_0 as caused by the negative charge accumulated in the central ion by the donation of electrons (with arbitrary symmetry: even γ_1, odd γ_4, etc.) from the ligands to the central ion by covalent bonding. Both mental pictures emphasize that the radial function of the partly filled shell is only changed by the presence of more negative charge, screening the nucleus from the partly filled shell, in the complex than in the gaseous ion.

If we re-write eqn. (39) with $V_0 = f(\eta)$ where η (in suitable units) represents the average value of r^{-1} for the radial function (thus decreasing for increasing average radius), and if we retain only two terms in a Taylor expansion $V_0 = v + w\eta$ (both v and w must be positive in any reasonable model of V_0), then

$$E = \eta^2 - 2\eta + k\eta + v + w\eta \tag{79}$$

having the minimum at

$$\frac{dE}{d\eta} = 2\eta - 2 + k + w = 0 \quad \text{for} \quad \eta = 1 - \frac{k}{2} - \frac{w}{2} \tag{80}$$

This value of η corresponds to an energy in eqn. (79)

---

* This word was proposed by Professor Kaj Barr, University of Copenhagen, and signifies "cloud-expanding" in Greek.

$$E = -1 + v + k + w - \frac{kw}{2} - \frac{k^2}{4} - \frac{w^2}{4} \qquad (81)$$

If we consider two different values, k_1 and k_2, of the coefficient k characterizing the interelectronic repulsion, the energy difference will be

$$k_2 - k_1 - \frac{(k_2 - k_1)w}{2} - \frac{k_2^2}{4} + \frac{k_1^2}{4} \qquad (82)$$

representing a nephelauxetic effect $-(k_2 - k_1)w/2$ relative to the gaseous ion (eqn. 40). This effect may be described in two different ways: we may consider it as a second-order perturbation (of the type produced by the non-diagonal elements $\int(3d)V_0(nd)$ with $n = 4, 5, \ldots$) expanding the radial function from its original value of $\eta = 1 - k/2$ to $\eta = 1 - (k + w)/2$. Since the interelectronic repulsion parameters are proportional to η, we would then expect an energy decrease $kw/2$, giving the contribution $-(k_2 - k_1)w/2$ to the energy difference between two terms.

Surprisingly enough, we may also consider the effect as different values of the first-order perturbation energy (of the type $\int(3d, L, S)V_0(3d, L, S))$ according to the different values of $\eta_1 = 1 - k_1/2$ and $\eta_2 = 1 - k_2/2$. Since this difference is $w(\eta_2 - \eta_1)$, we also get $-(k_2 - k_1)w/2$ with this explanation.

Though these calculations are qualitatively illustrative, we cannot assume that the simplifications involved (e.g. the identical shape of the radial function by pure scaling) are quantitatively correct. However, they show a very important feature of the self-consistent orbitals: that they take account of the main part of the interelectronic repulsion as a contribution to the core field. The quantities involved in the M.O. J-integrals discussed above seem enormous. However, the main part of it is common to each configuration and does not participate in the separation of the terms characterized by S and Γ_n. Thus, for pure d-electrons, all the J-integrals in eqn. (75) contain the common quantity A, consisting mainly of the integral F^0. It can be estimated that A is approximately equal to $100B$ or $25C$, making the deviations of the J-integrals from their average value of the same order of magnitude as the K-integrals, some 4 to $15B$.

It is customary in atomic spectroscopy to define *empirical orbital energies* from the *baricentres* of a given configuration (the average of the energy levels, weighted by their degeneracy number e). These empirical orbital energies are ill-defined mixtures of one-electron and two-electron operator quantities. Actually, they are not strictly additive in different configurations. It is too primitive to say that the empirical orbital energies contain simply the interelectronic repulsion energy for a spherical symmetrical wave function. Thus, the energies in p^2 are

$$
\left.
\begin{array}{lll}
{}^3P & F^0 - \tfrac{1}{5}F^2 & \text{baricentre: } F^0 - \tfrac{2}{25}F^2 \\
{}^1D & F^0 + \tfrac{1}{25}F^2 & J(xy) = F^0 - \tfrac{2}{25}F^2 \\
{}^1S & F^0 + \tfrac{2}{5}F^2 & K(xy) = \tfrac{3}{25}F^2
\end{array}
\right\}
\tag{83}
$$

The coincidence between the baricentre and the J-integral between two different p-orbitals occurs for all p^n (compare eqn. (69) with (70)), while similar relations do not occur for d^n. A spherical symmetrical distribution $x^2 + y^2 + z^2$ has the J-integral $3J(xx) + 6J(xy)$ with itself, giving an average value of $J(xy) + \tfrac{2}{3}K(xy) = F^0$ by division with 9. Hence, the antisymmetrization, producing another second-order density matrix (p. 21) than derived from a pure product function of the first-order density matrix, stabilizes the baricentres of configurations slightly, compared to a completely spherical wave function without internal correlations.

However, the errors introduced by defining the baricentre energy from one or another of reasonable charge distributions are small, compared to the large contributions to the orbital energy in the empirical sense, derived from the interelectronic repulsion parameters. Thus, the orbital energy difference Δ between the pure sub-shell configurations ${}^4\Gamma_2$ of $\gamma_5{}^3$ and ${}^4\Gamma_5$ of $\gamma_5{}^2\gamma_3$ contains a contribution 2Θ in the general M.O. theory, where

$$
\Theta = J(3, 5) - K(3, 5) - J(4, 5) + K(4, 5) \tag{84}
$$

is zero for $l = 2$ (eqns. 74 and 75). If it is assumed that the interaction between a γ_3- and a γ_5-electron is decreased by ϑ and between two γ_3-electrons decreased by 2ϑ, relative to the interaction between two γ_5-electrons (as seems reasonable because the σ-antibonding γ_3-electrons are more delocalized than the γ_5-electrons), the value of Δ is always decreased by $(n - 1)\vartheta$ in a d^n-system, independent of whether we consider one-electron transitions or two-electron transitions (the latter occurring between $\gamma_5{}^a\gamma_3{}^b$ and $\gamma_5{}^{a-2}\gamma_3{}^{b+2}$).

Since ϑ might very well be several per cent of $F^0 \sim 80$ kK, this contribution may decrease Δ by a factor 2 or more. However, this is not an effect of the interelectronic repulsion to be taken as seriously as the observable energy differences between terms, as expressed by multiples of, for example, a K-integral. It is rather a common feature of the orbital energy differences such as Δ to contain huge, formal contributions from two-electron operators.

The excited states of electron transfer spectra have not very large values of the *squared overlap* $\int \psi_1{}^2 \psi_2{}^2$ between the orbital (mainly concentrated in the ligands) which has lost an electron, and the acceptor orbital (mainly concentrated in the central ion). Consequently, the K-integrals are small, in addition to the fact that the average values of r^{-1} occurring are smaller than for the ligand field transitions. Actually, the splitting of electron transfer configurations is surprisingly small, resembling the naïve M.O. picture, as will be discussed below. On the other hand, large splittings might occur if determined by differences between J(aa) of a doubly occupied orbital and J(ab) of two singly occupied orbitals. The degeneracy between $^2\Gamma_3$ and $^2\Gamma_4$ of $\gamma_5{}^3$ is not removed to any large extent in d^3-complexes, suggesting that their γ_5-electrons probably have a rather well-defined $l = 2$ (cf. eqn. 68).

Even the Slater ratio 5:3 between the excitation energy of $^2\Gamma_5$ and the degenerate $^2\Gamma_3$, $^2\Gamma_4$ (eqn. 77) is not much disturbed in complexes of Cr(III), Mo(III) and Re(IV), where effects of intermediate coupling and sub-shell configuration intermixing are not completely negligible. Among physicists, the best known fact about Slater's theory of partly filled shells is that the intervals 2:3 in eqn. (77) have the observed ratio $1:1\cdot14$ in $2p^2$ and $2p^4$ systems (but are nearly correct in $3p^2$ and $3p^4$), while the ratio is $1:2\cdot01$ in $2p^3$. This phenomenon is probably not caused by an expansion of the radial function of the type eqn. (40), since the interval is remarkably constant as a function of ionic charge ($1\cdot137$ for C, $1\cdot143$ O^{2+}, $1\cdot140$ F^{3+}, $1\cdot142$ Na^{5+}; $1\cdot135$ O, $1\cdot172$ Ne^{2+}, $1\cdot169$ Al^{5+}) where the relative importance of diagonal elements of interelectronic repulsion, compared to the orbital energies, decrease. This deviation from Slater's theory may be a rather special case, favouring angular correlation by interaction between the configurations $2s^2 2p^n$ and $2p^{n+2}$, as discussed on p. 47. The diagonal elements of Table 8 are those of a *strong-field determinant* where the non-diagonal elements contain only multiples of the interelectronic repulsion parameters B and C. For Δ increasing

towards large values, compared to B and C, the energy levels therefore approach asymptotically the diagonal elements, containing the
term $(-0.4a + 0.6b)\Delta$ corresponding to the pure sub-shell configurations $\gamma_5{}^a\gamma_3{}^b$. Only the intermixing, due to the possible presence
of more $^{2S+1}\Gamma_n$ of the same sub-shell configurations persist for $\Delta \to \infty$.

The *weak-field determinants* have also diagonal elements, being a
sum of a (not necessarily integral) multiple of Δ and of B and C, but
non-diagonal elements, containing only Δ and not interelectronic
repulsion parameters. This description was much used in the early
days of ligand field theory, since it was assumed that Δ was only a
small perturbation (perhaps 2 kK) on the multiplet terms of atomic
spectroscopy. The diagonal elements of weak-field determinants
correspond to wave functions with well-defined L. The two types of
determinants, taking also the non-diagonal elements into account, are
of course completely equivalent, producing the same polynomium
$E_p(E, \Delta, B, C)$. It is only a calculational convenience that it may be
quicker to find the eigenvalues in one of the two representations. If
the determinants are of first degree (as $^4\Gamma_2$ and $^4\Gamma_5$ of d^3 and d^7, or
$^5\Gamma_5$ and $^5\Gamma_3$ of d^4 and d^6), the wave function has for any value of Δ
both well-defined L and well-defined sub-shell configurations. In
the *Orgel diagram* which is the plot of the energy levels as function
of Δ, assuming fixed values of B and C (e.g. those of the gaseous
ion) or in the *Tanabe–Sugano diagram* as function of Δ/B (assuming
a fixed value, e.g. $C/B = 4$), these special levels are depicted by
straight lines.

If two levels with the same S and Γ_n interact, we get a hyberbola
in the Orgel or the Tanabe–Sugano diagram. An important example
is provided by the two levels Γ_4 with maximum value of S of d^2,
d^3, d^7 and d^8, having the secular determinants

$$\begin{vmatrix} 15B - E & 0.4\Delta \\ 0.4\Delta & 0.6\Delta - E \end{vmatrix}$$

$$= \begin{vmatrix} 0.8\Delta + 3B - E & 6B \\ 6B & -0.2\Delta + 12B - E \end{vmatrix} = 0 \quad (85)$$

(the value of Δ given has its sign inverted for d^2 and d^7). It is easy
to give the general transformation (the *re-diagonalization*) of a
second-degree weak-field determinant

$$\begin{vmatrix} A_1 + B_1 - E & A_{12} \\ A_{12} & A_2 + B_2 - E \end{vmatrix} = 0 \quad (86)$$

to the corresponding strong-field determinant

$$\begin{vmatrix} A_{11} + B_{11} - E & B_{12} \\ B_{12} & A_{22} + B_{22} - E \end{vmatrix} = 0 \qquad (87)$$

where A may represent the orbital energy difference Δ and B the parameters of interelectronic repulsion. Moreover, the results may be applied to many other types of variables. Since eqns. (86) and (87) represent the same polynomium E_p, we may identify the terms of each degree in E:

$$\begin{aligned} E^2: \quad & 1 = 1 \\ AE: \quad & -A_1 - A_2 = -A_{11} - A_{22} \\ BE: \quad & -B_1 - B_2 = -B_{11} - B_{22} \\ A^2: \quad & A_1 A_2 - A_{12}^2 = A_{11} A_{22} \\ AB: \quad & A_1 B_2 + B_1 A_2 = A_{11} B_{22} + B_{11} A_{22} \\ B^2: \quad & B_1 B_2 = B_{11} B_{22} - B_{12}^2 \end{aligned} \qquad (88)$$

The second and third lines of eqn. (88) represent the diagonal sum rule. The five relations between the ten parameters generally allow their full determination, if five of them are known (except for the sign of the non-diagonal elements, which is of no consequence for the energy levels). It is possible to characterize the transformation from eqn. (86) to eqn. (87) by a single parameter x, defined by

$$\begin{aligned} B_{11} &= xB_1 + (1-x)B_2 & B_{22} &= (1-x)B_1 + xB_2 \\ A_1 &= xA_{11} + (1-x)A_{22} & A_2 &= (1-x)A_{11} + xA_{22} \end{aligned} \qquad (89)$$

in the interval $0 \leqslant x \leqslant 1$. It can be found by the transformation of the weak-field wave functions Ψ_1 and Ψ_2 to the strong-field wave functions Ψ_{11} and Ψ_{22}

$$\begin{aligned} \Psi_1 &= a\Psi_{11} + b\Psi_{22} \\ \Psi_2 &= b\Psi_{11} - a\Psi_{22} & a^2 &= x \\ \Psi_{11} &= a\Psi_1 + b\Psi_2 & b^2 &= 1 - x \\ \Psi_{22} &= b\Psi_1 - a\Psi_2 \end{aligned} \qquad (90)$$

giving eqn. (89) on insertion of the symbols from eqn. (86) and eqn. (87). Actually, x is the square of the cosine to the angle in the Hilbert space between the wave functions Ψ_1 and Ψ_{11}. For the case eqn. (85), $x = 0.2$, indicating that the lower energy level Γ_4 contains from 20 to 100 per cent of the lowest strong-field wave function by going from $\Delta = 0$ to $\Delta = \infty$, and from 100 to 20 per cent of the lowest weak-field wave function ($L = 3$) in the same range. (For d^2 and d^7, x is 0·8.) However, for intermediate values of Δ/B, the coefficients relating the actual wave function to those of the weak-field wave functions are not the same as those relating it to the strong-field wave function, since

$$\Psi_x = c\Psi_1 + d\Psi_2 = (ac + bd)\Psi_{11} + (bc - ad)\Psi_{22} \quad (91)$$

The three other second-order determinants of d^2 and d^8 are also fully described by their values of x. Thus, for d^8, $^1\Gamma_5$ and $^1\Gamma_3$ have both $x = \frac{4}{7}$ and $^1\Gamma_1$ has $x = \frac{3}{5}$. The non-diagonal elements in a given representation can be found from

$$A_{12}^2 = x(1 - x)(A_{11} - A_{22})^2 \text{ and } B_{12}^2 = x(1 - x)(B_1 - B_2)^2 \quad (92)$$

For determinants of arbitarily high degree, the polynomiae of type eqn. (89) can be studied. It is especially useful to remember that, in addition to the diagonal sum rule, the squared sum of contributions from a given parameter B (the non-diagonal elements counted twice)

$$\sum_n B_{nn}^2 + 2 \sum_{m \neq n} B_{mn}^2 \quad (93)$$

is invariant by going from one determinant to another, representing the same energy levels.

Among secular determinants of third or higher degree in ligand field theory, it is useful to compare Finkelstein and Van Vleck's weak-field determinants for all levels of d^3 in O_h with the strong-field determinants, given by Tanabe and Sugano. In the case of two interacting energy levels, the description demands just two parameters, e.g. Δ and B in eqn. (85). In the case of three or more interacting levels, we may meet a problem by comparison with experiments. Thus, the three levels $^4\Gamma_4$ of d^5 have the determinants

$$\begin{vmatrix} ^4F - E & 0 & 2\Delta/\sqrt{5} \\ 0 & ^4P - E & 4\Delta/\sqrt{5} \\ 2\Delta/\sqrt{5} & 4\Delta/\sqrt{5} & ^4G - E \end{vmatrix}$$
$$= \begin{vmatrix} \Delta + 10B + 6C - E & -3\sqrt{2}\,B & C \\ -3\sqrt{2}\,B & 19B + 7C - E & -3\sqrt{2}\,B \\ C & -3\sqrt{2}\,B & -\Delta + 10B + 6C - E \end{vmatrix} = 0$$
$$(94)$$

and the three levels $^4\Gamma_5$ of d^5 the determinants

$$
\begin{vmatrix}
^4F - E & 2\Delta/\sqrt{7} & \sqrt{3}\Delta/\sqrt{7} \\
2\Delta/\sqrt{7} & ^4D - E & 0 \\
\sqrt{3}\Delta/\sqrt{7} & 0 & ^4G - E
\end{vmatrix}
$$

$$
= \begin{vmatrix}
\Delta + 18B + 6C - E & -\sqrt{6}\,B & 4B + C \\
-\sqrt{6}\,B & 13B + 5C - E & \sqrt{6}\,B \\
4B + C & \sqrt{6}\,B & -\Delta + 18B + 6C - E
\end{vmatrix} = 0
$$

(95)

Since Slater's theory in terms of B and C (Table 2) does not exactly agree with the distribution of the multiplets ^4G, ^4P, ^4D and ^4F in gaseous Mn^{2+}, we may prefer to use Orgel's weak-field determinants at the left-hand side, inserting the energy levels known from atomic spectroscopy. In these cases, the determinants are not equivalent, because we recognize the deviation from the Slater model with $l = 2$ in the gaseous ion. Since similar corrections presumably occur in

TABLE 9. THE VALUES OF LANDÉ PARAMETERS
IN kK DERIVED FROM SPIN-FORBIDDEN ABSORP-
TION BANDS AND FROM ATOMIC SPECTROSCOPY

		$k\zeta_{nd}$	ζ_{nd} for the gaseous ion
$3d^3$	Cr(III)	0·2–0·3	0·275
$3d^6$	Fe(II)	0·6	0·408
$3d^6$	Co(III)	0·5	0·6
$3d^8$	Ni(II)	0·5–0·8	0·63
$4d^3$	Mo(III)	0·6–0·7	0·80
$4d^6$	Rh(III)	0·9	$\sim$1·4
$4d^8$	Pd(II)	1	1·5
$5d^1$	Re(VI)	3·4 (k=1)	$\sim$4
$5d^3$	Ir(VI)	3·4 (k=1)	$\sim$6
$5d^6$	Ir(III)	2	$\sim$4
$5d^8$	Pt(II)	2	$\sim$4

complexes, it must not be taken *too* seriously to give definite values of B and C by adjusting a Tanabe–Sugano diagram to an experimental material of band wave numbers.

CHAPTER 6

CHARACTERISTICS OF ABSORPTION BANDS

THE absorption bands to be discussed in this book occur mainly in the near infra-red, in the visible and in the near ultra-violet regions. In the literature, most bands are characterized by their wave length λ, which can be measured in units of mμ, where 1 millimicron = 10^{-7} cm, or Å, where 1 ångström unit = 10^{-8} cm. However, we are more interested in the reciprocal of the wave length, the *wave number* σ as measured in cm^{-1}. This unit is now called kayser, K, and 1000 cm^{-1} is the kilokayser kK. The wave number σ is related to the energy difference $E_1 - E_2$ between two energy levels by the equation

$$E_1 - E_2 = hc\sigma \qquad (96)$$

according to N. Bohr's postulates from 1913, where h is Planck's constant and c the velocity of light. Some authors, especially in Japan, often indicate the *frequency* ν of a band in sec^{-1} (hertz). Since $\nu = c\sigma$, this quantity is 3×10^{10} times larger than the wave number. It is interesting for chemists to compare the wave number unit kK (which is actually an energy unit, because eqn. (96) is tacitly assumed) with other energy units, e.g.

$$
\begin{aligned}
1 \text{ kK} &= 0 \cdot 1239 \text{ eV (electron volt)} \\
&= 2 \cdot 85 \quad \text{kcal/mole} \\
&= 2 \cdot 1 \quad \text{powers of ten in equilibria constants} \\
&\qquad K_{equ} \text{ (at 25°C), related to free energy G} \\
&\qquad \text{by the equation } \Delta G = - RT \ln K_{equ} \\
&= 11,900 \text{ joules/mole}
\end{aligned} \qquad (97)
$$

The visible part of the electromagnetic wave spectrum does not have very sharp definition, but most people have no visual impression of radiation outside the range 12 to 26 kK. At ordinary light intensities, the maximum of eye sensitivity occurs at 18 kK, and the sensitivity vanishes very quickly below 15 kK and above 22 kK. Since the maximum of radiation from a "black body" occurs at

$T/(300°)$ kK, the sun, with a surface temperature T of some $6000°K$ (degrees Kelvin $= °K$), emits most light near to the sensitivity maximum of the eye.

The different colour sensations are distributed in a rather peculiar way on the wave number range. The red is very broad, going up to 16 kK, while the orange 16–17 and the yellow 17–17·5 kK are very narrow ranges. The green (with some shades) has a larger interval, 17·5–21 kK. It is customary to divide the following region in a clear blue, an "indigo blue" and a violet region. It is often assumed that the colours of monochromatic light (with a definite λ and σ) can be described in terms of varying portions of three trichromatic stimuli, having their maxima corresponding to a red, a yellow-green and a blue colour. With this definition, it is interesting to observe that the red stimulus re-appears above 22 kK, changing the blue to different violet and purple shades.

Any mixture of light observed by us seems to have a colour roughly corresponding to the impression of some monochromatic colour (some reddish purple shades, however, do not exist as such) mixed with a small or large amount of neutral, white light. Thus, we weigh the different proportions of monochromatic colours according to their absolute intensity, multiplied by our sensitivity, and obtain by direct inspection the impression of only one colour. It is well known that the absence of one narrow spectral range gives the impression of the *complementary colour* in the remaining light. Thus, the transmission through a transparent substance, having an absorption band in the green, or the reflection of light from such a substance, gives us the impression of a red colour. In this way, a biunique correspondence can be obtained between definite shades, e.g., red–green, orange–blue, and yellow–violet. However, if more than one absorption band occurs, we may have impressions of nearly neutral grey colours, which are easily shifted in one of the directions mentioned above. Thus, according to the concentration and layer thickness, and according to the illumination (daylight or tungsten lamps), $Cr(H_2O)_6^{3+}$ appears either red-violet or bluish green, and $RhBr_6^{3-}$ either dark cherry-red or greyish green. Though the eye has not a definite spectroscopic sense, these peculiarities of colours may sometimes indicate to us the absorption band positions. Thus, the pink colour desired by cosmetic experts occurs exactly in powders of neodymium salts, having a narrow absorption band in the yellow, as have also the porphyrine rings of the haemoglobin molecules, which can be clearly observed in the reflected light from the skin.

The most conspicuous spectroscopic phenomenon exhibited in Nature is the rainbow, which was explained by Descartes as the refraction effects in drops of water. Newton used a glass prism for a study of the decomposition of white light into its coloured components. However, a clear separation into monochromatic portions was not obtained before Fraunhofer used a very narrow slit, parallel to the prism, for the entrance of the light. Thus he discovered the many dark, narrow lines in the spectrum of the sun (and the reflected daylight) which were later identified by Bunsen and Kirchhoff as the wave numbers missing in the radiation from the hot, opaque gases below the solar surface, because cooler vapours of different elements in gaseous state absorb selectively some of the spectral lines that they also emit by excitation in a Bunsen gas burner or an electrical arc.

Our knowledge of spectroscopy is solely built on the use of such apparatus and it is very seldom that we observe absorption bands without a spectroscope. The present writer once observed narrow, dark bands in the rainbow-coloured, refracted light from a nearly point-source lamp, seen through a round flask, containing a solution of rare earths. Perhaps the red and blue separate images of lamps, which may sometimes be observed in a glass of red wine, absorbing in the middle of the visible spectrum, are due to the same cause.

Around 1900, mainly visual observations with spectrometers were made of the positions of absorption bands of coloured solutions, especially of the synthetic, organic dye materials. Formanék described many such measurements, but the broad bands of transition group complexes were generally not easily observed. In a few cases, such as the vibrational structure of the bands of MnO_4^- or UO_2^{2+} and in the rare earths, narrow bands were observed. As will be discussed below, this phenomenon can also be observed with a pocket spectroscope in the case of some extremely narrow ligand field transitions, e.g. of $ReCl_6^{2-}$ and $OsCl_6^{2-}$.

Around 1910, Jones and several other authors photographed spectra of solutions, making it possible to study a wider range of wave numbers than with visible light. However, water is only fairly transparent in the range 8 to 55 kK. In the infra-red the vibrational bands of O—H stretching are widened so much in the liquid as to block out all radiation, and the overtones can still be observed as weak bands at 13 kK. In the far ultra-violet, almost all materials have a high absorption, and it has become customary to call the region above the position of the strong bands of oxygen molecules the *Schumann violet*. It is necessary to work with *vacuo* spectrographs in this region, and it is very difficult to provide a source of continuous radiation without line structure. Above 90 kK it is necessary to use grating spectrographs, since even the most transparent optical materials, such as LiF and Al_2O_3 can no longer be used. The ordinary materials begin to be transparent again for radiations above 1000 kK, belonging to the X-ray region. (The wavelength unit 1 X is very nearly 10^{-11} cm, and most X-ray studies stop below 10 kX.) In this region, it is possible to observe interesting line spectra of the elements by excitation with anticathodes and the compounds have slightly broader absorption bands. These features, characteristic for each element, continue up to a wave number of order of magnitude Z^2 ry which is some 900,000 kK for uranium. The chemical bonding has a very small, but observable shifting effect on these bands, as will be discussed later on. For increasing wave number, all matter shows a general absorption, ionizing electrons away from the atoms, until the threshold limit for emission of neutrons from the nuclei is reached. This is some 10^7 kK for the most easily decomposed nuclei, deuterium and beryllium 9, and 6×10^7 kK for most other nuclei, indicating the binding energy per nucleon. When the wave number 4×137^2 ry $\sim 8 \times 10^6$ kK is reached, the light quanta contain sufficient energy to create an electron and a positron.

It is very difficult to measure the extinction coefficients defined below exactly by photographing spectra. One of the difficulties is that the sensitivity of most plates varies rather strongly with σ, and often exhibits secondary maxima for the sensitized plates used below 20 kK. For some years, visual *spectrophotometers* were much used, utilizing well-known techniques of visual comparison of light intensities. However, they could only be used significantly in the narrow range 15–23 kK. Before 1939, several laboratories constructed photoelectrical devices for spectroscopical measurements, but it is only since 1946 that such instruments, as those made by Beckman, Cary and Unicam, have become commercially available; these instruments have now become the standard equipment of many laboratories. It may safely be concluded that the recently evolved interest in the spectral range 10–50 kK is certainly connected with this availability of versatile instruments, without which the investi-gator must spend many years of constructional effort, and he generally ends up by becoming more interested in electronics than in spectra.

With a photoelectric spectrophotometer, we can measure the light absorption of a coloured solution as a function of λ (most instruments have their scale divided into wave length units), either at definite points, where λ in most cases can be reproduced within $\frac{1}{2}$ mμ, or continuously by a recording instrument on a paper. Generally, the intensity I of a beam of monochromatic light passing through an absorption cell, with parallel windows (spaced 1 cm apart) containing the coloured solution, is compared to the light intensity I_0 traversing a comparison cell filled with water (or another solvent, without the coloured substance) under similar circumstances. We define an *optical density* (*absorbancy*) D

$$D = \log_{10}(I_0/I) = \epsilon cl \qquad (98)$$

as the decrease of the (decadic) logarithm of the light intensity in the coloured solution. It is obvious that *Lambert's law*, i.e. D is propor-tional to l, is valid, if the monochromatic light retains its properties (if a cell absorbs half the monochromatic light, then two such cells, one after the other, absorb $\frac{1}{2} + \frac{1}{2} \times \frac{1}{2} = \frac{3}{4}$ of the light). We define the *molar extinction coefficient* ϵ from eqn. (98), where only one coloured substance is considered. If ϵ does not vary with the molar concentration c in moles per litre, *Beer's law* is obeyed. This is often obeyed with great accuracy (within 1 per cent) if the coloured substance does not dissociate or react with the solvent by dilution. If more coloured substances are present, we may rewrite eqn. (98)

$$D = l(\epsilon_1 c_1 + \epsilon_2 c_2 + \ldots + \epsilon_n c_n) \qquad (99)$$

In the principle of *colorimetric analysis* eqn. (99) must be applied, and the molar concentrations $c_1, c_2, \ldots$ may be found if the extinction coefficients $\epsilon_1, \epsilon_2, \ldots$ are previously known from experiments on solutions containing only one coloured component. Instrumental deviations from Beer's law may occur, caused by the light beam not being completely monochromatic. In that case D generally increases too slowly with increasing c_n, since some of the light has a λ, corre-sponding to a smaller ϵ than the average λ. It may even be possible to get entirely false results from parasite stray light. Thus, the slit widths increase in most photoelectric spectrophotometers near to the limits of their applicability, in the infra-red because the photocells grow too insensitive, and in the ultra-violet because the light-source emits too little radiation. In double-beam apparatus solvents in the comparison cell may show high absorption. In all these cases of excessive slit width the false light, which is only a fractional amount of the light intensity in ordinary measurements, becomes the main source of the photoelectric impulse. Thus, many spectrophotometers produce readings in the range 2000 to 1300 $m\mu$, which are sometimes given in literature as broad absorption bands. Since 1 cm of water is completely opaque to this radiation, it does not correspond to a real band there.

The colorimetric analysis method is in principle a very informative one, since q different coloured substances can be determined by measurements at q different wave lengths, forming a set of q linear equations of the form eqn. (99). It can further be shown that no more than these q substances occur by measuring at more than q wave lengths. However, the experimental uncertainties may be too large to make possible the full exploitation of this mathematical technique, especially if one of the spectra is almost a linear combination of two other spectra. If this happens exactly, our linear equations (99) cannot separate this species out from the appropriate mixture of the two other species. It is often important for a chemist to demonstrate that a given set of solutions does only contain two coloured species in measurable and varying concentrations. For this purpose, the occurrence of *isosbestic* points (with $\epsilon_1 = \epsilon_2$) has often been studied in literature, and it is argued that a third ϵ_3 would probably be different at that λ and would therefore disturb the regular crossing of the set of curves. However, this is only a special case of the general investigation, which we always may make into such a set of curves, as to whether all of them can be represented as a linear combination $c_1\epsilon_1 + c_2\epsilon_2$ (and, if we know the total concentration of the two coloured species from preparation of the solution or chemical analysis, as to whether $c_1 + c_2$ agrees with this value). For this type of study, it is very important to avoid or to correct for *dust errors*. If the solutions are not completely limpid (and even hard filters may sometimes fail for this purpose, especially for large l) the loss of scattered light corresponds to a weak absorption, increasing slowly with decreasing wave length. Often, this effect may be estimated from places in the spectrum, where ϵ of all the species are known to be very small.

From the study of chemical equilibria between complexes in solution, it may sometimes happen that the concentrations $c_1, c_2, \ldots$ in eqn. (99) are known, and then the linear equations for several wave

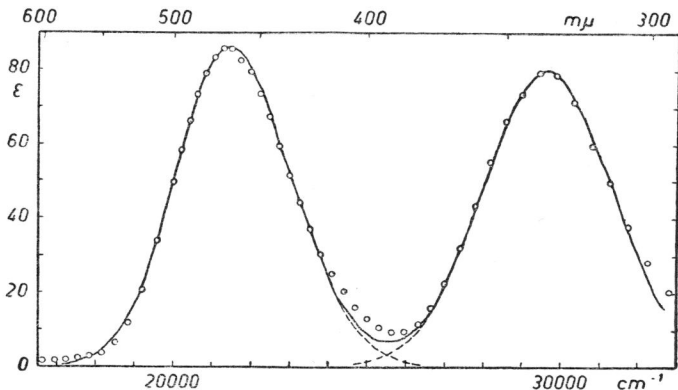

FIG. 3. The absorption spectrum of the tris (ethylenediamine) cobalt (III) ion Co en_3^{+3}. The curves drawn are Gaussian curves with parameters from eqn. (100)

σ_0	ϵ_0	$\delta(-)$	$\delta(+)$
21·4 kK	86	1·6 kK	1·9kK
29·6	80	1·9	2·0

Further, a spin-forbidden band can be seen in the red (σ_0 13·7 kK, ϵ_0 0·33, δ (–) 1·7 kK). The Figure is reproduced with permission from *Acta Chem. Scand.* (ref. 264).

lengths may be used for determining the extinction coefficients $\epsilon_1, \epsilon_2, \ldots$ This is especially important in the cases studied by J. Bjerrum (since 1929) of complexes such as $Ni(NH_3)_n(H_2O)_{6-n}^{2+}$ with $n = 1, 2, 3, 4, 5$, (which cannot be made the only coloured species in any solvent) by varying the ammonia concentration. In the chapter on identification of complexes and determination of their equilibrium formation constants, we shall return to some of these applications of absorption spectra. We may even draw some conclusions in the cases where neither ϵ_n nor c_n in eqn. (99) are known previously.

We may plot the absorption spectrum as ϵ as a function of λ or of σ. It is not reasonable (for coloured solutions) to indicate the transmission T, which is the percentage ratio between I and I_0, since the behaviour as a function of varying concentration gives highly different pictures of T. Sometimes, $\log_{10}\epsilon$ as a function of λ or σ is indicated in literature. This has the advantage that very small and very high bands can conveniently be represented on the same diagram without changing the scale of ϵ; on the other hand, the absorption of two species is no longer directly additive on these diagrams, and fine details, corresponding to a variation of ϵ some 10 per cent are lost, being only a change in $\log_{10}\epsilon$ of some 0·04.

The broad absorption bands of complexes generally have the shape of *Gaussian error-curves*, and often a region of the spectrum can be successfully represented as a linear combination of two or three such Gaussian curves (Fig. 3). Kuhn and Braun proposed (1930) the Gaussian curve in wave number σ:

$$\epsilon = \epsilon_0 \exp\left(-(\sigma - \sigma_0)^2/\vartheta^2\right) = \epsilon_0 2^{-(\sigma-\sigma_0)^2/\delta^2} \tag{100}$$

with ϵ_0 of the maximum at σ_0 and the half width δ (since $\epsilon = \epsilon_0/2$ for $\sigma = \sigma_0 \pm \delta$) being $\vartheta\sqrt{(\ln 2)}$. The area of a band in this representation is proportional to the product of δ and ϵ_0, since for $\sigma_0 \gg \delta$

$$\int_0^\infty \epsilon d\sigma = 2\epsilon_0 \int_0^\infty \left[\exp\left(-(\sigma - \sigma_0)^2 \ln 2/\delta^2\right)\right] d(\sigma - \sigma_0)$$

$$= \frac{\sqrt{(\pi)}}{\sqrt{(\ln 2)}} \epsilon_0\delta = 2 \cdot 1289\ \epsilon_0\delta \tag{101}$$

This area is proportional to the *oscillator strength P* of the transition, being

$$P = \frac{1000}{N} \cdot \frac{mc^2}{\pi e^2} \cdot 2 \cdot 30 \int \epsilon d\sigma = 4 \cdot 32 \times 10^{-9} \int \epsilon d\sigma$$

$$= 9 \cdot 20 \times 10^{-9}\epsilon_0\delta \tag{102}$$

where N is Avogadro's number, $-e$ and m the charge and mass of the electron, and c the velocity of light.

Lowry and Hudson (1933) suggested a Gaussian shape with λ as independent variable, and Mead showed that this agreed better with the spectra of several chromium (III) and cobalt (III) complexes, as Claus E. Schäffer will discuss later. For a classification of absorption bands, the present writer (1954) suggested a Gaussian curve eqn. (100) with different half widths, $\delta(-)$ and $\delta(+)$, towards smaller and larger wave number σ. Shimura has defined an inclination parameter δ by

$$\delta = \{\delta(+) - \delta(-)\}/2 \qquad (103)$$

calling the double half-width $1 = \delta(-) + \delta(+)$.

All these three proposals (with $\delta(+)$ generally some 10 per cent larger than $\delta(-)$) are in consequence nearly coincident with Henri's proposal (1913)

$$\epsilon = a\sigma \exp \{-b(\sigma - \sigma_0)^2\} \qquad (104)$$

actually being more related to quantum mechanics than the other proposals, as shall be discussed below. However, the introduction of a factor σ has rather complicated mathematical consequences. Thus, the maximum value of ϵ no longer occurs at σ_0, but is displaced to some $\sigma_0 + \delta^2/(\sigma_0 2 \ln 2)$, and the ratio $\delta(+)/\delta(-)$ is approximately $1 + \delta/\sigma_0 \ln 2$. With the usual conditions of $\delta \sim 1\cdot 5$ kK and $\sigma_0 \sim 20$ kK, this agrees excellently with the experiment. However, for most practical purposes, the simpler eqn. (100), possibly with different $\delta(-)$ and $\delta(+)$, is just as useful, especially since small deviations from eqn. (104) must occur for several reasons. In practice, the maximum is not determined simply by choosing the highest optical density of a given measured band, D. As recommended by Scheibe, a much more accurate method is to find the limiting value of the mean wave lengths $(\lambda_1 + \lambda_2)/2$ where D is the same for λ_1 and λ_2, for $D \to D_0$. After some training, this limiting value can be estimated directly from the behaviour of D in the interval $0\cdot 7$–$1\cdot 0$ D_0, and the maximum λ_0 determined at least five times more accurately than simply by looking for the highest point of the curve.

Orgel (1955) explained the width of the absorption bands from the *Franck–Condon principle*, using the concept of *potential curves* as function of internuclear distances. Molecular spectroscopy has a set of variables not known in atomic spectroscopy: the energy of the *electronic states*, the main subject of this book, depends in a continuous way on the internuclear distances. These produce a high number of variables in a polyatomic molecule, and it is convenient to classify them according to *normal modes of vibration*, having the same group-theoretical quantum numbers Γ_n (and parity) as the electronic states. If we neglect (the sometimes considerable) inter-mixing of normal modes, having the same Γ_n and parity, we may divide the normal modes into *stretching* modes, changing the distances from the central atom, but not the angles between the ligands,

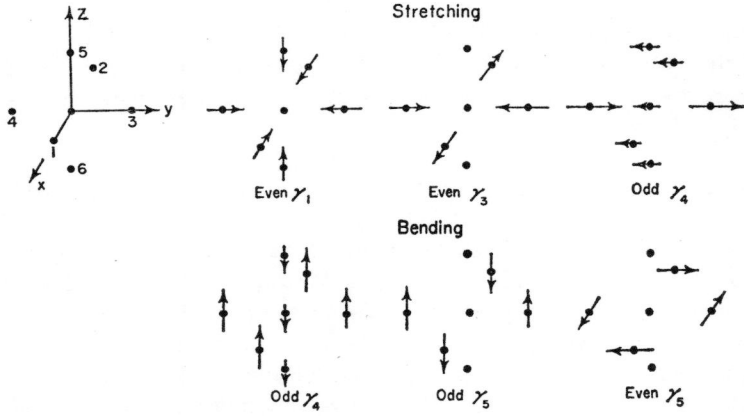

FIG. 4. The normal modes of vibration for an octahedral complex. The co-ordinate system as in Fig. 2. Linear combinations can be formed of the three vibrations having the same Γ_n, odd γ_4, viz. the stretching and the bending modes shown on the Figure and the pure translation of all seven atoms.

and *bending* modes, changing the mutual distances between the ligands without affecting the length of the bonds to the central atom. For an octahedral complex, the modes are (compare Fig. 4 with the orbitals of Fig. 2):

$$\text{stretching: even } \Gamma_1, \text{ odd } \Gamma_4, \text{ even } \Gamma_3$$
$$\text{bending: odd } \Gamma_4, \text{ even } \Gamma_5, \text{ odd } \Gamma_5 \qquad (105)$$

The sum of degeneracy numbers e according to eqn. (43) is 15, representing the main part of the twenty-one degrees of freedom necessary for describing the motion of seven particles in ML_6. The rest is the three degrees of pure translation of the whole complex and the three degrees of pure rotation. It is seen that a formal correspondence exists between the stretching modes of eqn. (105) and the σ-orbitals of Table 4. In the same way, the bending modes correspond to the π-bonding orbitals of Table 4, except that even Γ_4 has disappeared as a normal mode of vibration, being used for the pure rotation. Each of the normal modes of vibration corresponds to the first approximation (as a Taylor expansion, where the term of first degree necessarily is absent near to the equilibrium) to a *harmonic oscillator*. The potential energy is a contribution of the form $U_0 + k_1(x - x_0)^2$ where x is the normal co-ordinate in the vibration, and k_1 the force constant. The eigenvalues of energy of vibration in a harmonic oscillator is

$$E = (n + \tfrac{1}{2})hc\sigma_c \quad \text{with} \quad \sigma_c = \frac{1}{2\pi c}\sqrt{\frac{k_1}{m}} \qquad (106)$$

n is an integer $= 0, 1, 2, \ldots$, h Planck's constant, m the reduced mass of the oscillating system and σ_c is the characteristic wave number of the vibration. In classical mechanics, this energy E would correspond to a maximum amplitude

$$\sqrt{(2n + 1)}x_c = |x - x_0| = \sqrt{\left\{\frac{(2n + 1)h}{4\pi}\right\}} \; \frac{1}{\sqrt[4]{(k_1 m)}} \qquad (107)$$

In quantum mechanics, the wave functions have roughly the same spatial extension, but are complicated polynomiae (having n nodes) multiplied by exponential functions. The simplest wave function, that of the ground state with $n = 0$, is a Gaussian function with the normalisation factor N_0

$$\Psi_0 = N_0 \exp\left(-\tfrac{1}{2}x^2/x^2{}_c\right) \qquad (108)$$

The Franck–Condon principle states that the oscillator strengths are proportional (by a factor containing σ) to the square of a matrix element (of electric dipole moment to be discussed below), containing the products of the ground state functions $\Psi_{elec}\Psi_{vib}$ with the excited state functions $\Psi'_{elec}\Psi'_{vib}$ at the *same internuclear distances*. We may interest ourselves mainly in the *totally symmetric* stretching modes (having the symmetry even Γ_1). In this normal mode, all (the identical) six distances M–L (nuclei) in the octahedral complex are changed by the same amount (at a given moment), retaining the regularly octahedral symmetry. If we do not have special effects of the Jahn–Teller instability, to be mentioned later on, the main deviation of the excited electronic state from the potential curve of the electronic ground state is caused by the totally symmetric stretching mode, the equilibrium internuclear distances (corresponding to the minimum of energy) being all multiplied by the same factor in the excited electronic state. This agrees well with the changes in *crystallographic ionic radii*, pointed out by van Santen and van Wieringen (1952) with the sub-shell configurations $\gamma_5{}^a\gamma_3{}^b$ in the first transition group, that γ_3-electrons expand the central ion by some 0·1 Å, compared to γ_5-electrons.

It might at first glance seem strange that the vibrations are sufficient for explaining the large half widths $\delta \sim 2$ kK occurring in many ligand field bands, especially since the vibrational excitation of the electronic ground state has the same order of magnitude as kT (which at room temperature is as small as 0·21 kK). Contrary to the case in atomic spectroscopy, we never observe absorption bands in complexes from excited electronic states, if their energy is higher than some 0·4 kK above the ground state (and nearly all states are at least above 10 kK). We know from the infra-red and Raman studies of many molecules that the vibrational wave numbers σ_e from eqn. (106) are nearly always below 1 kK, except the stretching vibrations of some multiple bonds, having large force constants k_1 and the stretching vibrations of molecules containing hydrogen (small m), which may have σ_e in C—H, N—H and O—H slightly above 3 kK. The latter vibrations can be identified by substituting hydrogen with deuterium, decreasing the wave numbers by a factor of about $1/\sqrt{2}$. Unfortunately, we do not know much about the vibrational wave numbers in complexes. In the tetrahedral $MnO_4{}^-$ and the linear $UO_2{}^{2+}$, the *vibrational fine-structure* of the absorption bands indicate accurately a stretching wave number 0·75 kK. However, this is connected with an unusually large force constant, corresponding to the high oxidation number of Mn and U. A large number of gaseous fluorides MF_6 are known, and their infra-red spectra have supplied valuable information about the vibrations of octahedral molecules. The stretching wave numbers vary between 0·5 and 0·6 kK, while the bending wave numbers are between 0·1 and 0·3 kK, Since the reduced mass m in eqn. (106) is simply the mass of the ligand for the totally symmetric mode of vibration even Γ_1, the constancy of the corresponding wave number in fluorides such as SF_6 and UF_6 implies a rather constant force constant. In the more familiar complexes of water and ammonia, the stretching wave numbers are probably quite small, from 0·2 to 0·4 kK, but it is very difficult to measure them and to distinguish them from other, bending wave numbers of hydrogen-containing complexes. The measurements of Dreisch and Trommer, showing the vibrational structure of some ligand field bands in the infra-red, have later been shown to be caused by an instrumental error, and the only known certain cases of vibrational structure of the bands of octahedral complexes in

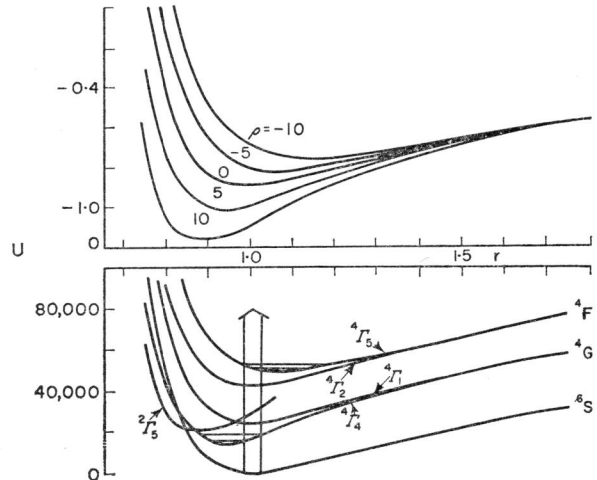

FIG. 5. The potential curve of eqn. (109) with different values of the ligand field stabilization parameter ρ. As an example, the energy levels of manganese (II) complexes are given below as function of the ligand distance in the octahedral symmetry. The mechanism of band widths is indicated by the vertical projection of the vibration of the electronic ground level on the potential curves of the excited levels. It is also seen how the low-spin levels (though not the actual ground level of Mn (II)) sometimes have smaller equilibrium internuclear distances than the high-spin levels. The Figure is reproduced, by permission from *Acta Chem. Scand.* (ref. 283).

solution are $ReCl_6^{2-}$ and $OsCl_6^{2-}$, having wave number differences of 0·17 kK between some five consecutive components, and some structure in FeF_6^{3-} of the same order of magnitude. However, it is difficult to identify these vibrations with definite normal modes.

The Franck–Condon principle, as discussed above, functions as a *projection* of the (squared) vibrational wave function of the electronic ground state upon the potential curve of the excited electronic state. Fig. 5 illustrates this effect, using as numerical example the potential U as a polynomium in r^{-1}:

$$U = \frac{1}{9r^9} - \frac{0 \cdot 02\rho}{r^5} - \frac{1}{r} \qquad (109)$$

where the unit of r is chosen such that for $\rho = 0$, the minimum of U occurs at $r = 1$. The first of the three parts of the expression represents the mutual repulsion between the closed shells, the second the ligand field effects (with $\rho = 2n_5 - 3n_3$, n_3 and n_5 being the numbers of γ_3- and γ_5-electrons) and the third the attraction of the ligand. The equilibrium distances r_L occur for:

$$
\begin{array}{cccccc}
\rho = & +10 & +5 & 0 & -5 & -10 \\
r_L = & 0 \cdot 887 & 0 \cdot 940 & 1 \cdot 000 & 1 \cdot 064 & 1 \cdot 128
\end{array} \qquad (110)
$$

showing the influence of ρ on the crystallographic radii. Even though the potential U is certainly not purely electrostatic, it has been assumed that the energy difference Δ between the γ_3- and γ_5-orbitals varies proportionally to r^{-5}, as given by the electrostatic model (eqn. 51). Experience seems indirectly to show that this is a fair approximation.

The square of the vibrational wave function Ψ_0^2 (eqn. 108) of the lowest level of a harmonic oscillator is a Gaussian curve with the half width $x_c \sqrt{(\ln 2)}$, producing by the projection an almost Gaussian-shaped absorption band (cf. eqn. 104)

$$\epsilon = \frac{\sigma}{\sigma_0} \, \epsilon_0 \cdot 2^{-(\sigma - \sigma_0)^2/\delta^2} \qquad (111)$$

with the half width δ of the Gaussian part

$$\delta = 0 \cdot 834 x_c \cdot \frac{dU}{dr} \qquad (112)$$

assuming that the potential curve U of the excited electronic state can be approximated by a straight line for values of r near to the equilibrium value r_L of the electronic ground state. With the assumption of Δ proportional to r^{-5}, this implies for small values of x_c

$$\delta = 4 \cdot 17 \frac{x_c}{r_L} \Delta \qquad (113)$$

without specific assumptions on the parts of U common to the excited and the ground state of the ligand field transition. Since x_c is proportional to $(k_1 m)^{-\frac{1}{4}}$ (eqn. 107), it is not expected to vary more than between 0·03 and 0·08 Å. Actually, manganese (II) and nickel (II) complexes have $\delta \sim 0.15\Delta$, while rhodium (III) has $\delta \sim 0.09\Delta$ and iridium (III) (mainly halide complexes are known) $\delta \sim 0.08\Delta$, corresponding to higher values of km in the platinum group complexes.

However, if σ_c is not much larger than 0·21 kK (the mean Boltzmann energy at room temperature), transitions will occur from the nth vibrational state of the ground level, having two strong maxima of the wave function near $r_L \pm x_c\sqrt{(2n + 1)}$. The feet of the absorption bands will thus be too high and show a strong increase with the absolute temperature T, while δ is approximately proportional to $\sqrt{T}$ for a small σ_c. On the other hand, all bands will vanish too quickly towards the low wave numbers, because dU/dr of eqn. (112) is not constant, but approaches 0 in the bottom of the potential hole.

Recently, Stolarczyk has shown that for high temperatures $(kT \gg hc\sigma_c)$ the absorption bands are again Gaussian shaped with δ proportional to $\sqrt{T}$. Since this is definitely not the case for intermediate values of σ_c, the general occurrence of Gaussian-shaped absorption bands might indicate that we are near to one of the limiting cases, either $kT \gg hc\sigma_c$ or $kT \ll hc\sigma_c$. Further experimental work is necessary, but since δ of several bands have been shown to increase some 10 per cent by going from 290°K to 360°K, the former possibility is perhaps the most probable. However, the presence of many normal modes of vibrations, many of which have very low σ_c, may be the explanation of the completely smooth Gaussian curves. There must be some reason for the single transitions, going from one definite vibrational state of the electronic ground state to another definite vibrational state of the excited electronic state, being blurred out to a width at least some 50 per cent of the average distance between the wave numbers of such transitions.*

The common Gaussian shape for most absorption bands makes it possible to concentrate on only three principal properties of each

---

* The situation is entirely different in gaseous 5d-hexafluorides and in many solids studied recently below 80° K. Thus, Co(II), Ni(II), Cu(II) imbedded in MgO, ZnO or CdS show generally one or a few, extremely narrow lines followed by a broader structure in each ligand field transition (private communications by Dr. Pappalardo, and by Dr. McClure and Dr. H. Weakliem).

band: the position, the half width, and the area or intensity. The *position* indicates the energy difference between the excited and the ground state at the minimum of the potential curve of the ground state. It is this quantity that the bulk of this book is concerned with, and we shall assume that the maximum value of ϵ represents this position, though eqn. (104) may indicate that this is a slight over-simplification. The *half width*, as discussed above, is a measure of the anti-bonding character of the excited state. Even though we cannot estimate the increase of internuclear equilibrium distance r_L from δ alone (see eqn. 112) without knowledge of the force constant of the excited electronic state (and its anharmonicity), we may certainly expect a monotonic increase of r_L with increasing δ in a given complex. In particular, the transitions within the same M.O. configuration, i.e. not changing the electron density in the molecule (but only the average mutual distance between the electrons in the partly filled M.O.), correspond to identical values of r_L in the excited and the ground state, and thus, to very narrow bands. Finally, the area or *intensity* shall be discussed below. According to eqn. (101), it is proportional to $\epsilon_0\delta$. There is one more valuable consequence of the near generality of Gaussian shape: that independent bands may be inferred from the presence of *shoulders* or weak *inflexions* in the absorption spectrum. These bands are separated by *curve analysis*. However, since the addition of two Gaussian curves with different values of σ_0 does not yield a simple expression, there is no general mathematical method available for the curve analysis, and it is mainly performed by trial and error. The resulting heights ϵ_0 and to a lesser extent, the half widths δ, of the component bands are often highly uncertain, in the worst cases by a factor 5. The individual maxima σ_0 are much better determined, often with an accuracy of some 0·1 kK. This is due to the fact that the differential quotient of a sum of two functions f and g equals the sum of their differential quotients $f' + g'$. Since the maximum of f occurs where $f' = 0$, we look for the point on the experimental absorption spectrum where the slope equals that of the isolated function g', reasonably estimated (thus, the maxima "attract" each other from their original positions, if two maxima can actually be observed: the latter occur where f' cancels g' with the opposite sign). This estimate is largely a matter of practice; two principal methods may assist in the choice of the component bands f and g: the *reflection in the median*, where one of the two sides of the band f is sufficiently free for other bands to allow an estimate of $\delta(-)$ or $\delta(+)$, and then extrapolating it with the same or a slightly altered δ on the opposite side. When this approximation is subtracted from the experimental spectrum, the other band should come free. This process may be iterated until both bands have a reasonable shape. The other method, the *subtraction of the extrapolated background* (which may be useful also for rescuing results disturbed by inevitable dust errors) is applied to small bands at the foot of a strongly increasing band. Here, it is necessary to get accustomed to the fading away of a band, when no other band is present, and if possible to study the same spectrum at several ranges of concentration c (or layer thickness l), since most observers sometimes believe that they can see weak shoulders at the feet of perfect Gaussian curves at certain slopes of the optical density D versus the wave length scale. For subtraction from the extrapolated high band, it may be useful to plot log ϵ as function of σ, since pure Gaussian curves are parabolae in this representation (cf. the large atlas in the Landolt–Börnstein tables, Volume 1, Division 3 of the 6th edition, 1951). However, the subtraction process cannot be performed directly on irregularities of these parabolae, but must be transformed to the ϵ-diagrams.

The intensity (proportional to the product $\epsilon_0\delta$) is probably the quantity most varying among the absorption bands known. Some organic molecules containing a large number of atoms may have values of ϵ_0 in the range up to 10^6. The highest electron transfer band as found in complexes has $\epsilon_0 = 70,000$ (PtBr$_6^{2-}$). On the other hand, very weak bands have been observed for other complexes, such as $\epsilon_0 = 0.014$ for one of the bands of $Mn(H_2O)_6^{2+}$. Obviously, the accurate study of such weak bands demands very severe conditions, and the possible presence of some parts per million of impurities having strong absorption bands in the same spectral range must be guarded against.

Within this range of ϵ_0, varying by a factor of five millions, the bands of transition group complexes tend definitely to fall within two classes: the *Laporte-allowed* bands with ϵ_0 generally above 2000, and the *Laporte-forbidden* bands with ϵ_0 generally below 200. The physical explanation of this grouping is the following: As pointed out by Van Vleck (1937) and Broer, Gorter, and Hoogschagen (1945), the only mechanism of light absorption, producing oscillator strengths P (eqn. 102) above 10^{-6}, is *electric dipole radiation*, while other mechanisms such as magnetic dipole and electric quadrupole

H

radiation at most are responsible for some very weak bands of europium (III) and other rare earths. However, the electric dipole moment transforms as odd Γ_4 in octahedral symmetry O_h and has odd parity in any symmetry containing a centre of inversion. Since transitions are only possible between two states, having the wave functions Ψ_a and Ψ_b, where the product

$$\Gamma_a \times \Gamma \text{ (electric dipole moment)} \times \Gamma_b \qquad (114)$$

contains the totally symmetric representation even Γ_1, we have the Laporte rule, originally discovered in atomic spectroscopy, that *transitions are only allowed between states of opposite parity*, because even $\times$ odd $\times$ even is odd and not even.

If we use the multiplication table for O_h, Table 5, we see that only some Laporte-allowed transitions are *symmetry-allowed*, viz.

$$\left.\begin{array}{ll} \Gamma_1 \rightarrow \Gamma_4 & \Gamma_4 \rightarrow \Gamma_1, \Gamma_3, \Gamma_4, \text{ and } \Gamma_5 \\ \Gamma_2 \rightarrow \Gamma_5 & \Gamma_5 \rightarrow \Gamma_2, \Gamma_3, \Gamma_4, \text{ and } \Gamma_5 \\ \Gamma_3 \rightarrow \Gamma_4 \text{ and } \Gamma_5 & \end{array}\right\} \qquad (115)$$

while the others are *symmetry-forbidden*.

However, an analogy may be drawn between the atoms and molecules, and human society, that some individuals generally find ways to transgress the existing laws. Actually, Laporte-forbidden transitions are also observed, but with an intensity only between 10^{-4} and 10^{-1} of the Laporte-allowed, symmetry-allowed transitions. This is described in quantum mechanics as second-order perturbations of operators of odd parity, mixing together states of opposite parity. We may visualize two such types of odd operators: *hemihedric* ligand fields (as contrasted with the *holohedric* parts of the ligand field, the latter being identical in the points (x, y, z) and ($-$x, $-$y, $-$z) of a Cartesian system with the centre of inversion at the origin) and *vibronic intermixing* of electronic states by odd vibrations.

The hemihedric ligand fields do not seem important for the main part of the intensity of the Laporte-forbidden ligand field bands, since the complexes without a centre of inversion, $Co(NH_3)_5Cl^{2+}$ and *cis*-$Co(NH_3)_4Cl_2^+$, do not have much more intense absorption bands in the visible region than the holohedric complexes $Co(NH_3)_6^{3+}$ and *trans*-$Co(NH_3)_4Cl_2^+$. This is probably caused by the partly filled shell still continuing to be mainly a d-shell with even parity, though the group theory no longer demands it to have a definite parity. This illustrates the fact that matrix elements, which are not required to vanish by group theory, may still approximately retain the value zero they would have in higher symmetries. It is sometimes argued that the high intensities of ligand field bands in tetrahedral $CoCl_4^{2-}$ are caused by the absence of a centre of inversion. This is undoubtedly a necessary (since no adjacent strong bands are observed), but not a sufficient condition, since tetrahedral blue-violet $NiCl_4^{2-}$ (in chloride melts) seems to have five times weaker ligand field bands.

As emphasized by Tanabe and Sugano and several other ligand field theoreticists, the odd vibrations (the odd Γ_4 and odd Γ_5 from eqn. (105)) are the main causes of intensity in the Laporte-forbidden bands of octahedral complexes. During the performance of these odd vibrations, the electronic wave functions of the complex are composed of much Ψ_{even} plus a small admixture of Ψ_{odd}, produced by the second-order perturbation, called *vibronic interaction*. It is an experimental fact that the ratio between the oscillator strength P_f of Laporte-forbidden and P_a of the nearest Laporte-allowed band is

$$\frac{P_f}{P_a} = \frac{\sigma_f}{\sigma_a} \cdot \frac{E_{af}^2}{(\sigma_a - \sigma_f)^2} \tag{116}$$

with E_{af} surprisingly constant between 1·5 and 2·5 kK in different complexes of the three ordinary transition groups and of the actinides. E_{af} is not an ordinary perturbation energy, but it is a complicated expression, involving the amplitudes of the odd vibration and the changes in the electronic wave function during the odd vibration. Liehr and Ballhausen have calculated the intensity of ligand field bands, e.g. of $Ti(H_2O)_6^{3+}$ and $Ni(H_2O)_6^{2+}$, guessing wave numbers σ_c of the odd vibrations (which are not known from experiment), assuming the odd, excited states to be intermixed to be $3d^{n-1}4p$ in $3d^n$, and assuming hydrogenic radial functions (eqn. 6). They obtained agreement with the experimental value of P within some 20 per cent and conclude that they have thus proved the correctness of vibronic interaction. This somewhat optimistic statement offends the logical theorem

$$(a \to b)(b) \not\to a \tag{117}$$

which is always troublesome to workers in natural sciences. When a quantitative model a (perhaps with many fine parameters, given with several significant figures) gives the result b, which is known from experiment to be true, it is not possible to imply in the opposite sense that a is true. It is easy for us to believe that a good explanation, giving accurate results, is the only correct one; but the history of theories of chemical bonding has already seen the decay of several such good explanations, when they were applied to new material.

Since the excited states of the Laporte-allowed transitions in eqn. (116) are almost certainly electron transfer states in most complexes, as shall be discussed below, it is not a good approximation to consider the odd states in the vibronic intermixing to be caused by an excitation of a 3d-electron to the 4p-shell. Actually, since the bonding electrons of the ligands resemble far more 4p-electrons in their spatial extension, it would be a slightly better approximation to characterize the excited states of the electron transfer bands by $3d^{n+1}4p^{-1}$, even though these configurations from spherical symmetry are not appropriate to the octahedral microsymmetry prevailing in the bond regions of such a complex.

The *selection rules* of the type eqn. (115) do not apply to Laporte-forbidden transitions in O_h, where we must multiply eqn. (114) also with odd Γ_4 or Γ_5, permitting all transitions to the same, small extent. In lower symmetries, such as the tetragonal in *trans*-$Co(NH_3)_4Cl_2^+$, there seem, according to Moffitt and Ballhausen, to be selection rules valid for polarized light. However, this subject is somewhat uncertain due to the fact that more than one odd normal mode of vibration may be excited in the ground state, if σ_c is small.

It is less useless to attempt to calculate the oscillator strength of Laporte-allowed transitions. The oscillator strength is connected with the square of the matrix element of electric dipole moment $\bar{r}$ between the ground state Ψ_1 and the excited state Ψ_2 with the degeneracy numbers e_1 and e_2

$$P \cong \frac{\sigma}{3\,\mathrm{ry}} \cdot \frac{e_2}{e_1} \cdot \frac{1}{a_0^2} \left[\int \Psi_1\, \bar{r}\, \Psi_2\, dr \right]^2 \tag{118}$$

where ry is the Rydberg constant $109{\cdot}7$ kK and a_0 the Bohr radius. If Ψ_1 and Ψ_2 can be considered as pure M.O. configurations, eqn. (118) is equivalent to a calculation of dipole moments between orbitals ψ_1 and ψ_2. Here, the selection rules eqn. (115) apply for the γ_n of these orbitals. Graphically, a qualitative picture can be obtained by drawing the overlap areas $\psi_1\psi_2$ and finding the dipole moment of the corresponding distribution of positive and negative charges (according to the sign of the product $\psi_1\psi_2$). However, when quantitative calculations are attempted, they are generally incorrect by a factor 2 or 5. The assumption of pure M.O. configurations produces the selection rule of *one-electron transitions*, making possible only transitions where only one electron changes its orbital.

The sum of oscillator strength P present in a molecule equals the number of electrons. Actually, most observed absorption spectra (up to some 45 kK) have values of P much below 1, since even a large electron transfer band with $\epsilon_0 = 20,000$ and $\delta = 3$ kK has only $P = 0.6$ (cf. eqn. 102). Thus, the absorption in the far ultraviolet accounts for nearly all the oscillator strength of an ordinary complex, and we have cases such as $Mn(H_2O)_6^{2+}$, where the sum of P of the observed bands is only 2×10^{-6}.

In addition to the Laporte- and symmetry-forbiddenness, we may have *spin-forbidden* transitions between states having a different value of S. Until now, no certain spin-forbidden electron transfer band has been observed, probably because it would always be very close to a broad, spin-allowed band. As discussed below, we have spin-forbidden transitions with moderate intensities ($P = 0.02$) in Sn(II) and Sb(III) complexes and large intensities ($P = 0.15$) in Pb(II) and Bi(III) complexes. In the latter cases, S is very strongly intermixed, and the corresponding spin-allowed transitions have $P = 0.8$. The main subject for the study of spin-forbidden transitions is that of the ligand field bands, showing very weak bands in $3d^n$-, weak bands in $4d^n$-, and comparatively strong bands in $5d^n$-complexes. Before the non-diagonal elements of intermediate coupling were known in O_h, empirical values of $k\zeta_{nl}$ (with k a constant about 1) were estimated from many complexes, as seen from Table 9. Linhard and Weigel (1957) reported that the spin-forbidden bands are stronger of $Co(NH_3)_5Br^{2+}$ and much stronger of $Co(NH_3)_5I^{2+}$ than for $Co(NH_3)_5Cl^{2+}$ or $Co(NH_3)_6^{3+}$. This must be interpreted as a delocalization effect, adding a part of the very large values of ζ_{np} in the heavy halogens (eqn. 151) to $k\zeta_{nl}$ of the central ion. Similar effects are known of organic compounds with extremely weak, spin-forbidden bands (ζ_{2p} is only 0.03 kK in neutral carbon atoms) where solvents such as C_2H_5I produce much higher intensities.

The values of $k\zeta_{nl}$ given are rather uncertain, since we do not know the exact mechanism of the small intensity induced by the vibronic interactions in the Laporte-forbidden ligand field bands. Table 9 gives the general impression that the values of ζ_{nl} are somewhat smaller in complexes than in the corresponding gaseous ions. More certain evidence for this decrease of ζ_{nl} was presented by Owen (1955) from paramagnetic resonance experiments, where the *gyromagnetic factor* g is determined. In octahedral and tetrahedral complexes, with $e = 1$, g should, according to the ligand field theory, be expressed as

$$g = 2\cdot00229 + k_g \frac{\zeta_{nl}}{\Delta} \tag{119}$$

where the (positive or negative) constant k_g is a function of the number of d-electrons only (being $-8/3$ for octahedral Cr(III) and tetrahedral Co(II), and $+4$ for octahedral Ni(II)). The deviation of g from the value given by quantum field theory for a free electron (2·00229) can be measured with high accuracy and eqn. (119) is found to agree with the values of Δ derived from absorption spectra only if the values of ζ_{nl} are decreased by about 10–30 per cent in the $3d^n$-hexa-aquo ions. However, the estimates of the decrease of inter-electronic repulsion parameters in complexes can be derived even more accurately from absorption spectra, as shall be discussed in the section on the nephelauxetic series.

The paramagnetic resonance experiments deliver actually, as discussed by Low (1959), four different kinds of evidence of covalent bonding: the orbital contribution to g for complexes with $e = 3$ is decreased (some 20 per cent for Fe(II) in MgO, 10 per cent for Fe(II) in ZnF_2, and 12 per cent for Co(II) in MgO); the second-order effect in eqn. (119); the decrease of the hyperfine-structure, caused by the nuclear magnetic moment of the central ion (cf. Van Wieringen, 1955); and the appearance of hyperfine-structure, caused by the nuclei of the ligands (cf. Owen and Stevens, 1953). The quantity measured by these four phenomena is not the same; the most direct way of measuring the presence of the partly filled shell in the ligands is the fourth effect. The density near to the nucleus is undoubtedly approximately proportional to the second and third effects, but it is difficult to give a quantitative treatment at present.

CHAPTER 7

THE SPECTROCHEMICAL SERIES

THE absorption bands of different complexes of a given lanthanide or actinide element with a given oxidation number have nearly the same positions, only dependent on the central ion. These narrow bands are identified as internal transitions in the partly filled 4f- or 5f-shell. Correspondingly, all complexes of a given central ion have the same colour (a remarkable exception is the pale blue Nd_2O_3, compared to the other, pink Nd(III) compounds, this being a case of a strong nephelauxetic effect acting on the multiplet term distances in $4f^3$, as shall be discussed in the next chapter).

The three ordinary transition groups exhibit much more varying colours. Many central ions, such as nickel (II) and cobalt (III), form complexes of any conceivable colour, green, blue, violet, red and yellow, arranged according to the complementary colour of one absorption band moving from low wave numbers in the red to high wave numbers in the ultra-violet. However, many regularities appear in this evolution. Werner remarked that the colour of cobalt (III) complexes is mainly dependent on the nature of the six ligand atoms in the first co-ordination sphere. Thus, six oxygen atoms produce green or blue colours, while two oxygen and four nitrogen give purple or cherry-red complexes and one oxygen and five nitrogen atoms brick-red colours. If the first co-ordination sphere is completely occupied by six nitrogen atoms, yellow complexes are formed. This was once an argument for distinguishing between the two isomers of $Co(NH_3)_5(NO_2)^{2+}$, the red nitrito complex being bound $Co—ONO$ and the yellow nitro complex $Co—NO_2$.

Around 1920, Shibata and later Tsuchida began a study of absorption spectra of complexes by photographic methods. They concluded that the spectra generally consist of the first and the second band (arranged according to increasing wave number) of low and comparable intensity, and the third band (and sometimes several more) of very high intensities in the ultra-violet. The third band occurs in complexes both outside and inside the transition groups ($PtCl_6^{2-}$, $HgCl_4^{2-}$, etc.) and is one of the Laporte-allowed electron transfer bands. The first and second band do only occur in the transition groups (actually, only d^3 and low-spin d^6 have the characteristic distribution of two Laporte-forbidden bands; thus, d^8 would show three such bands). Tsuchida assumed a different origin of the first and the second band, the former being caused by an internal transition in the d-shell and the latter by an excitation of the electrons, responsible for the covalent bond. According to ligand field theory,

TABLE 10. ABBREVIATIONS OF LIGANDS AND NUMBER OF ATOMS UTILIZED IN CO-ORDINATION

urea	$(NH)_2CO$	one O
ox^{2-}	oxalate	two O
mal^{2-}	malonate	two O
aca^-	acetylacetonate	two O
gly^-	aminoacetate=glycinate	one N, one O
ata^{3-}	nitrogentriacetate=nitrilotriacetate	one N, three O
$enta^{4-}$	ethylenediaminetetra-acetate	two N, four O
entol	tetra(hydroxyethyl)ethylenediamine	two N, four (?) O
py	pyridine	one N (heterocyclic)
en	ethylenediamine	two N
tn	trimethylenediamine	two N
den	diethylenetriamine	three N
trien	triethylenetetramine	four N
tren	tris(aminoethyl)amine	four N
dip	$a:a'$-dipyridyl	two N (heterocyclic)
phen	o-phenanthroline	two N (heterocyclic)
aes^-	mercaptoethylamine	one S, one N
cys^{2-}	cysteinate	one S, one N
$exan^-$	ethylxanthate	two S
dtp^-	diethyldithiophosphate	two S

this distinction cannot be supported; there remains one interesting feature of it that the wave number of the first band is essentially an expression of the orbital energy difference Δ to be discussed in this chapter, while the distance between the first and the second band is mainly a multiple of interelectronic repulsion parameters such as B in eqn. (35) and (75). Since the latter parameters are decreased by covalent bonding as discussed in the next chapter, we actually get an impression of two different properties of the complex by considering the two ligand field bands.

Tsuchida developed Werner's concept of the variation of visual colours to the *spectrochemical series*, arranging the ligands according to their shift-effects on the Laporte-forbidden bands of a given central ion. The series is, underlining the atom directly bound to the central ion in cases where the ligand can be bound at different atoms:

$$I^- < Br^- < Cl^- \sim \underline{S}CN^- < dtp^- < F^- \sim urea \sim OH^- \sim$$

$$\sim N\underline{O}_2^- \sim HCOO^- < ox^{2-} < H_2O \sim mal^{2-} < SC\underline{N}^- <$$

$$< gly^- \sim enta^{4-} < py \sim NH_3 < en \sim den \sim tren <$$

$$< \underline{S}O_3^{--} < dip < phen < \underline{N}O_2^- \lll \underline{C}N^- \qquad (120)$$

(for the abbreviation of the ligands, see Table 10). Actually, Tsuchida did not derive the spectrochemical series mainly from complexes, where the only ligands are those of eqn. (120), but implied most of the order from the spectral shifts observed in $Co(NH_3)_5X$-complexes. Thus, he applied the rule of *average environment* stating that the wave numbers of the bands of an octahedral complex MA_nB_{6-n} approximately are situated at

$$\sigma(MA_nB_{6-n}) \cong \frac{n}{6}\,\sigma(MA_6) + \frac{6-n}{6}\,\sigma(MB_6) \qquad (121)$$

interpolated between the wave numbers of the pure complexes MA_6 and MB_6. The spectrochemical series eqn. (120) can be arranged mainly according to the nature of the ligand atom "touching" the central ion:

$$I < Br < Cl < S < F < O < N < C \qquad (122)$$

This is the series of decreasing radii of the ligand atoms.* While this agrees with an electrostatic interpretation of the spectrochemical series, it is hardly conceivable in this model (eqn. 50) that negative ions such as OH^- and oxalate have a slightly earlier place in the series than water in eqn. (120). The series of atoms in eqn. (122), at least in the beginning, very much resembles a series Fajans con-constructed from the *bathochromic* influence (shift wave numbers towards smaller values, while a shift towards larger wave numbers is

---

* The position of sulphur in the series 122 is estimated from the spectra of mercaptoethylamine complexes of Co(III) (Gorin, Spessard, Wessler and Oliver, 1959) and of ethylxanthate complexes of Cr(III) and Co(III) (private communication from Dr. Claus Schäffer). From rather small spectral shifts in square-planar Pt(II) complexes, Chatt, Gamlen and Orgel (1959) suggest the order $R_2Te < R_2Se < R_2S < R_3As < R_3N < R_3P < (RO)_3P$ with Cl^- perhaps slightly before R_2Te. However, as Dr. Chatt says, this order may depend on the oxidation number and be different in octahedral complexes. A quite curious phenomenon is that SO_3^{2-} stands between the amines and NO_2^- according to Tsuchida. Since this ligand has only one lone-pair, no contribution $-\pi(L{\to}M)$ of eqn. (134) is expected, while S^{--} and dtp^- may form strong π bonds with their several lone-pairs.

TABLE 11. Δ AND B FOR OCTAHEDRAL AND TETRAHEDRAL COMPLEXES

				Δ (kK)	B(K)	β
3d	$^2\Gamma_5$	Ti(III)	6 H_2O	(20·3)	—	—
3d^2	$^3\Gamma_4$	V(III)	6 H_2O	(18·4)	620	0·72
			6 urea	(17·4)	590	0·69
			3 ox^{2-}	(17·8)	550	0·64
			6 O^{2-} (Al$_2$O$_3$)	(18·6)	540	0·63
3d^3	$^4\Gamma_2$	V(II)	6 H_2O	12·4	690	0·93
		Cr(III)	6 Cl$^-$	13·8	510	0·56
			3 dtp$^-$	14·4	410	0·45
			6 F$^-$	15·2	820	0·89
			3 exan$^-$	16·0	390	0·43
			6 urea	16·0	660	0·72
			3 ox^{2-}	17·5	620	0·68
			3 mal^{2-}	17·5	610	0·67
			6 H_2O	17·4	725	0·79
			6 O^{2-}	16·6	480	0·52
			6 O^{2-} (Al$_2$O$_3$)	18·1	630	0·69
			6 NCS$^-$	17·8	570	0·62
			6 NH$_3$	21·6	650	0·71
			3 en	21·9	620	0·67
			6 CN$^-$	26·7	530	0·58
		Mn(IV)	6 F$^-$	21·8	600	0·56
3d^5	$^6\Gamma_1$	Mn(II)	6 H_2O	8·5	835	0·93
			6 NCS$^-$	8·8	800	0·90
			3 en	10·1	785	0·88
			6 O^{--}	9·4	795	0·89
			6 F$^-$	8·4	845	0·94
			6 Cl$^-$	7·5	785	0·88
			4 Br$^-$	−2	740	0·83
			6 S^{--}	7·1	720	0·81
		Fe(III)	6 F$^-$	14·0	845	0·78
			6 urea	13·2	775	0·72
			3 ox^{2-}	13·7	740	0·68
			3 mal^{2-}	14·1	760	0·70
			6 H_2O	14·3	815	0·75
			4 Cl$^-$	−5	625	0·58
3d^6	$^5\Gamma_5$	Fe(II)	6 H_2O	(10·4)	—	—
	$^1\Gamma_1$		6 CN$^-$	31·4	(400)	(0·45)
		Co(III)	3 CO$_3$$^{2-}$	17·3	540	0·49
			3 ox^{2-}	18·0	540	0·49
			6 H_2O	18·2	670	0·61
			6 NH$_3$	22·9	620	0·56
			3 en	23·2	590	0·53
			6 CN$^-$	33·5	460	0·42
3d^7	$^4\Gamma_4$	Co(II)	6 H_2O	9·3	—	—
			6 NH$_3$	10·1	—	—
	$^4\Gamma_2$		4 I$^-$	−2·8	600	0·62
			4 Br$^-$	−2·9	660	0·68
			4 Cl$^-$	−3·7	670	0·69

				Δ (kK)	B(K)	β
			8 F⁻(CaF₂)	−3·4	930	0·94
			4 O²⁻(ZnO)	−3·8	700	0·72
			4 NCS⁻	−4·9	670	0·69
			4 S²⁻ (CdS)	−3·2	660	0·68
3d⁸	³Γ₂	Ni(II)	6 Br⁻	7·0	740	0·70
			6 Cl⁻	7·2	760	0·72
			6 F⁻	7·3	960	0·91
			6 H₂O	8·5	940	0·89
			3 gly⁻	10·1	930	0·88
			6 NH₃	10·8	890	0·84
			3 en	11·5	850	0·81
			3 phen	12·2	—	—
4d³	⁴Γ₂	Mo(III)	6 Cl⁻	19·2	440	0·73
		Tc(IV)	6 Br⁻	—	390	0·56
			6 Cl⁻	—	410	0·59
4d⁵	²Γ₅	Ru(III)	6 Cl⁻	21·6	—	—
4d⁶	¹Γ₁	Rh(III)	6 Br⁻	19·0	280	0·39
			6 Cl⁻	20·3	350	0·48
			3 dtp⁻	22·0	210	0·29
			3 ox²⁻	26·4	—	—
			3 mal²⁻	27·0	—	—
			6 H₂O	27·0	510	0·71
			6 NH₃	34·1	430	0·60
			3 en	34·6	420	0·59
		Pd(IV)	6 Cl⁻	22	—	—
5d²	³Γ₄	Os(VI)	6 F⁻	—	370	0·47
5d³	⁴Γ₂	Re(IV)	6 Br⁻	—	380	0·58
			6 Cl⁻	27·5	400	0·62
			6 F⁻	32	490	0·78
		Ir(VI)	6 F⁻	—	310	0·38
5d⁴	³Γ₄	Os(IV)	6 Br⁻	—	330	0·48
			6 Cl⁻	—	360	0·53
		Pt(VI)	6 F⁻	—	260	0·30
5d⁶	¹Γ₁	Ir(III)	6 Br⁻	23·1	250	0·38
			6 Cl⁻	25·0	300	0·46
			3 dtp⁻	26·6	160	0·24
			3 en	41·4	—	—
		Pt(IV)	6 Cl⁻	29	—	—
			6 F⁻	33·0	380	0·53

Values given in parentheses may be influenced by Jahn–Teller effect. B is from spin-allowed bands, except for Mn(II) and Fe(III), where the distance $^6S - {}^4G = 10B + 5C$ is identified with $30B$, and for Tc(IV), Os(VI), Re(IV), Ir(VI), Os(IV) and Pt(VI), where $3B + C$ is put equal to $7B$. Generally, the Tanabe–Sugano determinants are used for $C = 4B$. Claus Schäffer has calculated the values of Δ and B for d⁶-systems from electronic computer results on such an assumption. β is often found by use of extrapolated values of B in the corresponding gaseous ion, notably Fe³⁺ (1090 K), Co³⁺ (1100 K) and the 4d⁹ (0·66 times the values of the corresponding 3d⁹ ion) and 5d⁹ (0·60 times 3d⁹) groups.

called a *hypsochromic* influence in the empirical description of colour shifts) of halogen atoms on the Laporte-allowed bands of complexes. This coincidence will be discussed in the chapter on electron transfer spectra.

With the arrival of the ligand field theory, it was obvious that the parameter describing the spectrochemical series is the energy differ-

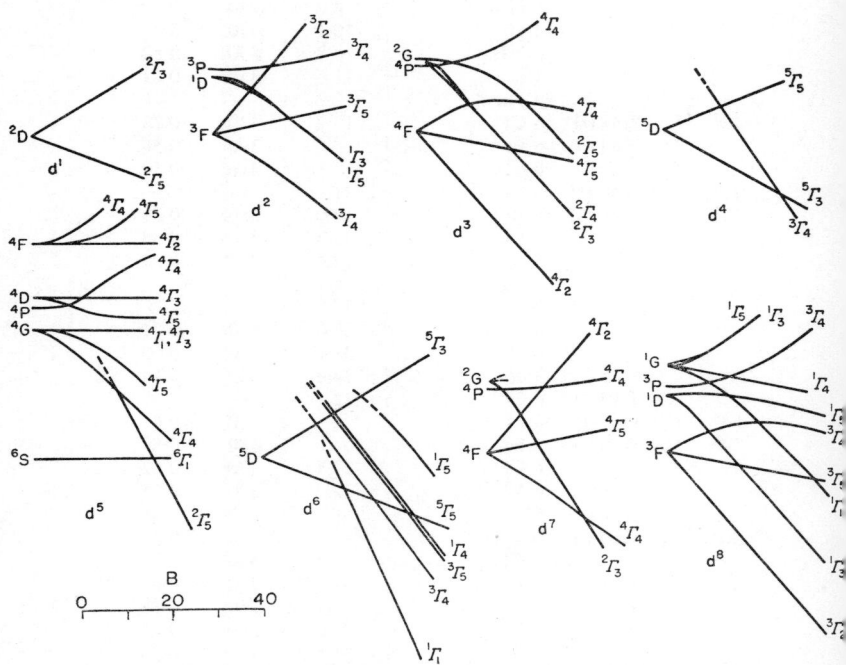

FIG. 6. Tanabe–Sugano diagrams for octahedral d^n-complexes. The free variable is Δ/B, assuming $C = 4B$, and the energy unit is B.

ence Δ between the two equivalent γ_3-orbitals and the three equivalent γ_5-orbitals in the partly filled shell. Actually, in octahedral complexes, assuming $l = 2$ (Table 8), the wave number of the first spin-allowed ligand field band is either exactly (d^3, high-spin d^6, d^8) or after some corrections for parameters of interelectronic repulsion, amounting at most to some 20 per cent of the wave number (d^2, low-spin d^6, high-spin d^7) a measure of Δ. With the *Orgel diagrams* (the energy levels as function of Δ with fixed values of the interelectronic repulsion parameters, e.g. those of a gaseous ion) or

Tanabe–Sugano diagrams (the energy levels as function of Δ/B, where C/B is assumed to have a fixed value, e.g. 4 in Fig. 6), the parameter Δ can be determined by fitting the observed wave numbers of absorption bands to the energy levels, cut by a vertical line in the diagrams. Table 11 indicates the values of Δ in a long series of different transition group complexes. The validity of the spectrochemical series as function of ligands (eqn. 120) is shown by the fact that as a rather good approximation, Δ can be written as a product of a function f of the ligands and g of the central ion:

$$\Delta = \text{f(ligands)} \cdot \text{g(central ion)} \qquad (123)$$

Representative values of the functions f and g are given in Table 12. Thus, for all hexachloro complexes, Δ is about 80 per cent of the value in the corresponding hexa-aqua ion, and for all hexammine and tris(ethylenediamine) complexes, Δ is about 120–130 per cent of the value of the hexa-aqua ion. We remember that it is an exceptional property of a function of two variables to be separable into the product of two functions of each one variable, such as eqn. (123).

TABLE 12. THE FUNCTIONS OF ONE VARIABLE FOR EXPRESSING THE SPECTROCHEMICAL SERIES (EQN. 123) AND THE NEPHELAUXETIC SERIES (EQN. 137)

	f	h		g (kK)	k
6 F$^-$	0·9	0·8	V(II)	12·3	0·08
6 H$_2$O	1·00	1·0	Cr(III)	17·4	0·21
6 urea	0·91	1·2	Mn(II)	8·0	0·07
6 NH$_3$	1·25	1·4	Mn(IV)	23	0·5
3 en	1·28	1·5	Fe(III)	14·0	0·24
3 ox^{2-}	0·98	1·5	Co(III)	19·0	0·35
6 Cl$^-$	0·80	2·0	Ni(II)	8·9	0·12
6 CN$^-$	1·7	2·0	Mo(III)	24·0	0·15
6 Br$^-$	0·76	2·3	Rh(III)	27·0	0·30
3 dtp$^-$	0·86	2·8	Re(IV)	35	0·2
			Ir(III)	32	0·3
			Pt(IV)	36	0·5

Actually, some irregularities occur: hexacyanide complexes have values of Δ between 150 and 300 per cent of the hexa-aqua ions; and ethylenediaminetetra-acetate may in some complexes even appear before water in eqn. (120). The latter phenomenon may be connected with a strong deviation from the octahedral symmetry in some of the complexes.

That no distinct difference between the results of the ligand field in complexes of anions and of neutral molecules exists, was demonstrated by Orgel (1955) from the close similarity of the spectra of solid MnF_2 and of $Mn(H_2O)_6^{2+}$ in solution. Actually, eqn. (120) contains anions and neutral ligands scattered in a rather accidental manner. It is also interesting to note that the function f (ligand) in eqn. (123) does not vary by more than a factor 2, apart from the cyanide complexes.

Also the function g (central ion) has a smaller range of variation than might perhaps be expected, resulting in a variation of all known values of Δ in octahedral complexes between 6 kK and 45 kK. We may define a *spectrochemical series of central ions* for the same ligand.

$$Mn(II) < Ni(II) < Co(II) < Fe(II) < V(II) < Fe(III) <$$
$$< Cr(III) \sim V(III) < Co(III) < Mn(IV) < Mo(III) <$$
$$< Rh(III) \sim Ru(III) < Pd(IV) < Ir(III) < Re(IV) <$$
$$< Pt(IV) \qquad (124)$$

showing a strict arrangement according to an increasing number of the transition group $3d^n < 4d^n < 5d^n$ (with relative values of the functions $1 : 1.45 : 1.75$) and according to increasing oxidation numbers $+2 < +3 < +4$ (with relative values of g approximately $1 : 1.6 : 1.9$). The strong increase of Δ from divalent to trivalent ions in the first transition group could hardly be explained by any reasonable electrostatic model.

It has often been written that the small variation of Δ for the same oxidation number in the same transition group for the same set of ligands is completely irregular, and an often cited example are the divalent hexa-aqua ions, having Δ in kK:

$$V(II)\ 12.3 \quad Cr(II)\ 13.9 \quad Mn(II)\ 8.3 \quad Fe(II)\ 10.4 \quad Co(II)\ 9.3$$
$$Ni(II)\ 8.5 \quad Cu(II)\ 12.6 \qquad (125)$$

However, we must consider the deviations from octahedral symmetry, caused by the Jahn–Teller effect, and the slight variation (see eqn. 130) of Δ with the ligand field stabilization parameter ρ of eqn. (110). Actually, the former deviations are so strong in Cr(II) and Cu(II) that it is certain that the value of Δ given in eqn. 125 does not have its usual meaning, but refers to an orbital energy difference in the tetragonal symmetry. It is argued by many authors that if d^4- and d^9-systems such as Cr(II) and Cu(II) had the octahedral symmetry O_h, then one spin-allowed transition would be observed, having (for $l = 2$) exactly the wave number Δ. Therefore, when $Cr(H_2O)_6^{2+}$ has a broad band at 13·9 kK and $Cu(H_2O)_6^{2+}$ a band (with a shoulder at 9·3 kK) with the maximum at 12·6 kK, these authors conclude that these wave numbers to some extent represent Δ. However, this is not realistic, as seen below, while this approximation is permissible, e.g. in the high-spin d^6-system $Fe(H_2O)_6^{2+}$ which is also Jahn–Teller unstable, but deviates much less from the octahedral symmetry than $Cu(H_2O)_6^{2+}$.

The *Jahn–Teller effect* is that the electronic ground state of a (non-linear) molecule must have the orbital degeneracy number $e = 1$. Should a molecule or complex happen to have a ground state with $e = 2$ or $e = 3$ in a high symmetry, it will spontaneously distort to a lower symmetry, separating out one or more states, forming a level with $e = 1$. If ζ_{nl} is large, we may also have a Jahn–Teller effect due to intermediate coupling, if the ground level values of Γ_J in high symmetry have e above 1 for an even number of electrons and e above 2 for an odd number of electrons (Kramer's doublets; for an odd number of electrons any energy level contains an even number of states).

The mechanism of the Jahn–Teller effect can be visualized as the deviation from, for example, octahedral symmetry. It is a negative first-order perturbation energy, being of first order in the displacement co-ordinate of some non-totally symmetric normal mode of vibration (cf. eqn. (105) and Fig. 4). Since the energy of excitation of such a displacement is always proportional to the square of the displacement co-ordinate near the equilibrium conditions, the distortion always proceeds to some extent, until the latter restoring forces makes it no longer energetically favourable. Thus, the presence of the Jahn–Teller effect is a purely qualitative theorem; the numerical extent is highly dependent on the details of the potential surfaces as function of internuclear distances.

The only levels in O_h, for which $e = 1$, are Γ_1 and Γ_2. Thus, according to Table 8, the only Jahn–Teller stable ground states of regularly octahedral transition group complexes are:

$$\left. \begin{array}{lll} d^3 & {}^4\Gamma_2 & \gamma_5{}^3 \\ \text{high-spin } d^5 & {}^6\Gamma_1 & \gamma_5{}^3\gamma_3{}^2 \\ \text{low-spin } d^6 & {}^1\Gamma_1 & \gamma_5{}^6 \\ \text{high-spin } d^8 & {}^3\Gamma_2 & \gamma_5{}^6\gamma_3{}^2 \\ d^{10} & {}^1\Gamma_1 & \gamma_5{}^6\gamma_3{}^4 \end{array} \right\} \quad (126)$$

The sub-shell configurations indicated cannot be intermixed within the d-shell in other cases than $^1\Gamma_1$ of d^6. If we consider the wave functions with the highest value of M_S (see p. 70), we have a clear picture of the physical reason for the Jahn–Teller effect. In the unstable configurations γ_5 and $\gamma_5{}^2$, the complex decreases its energy by

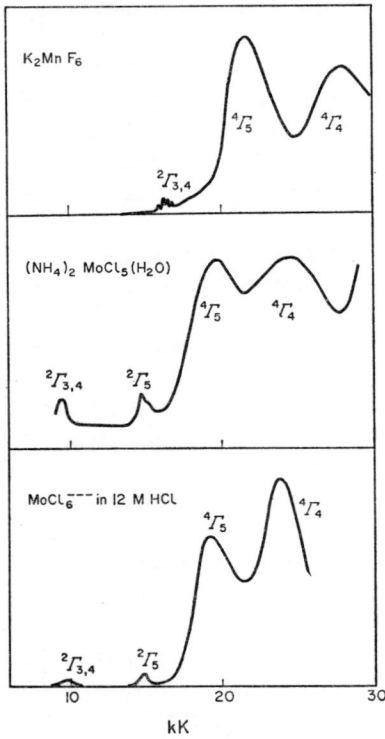

FIG. 7. The reflection and absorption spectra of some d^3-systems. The narrow, spin-forbidden transitions from the ground state $^4\Gamma_2$ to $^2\Gamma_3, {}^2\Gamma_4$ (nearly degenerate) and to $^2\Gamma_5$ are the "ruby lines", belonging to the sub-shell configuration $\gamma_5{}^3$, while the two broad bands have $^4\Gamma_5$ and $^4\Gamma_4$ of $\gamma_5{}^2\gamma_3$ as excited states (cf. Fig. 6).

a splitting of the three equivalent orbitals γ_5, letting one or two have lower energy and the rest higher energy. In the stable $\gamma_5{}^3$, this would lead to no net stabilization, one electron with the same spin direction m_s necessarily being in each orbital. However, in high-spin

$\gamma_5^3\gamma_3$ and low-spin γ_5^4 and γ_5^5, the arguments above are again valid, the complex preferring less equivalency among the orbitals than found in the octahedral symmetry. High-spin $d^5(\gamma_5^3\gamma_3^2)$ have one electron in each available orbital and low-spin $d^6(\gamma_5^6)$ two electrons in each of the three γ_5-orbitals; in both cases no advantage can be obtained by deviations from the octahedral symmetry. These arguments are easily continued for d^7 and d^9.

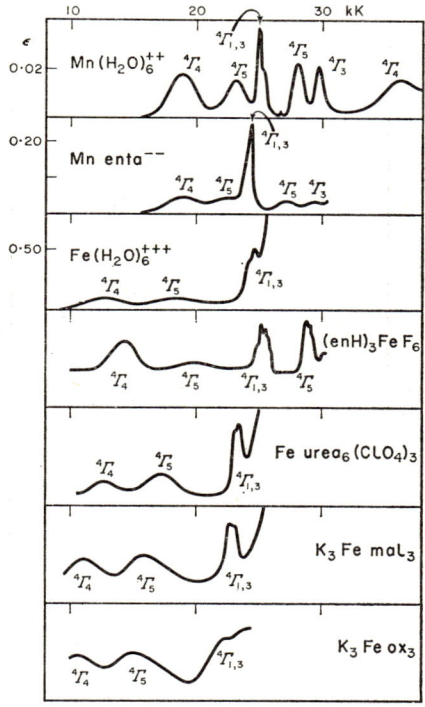

FIG. 8. The absorption and reflection spectra of high-spin d^5-systems. The reflection spectra of (en H)$_3$FeF$_6$; Fe(urea)$_6$(ClO$_4$)$_3$; K$_3$Fe ox$_3$, 3H$_2$O; and K$_3$Fe mal$_3$ are given. Experimental details are later to be published in a paper together with those of Dr. Claus Schäffer.

However, experience shows that the Jahn–Teller distortion in most cases is rather negligible in the ground states with $e = 3$ (d^1, d^2, low-spin d^4 and d^5, high-spin d^6 and d^7) while it is extremely important for the ground states Γ_3 with $e = 2$ (high-spin d^4, low-spin d^7 and d^9). The latter type of complexes (as also low-spin d^8) evolve tetragonal symmetry to so large an extent that most complexes appear

I

square-planar with only four ligand atoms co-ordinated. The explanation of the difference between small Jahn–Teller effect for $e = 3$ and large for $e = 2$ is that the sub-shell configurations in the former case evolve a non-equivalency between the three γ_5-orbitals,

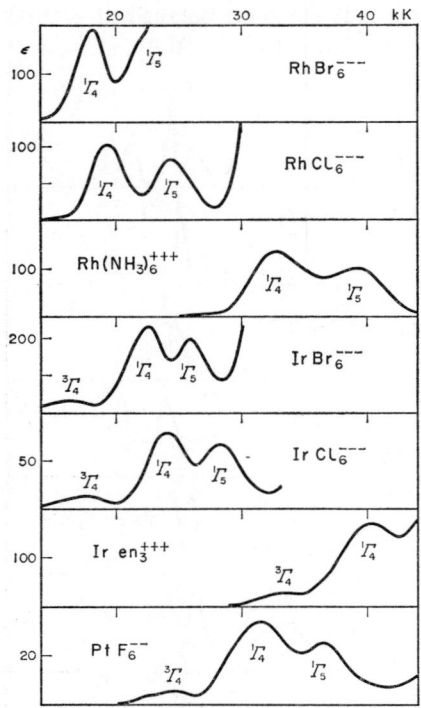

FIG. 9. The absorption spectra of low-spin d^6-systems with six identical ligand atoms. The spin-forbidden transitions to $^3\Gamma_4$ and $^3\Gamma_5$ of $\gamma_5^5\gamma_3$ at lower wave number than the spin-allowed transitions to $^1\Gamma_4$ and $^1\Gamma_5$ of the same sub-shell configuration are seen to increase in intensity 3d < 4d < 5d (cf. Fig. 3). The spectra have previously been measured by the present writer (ref. 276) except that of PtF_6^{2-}, kindly communicated by Dr. Th. Perros (cf. ref. 610).

while the latter case involves the two γ_3-orbitals. Since the γ_5-orbitals are only π-bonding (or π-antibonding) according to eqn. (53), while γ_3-orbitals are σ-antibonding, the latter interact much more with the ligands and are especially important for the distances between the central ion and the ligands.

Tetragonal distortions from octahedral symmetry may occur with two *signs*: one approaching square-planar co-ordination with weaker bonds and longer distances to the two perpendicular ligands; and one approaching linear co-ordination with weak bonds to the four ligands in the plane. It may be asked why the Jahn–Teller unstable complexes with $e = 2$ always choose the former possibility. Actually, for harmonic oscillator potentials of the normal mode of vibration,

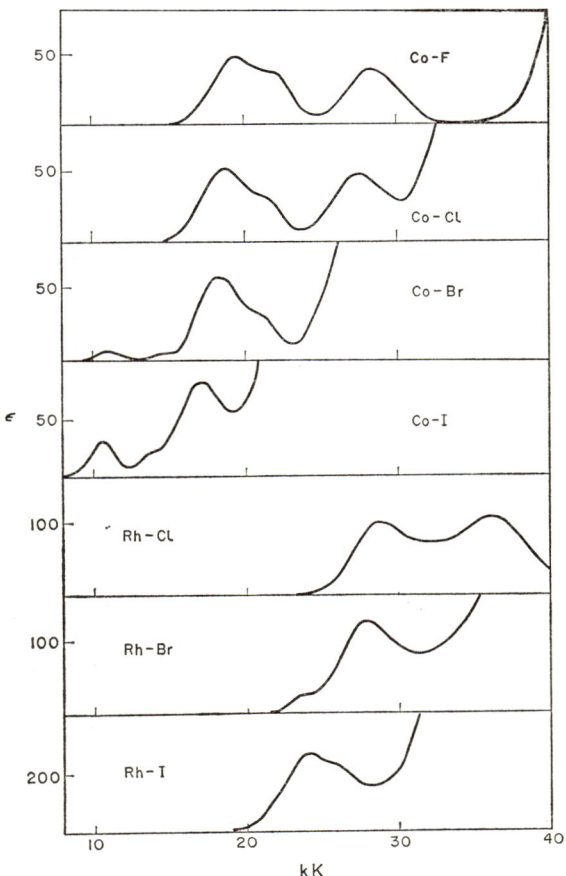

Fig. 10. The absorption spectra of halide-pentammine d^6-systems. The measurements of $Co(NH_3)_5X^{2+}$ are by Linhard and Weigel (ref. 366) and of $Rh(NH_3)_5X^{2+}$ by the present writer (ref. 276).

changing the distance of the ligands on the the z-axis (Fig. 2) from r_L to $r_L + 2x$ and the ligands on the x- and y-axes to $r_L - x$, the energy difference is identical for $\pm x$. If the baricentre rule applies to the energy of the two separated γ_3-orbitals (see eqn. 66), this

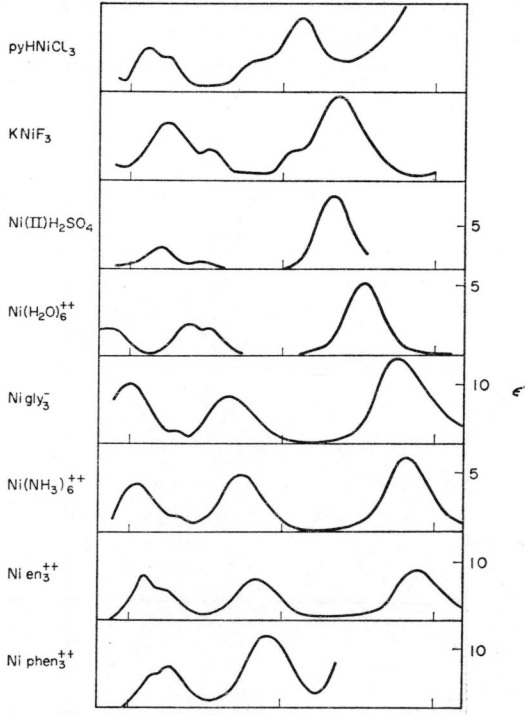

FIG. 11. The absorption and reflection spectra of high-spin d^8-systems. The solid pyHNiCl$_3$ has been measured by Asmussen and Bostrup (ref. 8), the solid KNiF$_3$ is to be described by Dr. Claus Schäffer and the writer. The identification of the individual bands can be made from Fig. 12. The environment of Ni(II) in the perovskites with three halide groups is actually octahedral, each halogen atom being bound to two nickel atoms.

assumption would lead to indifference with regard to the two signs of tetragonal distortion. Since the formation of the M.O. does not respect the baricentre rule, some particular orbital of the central ion may be responsible for the choice of $(x^2 - y^2)$ as the empty orbital. It can hardly be an interaction between the s-electrons and $z^2 - \frac{1}{2}(x^2 + y^2)$ alone, being both totally symmetric (γ_{t_1}) in the tetragonal symmetry, since d^{10}-complexes prefer linear symmetry in cases of small energy difference between nd and $(n + 1)$s. However, Öpik and Pryce (1957) point out in the more electrostatic picture of $Cu(H_2O)_6^{2+}$ that the anharmonicity of the potential curves tends to favour larger rather than smaller distances, compared to the equilibrium value r_L. Since such

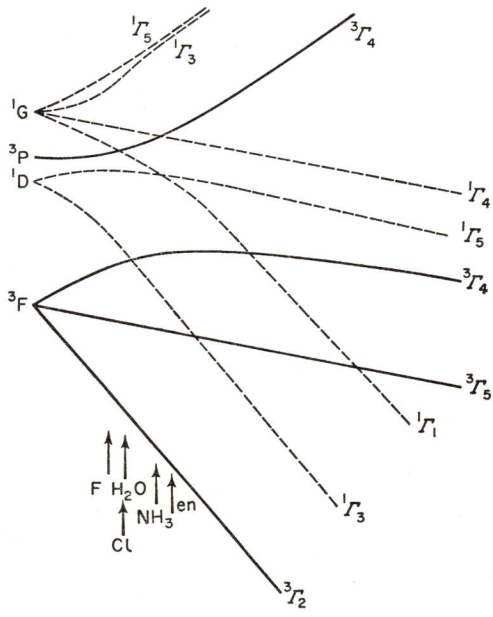

FIG. 12. Tanabe–Sugano diagram for octahedral d^8-systems. Though the perovskite pyHNiCl$_3$ has a much lower Δ than Ni(H$_2$O)$_6^{2+}$, the value of Δ/B is nearly the same, due to the larger nephelauxetic effect in the chloride.

an anharmonicity correction of the form $-x^3k_2$ gives the contribution $-12k_2x^3$ in the example mentioned above, the square-planar type with positive x would presumably have the lower energy.

The *spectrochemical series of tetragonal complexes* has, in the case low-spin d^8 (as found in square-planar palladium (II) and platinum (II) complexes) the same order of the ligands as eqn. (120). In these systems, one spin-allowed band is observed for complexes of neutral molecules, especially amines with only one lone pair used mainly for σ-bonding. The halide complexes, where the ligands form π-bonds also, have three spin-allowed ligand field bands. In addition, at least one spin-forbidden band is observed in Pd(II) and two in Pt(II).

The distribution of these bands is remarkably well described by the M.O. theory of covalent bonding without consideration of the electrostatic first-order perturbations. The order of the orbitals of the partly filled shell in tetragonal symmetry is, assuming the z-axis to be the tetragonal axis of symmetry (without ligands)

$$
\left.
\begin{array}{lll}
t3 & x^2 - y^2 & \sigma\text{-antibonding} \\
t4 & xy & \text{strongly } \pi\text{-antibonding} \\
t5 & xz, yz & \text{weakly } \pi\text{-antibonding} \\
t1 & z^2 - \frac{1}{2}(x^2 + y^2) &
\end{array}
\right\} \quad (127)
$$

The ground state of the complexes is the closed shell configuration $^1\Gamma_{t1}$, leaving the highest orbital γ_{t3} empty. The three spin-allowed absorption bands, as indicated by Chatt, Gamlen and Orgel (1958) are caused by the jump of an electron from the three lower sets of orbitals to the empty one. Each such configuration corresponds to a triplet and a singlet level according to a Heisenberg determinant

TABLE 13. THE RATIO σ_{Pd}/σ_{Rh} ($= \lambda_{Rh}/\lambda_{Pd}$) BETWEEN THE WAVE NUMBERS OF THE FIRST SPIN-ALLOWED LIGAND FIELD BAND OF CORRESPONDING PALLADIUM (II) AND RHODIUM (III) COMPLEXES. IT IS NOT CERTAIN THAT THE SPECIES, OCCURRING IN 2M NaOH, ARE THE MONOMERIC HYDROXO COMPLEXES GIVEN IN THE TABLE.

$RhBr_6^{3-}$	553 mμ	$PdBr_4^{2-}$	504 mμ	1·097
$RhCl_6^{3-}$	518	$PdCl_4^{2-}$	474	1·093
$Rh(SCN)_6^{3-}$	(516)	$Pd(SCN)_4^{2-}$	500	1·032
Rh dtp$_3$	469	Pd dtp$_2$	460	1·020
$Rh(H_2O)_6^{3+}$	393	$Pd(H_2O)_4^{2+}$	379	1·037
$Rh(OH)_6^{3-}$(?)	418	$Pd(OH)_4^{2-}$(?)	368	1·136
Rh ox$_3^{3-}$	398	Pd ox$_2^{2-}$	383	1·039
Rh mal$_3^{3-}$	390	Pd mal$_2^{2-}$	372	1·043
Rh enta$^-$	353	Pd enta^{2-}	337	1·047
Rh (NH$_3$)$_6^{3+}$	306	Pd(NH$_3$)$_4^{2+}$	296	1·034
Rh en$_3^{3+}$	301	Pd en$_2^{2+}$	287·5	1·044

(eqn. 64). Group theory allows both xy and the set xz, yz to interact with π-orbitals of the ligands, but since the former orbital has twice as large an overlap integral with the π-orbitals as one of the orbitals xz or yz, the latter orbitals are expected to be less affected by the π-bonding,

The reasons that complexes such as $Pd(NH_3)_4^{2+}$ (and also $Pd(H_2O)_4^{2+}$) have only one spin-allowed ligand field band are intelligible, since the filled orbitals of the ground sub-shell configuration have the same energy, if only σ-bonding is considered in eqn. (127). However, since the influence of interelectronic repulsion parameters (eqn. 75) is slightly different in the three excited sub-shell configurations, three adjacent bands were expected. This difficulty cannot be removed by assuming a strong delocalization of, for example, the orbital $(x^2 - y^2)$ so that the K-integrals with the other orbitals are negligible, since this assumption would make the singlet and triplet levels coincide. Probably, one of the three spin-allowed bands has a much higher intensity than the two others so that a curve analysis is not possible.

A much more surprising fact about the (first) spin-allowed ligand field band of Pd(II) with the wave number σ_{Pd} is that the ratio σ_{Pd}/σ_{Rh}, with the wave number σ_{Rh} of the first spin-allowed transition $^1\Gamma_1 \rightarrow {}^1\Gamma_4$ in the octahedral Rh(III) complex with the same ligands, is nearly constant in the range from 1·02 to 1·10. This similarity between a long series of square-planar Pd(II) and octahedral Rh(III) complexes (Table 13) is almost incredible. The only plausible explanation is that Δ in the octahedral complexes (as expressed in eqn. (134) is the difference between the σ-antibonding and the π-antibonding effects on the partly filled shell in the same way as is the orbital energy difference between γ_{t_3} and γ_{t_4} in eqn. (127). Similar effects occur in the low-spin, square-planar Ni(II) complexes compared to the octahedral, low-spin Co(III). Thus σ_{Ni}/σ_{Co} is about 1·1 for the band of the yellow amine complexes (cf. Jørgensen, 1957) and for the first of the two ligand field bands of the greenish-brown complexes of sulphur-containing ligands (Leussing, 1958–1959).

The d^9-systems (of which copper (II) complexes are the only well-known examples) pose the same problems as the diamagnetic d^8-systems, having the same order of orbitals (eqn. 127) in the M.O. description of covalent bonding, and one hole in the highest orbitals. We may define a ratio σ_{Cu}/σ_{Ni} (Table 14) between the wave number σ_{Cu} of the principal band of a copper (II) complex and σ_{Ni}, the wave number of the first, spin-allowed band of the high-spin, octahedral nickel (II) complex, with the same ligands (when corrected for effects of intermediate coupling, $\sigma_{Ni} = \Delta$). While σ_{Rh}/σ_{Pd} is almost constant (and corresponds to a value of σ_{Cu}/σ_{Ni} around 1·8, because we are comparing a trivalent with a divalent central ion), the copper (II) complexes fall into three classes: the highly tetragonal with σ_{Cu}/σ_{Ni} = from 1·6 to 1·8, having only one absorption band (probably a coincidence of the three bands, as discussed above),

TABLE 14. THE RATIO σ_{Cu}/σ_{Ni} BETWEEN THE WAVE NUMBER OF THE PRINCIPAL LIGAND FIELD BAND OF A COPPER (II) COMPLEX AND THE WAVE NUMBER OF THE FIRST SPIN-ALLOWED LIGAND FIELD BAND (AS CORRECTED FOR INTERMEDIATE COUPLING), i.e. Δ, OF THE CORRESPONDING NICKEL (II) COMPLEX

Ligands	σ_{Cu} (kK)	σ_{Ni} (kK)	σ_{Cu}/σ_{Ni}
tren, $2H_2O$	11·6	10·8	1·07
tren, $2NH_3$	12·7	11·4	1·11
2 dip, $2H_2O$	10·5, 13·9	—	1·1
3 dip	14·7	12·2	1 20
2 phen, $2H_2O$	10·2, 13·3	—	1·1
3 phen	14·7	12·25	1·20
ata^{3-}	11·4, 12·9	9·5	1·24
2 ata^{3-}	15·2	10·4	1·46
$enta^{4-}$, H_2O	13·6	10·1	1·35
$enta^{4-}$, NH_3	13·8	10·2	1·36
2 gly^-, $2H_2O$	15·8	9·8	1·60
3 gly^-	15·1	10·1	1·50
den, $3H_2O$	16·3	10	1·6
2 en, $2H_2O$	18·2	10·5	1·73
3 en	16·4	11·4	1·44
2 den	15·9	11·6	1·37
4 NH_3, $2H_2O$	16·9	10	1·7
6 NH_3	15·6	10·8	1·45
6 H_2O	12·6	8·5	1·48

the complexes of six identical ligands with σ_{Cu}/σ_{Ni} = 1·4, having a shoulder at lower wave number than σ_{Cu}. This shoulder represents perhaps a transition from γ_{t1}, being the σ-antibonding from the weakly bound ligands on the z-axis, to the hole in γ_{t3}, while the three other orbitals are nearly coincident. Finally, copper (II) complexes of low tetragonality, σ_{Cu}/σ_{Ni} = 1·1, have two or three absorption bands. This type of complex is formed with rather special ligands, such as three or two molecules of the heterocyclic di-imines (the latter seem to prefer the *cis*-configuration, contrary to all the tetra-co-ordinated *trans*-complexes of aliphatic amines) or the tetradentate amine "tren", where sterical hindrance prevents the formation of a square-planar complex. Corresponding to the varying values of σ_{Cu}/σ_{Ni}, the rule of average environment is invalidated in Cu(II) by J. Bjerrum's *pentammine effect*, the rather unusual phenomenon that the absorption band of $Cu(NH_3)_5{}^{2+}$ has a lower wave number than of $Cu(NH_3)_4{}^{2+}$ (the water ligands are not indicated), while the evolution in the series $Cu(NH_3)_n{}^{2+}$ with n = 0, 1, 2, 3, 4 is, as in octahedral complexes, a shift towards the higher wave numbers.

Linn Belford discussed copper (II) complexes, based on the M.O. theory, and in particular the acetylacetonates $Cuaca_2$ with perpendicular solvent molecules. The present writer (1957) has suggested the order of orbitals (see eqn. 127) t1 > t3 $\sim$ t5 > t4 in vanadyl complexes, containing the group VO^{2+}. This can be explained by the strong σ- (increasing the energy of $z^2 - \frac{1}{2}(x^2 + y^2)$) and π- (increasing the energy of the degenerate set xz, yz) bonding of the oxygen atom (here placed on the z-axis). However, in the special case of $VOaca_2$, Dr. Robert Feltham suggests a tetragonality with the same sign as that in Cu(II), i.e. with xy above xz, yz.

The electrostatic model for tetragonal complexes such as trans-$Co(NH_3)_4Cl_2{}^+$ gives the order of the orbitals t3 > t1 > t4 > t5 (eqn. 127). However, Griffith and Orgel (1956) and Yamatera (1958) pointed out that the splitting of $^1\Gamma_4$ from the octahedral symmetry to $^1\Gamma_{t5}$ at lower and $^1\Gamma_{t2}$ at higher energy, connected with the absence of measurable splitting of $^1\Gamma_5$ (cf. Linhard and Weigel, 1951) implies that the order of the orbitals is t3 > t1 > t5 > t4. The latter inversion is caused by the π-bonding effect of the two chloride ligands on the z-axis. In the case of Jahn–Teller stable, octahedral complexes, where the tetragonal symmetry is introduced by differences between the ligands, the splitting of the levels known from octahedral symmetry is often absent or surprisingly small (as known from a large number of nickel (II) complexes, studied by Ole Bostrup and the present writer, and from the chromium (III) and cobalt (III) complexes studied by Claus Schäffer) and the rule of average environment is valid with good approximation. This is also true for the baricentres of the band splittings observed in trans-$Rhpy_2Cl_4{}^-$ and trans-$Rhox_2Cl_2{}^{3-}$, which have nearly the same wave numbers as the actual bands of the corresponding cis-isomers. (These and many other salts were prepared by Marcel Delépine and the spectra measured and published 1957).

The *ligand field stabilization parameter* $\rho = 2n_5 - 3n_3$, where n_5 is the number of γ_5- and n_3 the number of γ_3-electrons, varies through a transition group for octahedral complexes, assuming pure sub-shell configurations. The values for strongly Jahn–Teller distorted complexes are given in parentheses.

$$
\begin{array}{lccccccccccc}
& d^1 & d^2 & d^3 & d^4 & d^5 & d^6 & d^7 & d^8 & d^9 & d^{10} \\
\text{high-spin} & 2 & 4 & 6 & (3) & 0 & 2 & 4 & 6 & (3) & 0 \\
\text{low-spin} & & & 8 & 10 & 12 & (9) & & & &
\end{array}
\qquad (128)
$$

If the baricentre rule was valid (as, for example, in the electrostatic and expanded radial function models), the presence of a partly filled shell produces exactly a ligand field stabilization $\rho\Delta/5$ compared to the energy of a similar complex with a spherically symmetrical central ion. This argument was applied by Orgel (1952) to explain the trends of hydration heats of gaseous ions in the first transition group and by J. Bjerrum and the present writer (1955) to explain the evolution of formation constants of complexes (see a later chapter) and especially the *Irving–Williams series* of complexity constants of a given ligand:

$$
\text{Cr(II)} > \text{Mn(II)} < \text{Fe(II)} < \text{Co(II)} < \text{Ni(II)} < \text{Cu(II)} \gg
$$
$$
\gg \text{Zn(II)} \qquad (129)
$$

Irving and Williams (1953) ascribed eqn. (129) to the variation of the electron affinity of the gaseous ions, e.g. Cu^{2+} being most interested in taking up an electron to complete the d-shell, while Zn^{2+} with a filled d-shell has a rather low electron affinity. As seen from eqn. (149), the hump at Mn(II) is caused by the half-filled shell d^5, changing the interelectronic repulsion energy (for the ground state with the maximum value of S) as function of the number of electrons in the d-shell. On the other hand, the presence of the M.O. energy difference must necessarily produce ligand field stabilizations of the type in eqn. (128), but their size is indeterminate, if the baricentre rule is not valid. For a very large number of complexes of the first transition group, the stabilization, as derived from the complex formation constants compared to a linear interpolation with Z between Mn(II) and Zn(II), is within about 20 per cent uncertainty (except for Cu(II), where it is too large by a factor 3 to 4) of that derived from the difference $0.2\rho\,(\Delta_{\text{complex}} - \Delta_{\text{hexa-aqua}})$ known from the absorption spectra. It must be remembered that this difference is rather small, compared to the absolute ligand field stabilization $0.2\rho\Delta_{\text{complex}}$.

Williams criticizes the application of the baricentre rule, pointing out that the formation constants of oxalates and acetylacetonates (always compared in aqueous solution to the stability of the hexa-aqua ions) also follows the Irving–Williams series eqn. (129), though the values of Δ for these and other oxygen anions are slightly below those of the aqua ions (eqn. 120). However, this may still be related to the Santen–Wieringen variation of the ionic radii and other aspects of the variation of ρ in eqn. (128). The high values of ρ may still furnish better conditions for formation of strong covalent bonds with other orbitals of the central ion and the ligands than those involved in the behaviour of the partly filled shell. Similar effects might cause the extraordinary stability of the strongly tetragonal Cu(II) complexes.

If the σ-antibonding effects in the M.O. theory are the only ones considered, high-spin octahedral complexes would exhibit two antibonding electrons in d^5, d^6, d^7 and d^8, three in d^9, and four in d^{10}. The deviation of Cu(II) in eqn. (129) might be explained by the presence of only one antibonding electron in square-planar d^9 (cf. eqn. 127). However, this explanation would raise the question why d^{10}-systems do not prefer square-planar or linear co-ordination, producing only two and not four antibonding electrons. Actually, it is remembered that Cu(I), Ag(I), Au(I) and Hg(II) prefer the latter possibility. The complete solution of this problem is not yet known.

At the present, it is impossible to predict, *a priori*, the total energy of a complex, or even the stabilization as compared to the hypothetical, analogous complex with a spherically symmetrical central ion. This does not affect the value of Δ found from absorption spectra, since the bonding and antibonding effects of orbitals of other symmetries are included in the terms of the potential U independent of ρ (eqn. 109). These effects, common to all sub-shell configurations of the partly filled shell, tend to decrease any ligand field stabilization directly calculated as $\rho\Delta/5$, because the value of r_L is decreased from the value for $\rho = 0$ (eqn. 110), increasing necessarily the part of the potential energy independent of ρ. However, since this deviation may be highly different in the hexa-aqua ion and in another complex, the ligand field stabilization in the complexity constants may alternatively be strongly increased or decreased, compared to the implication from absorption spectra and the baricentre rule.

We may ask about the dependence of Δ on r_L and thus, on $\rho\Delta$

itself. Actually, eqn. (109) suggests a value of Δ increasing about 30 per cent by going from $\gamma_5{}^a\gamma_3{}^b$ to $\gamma_5{}^{a+1}\gamma_3{}^{b-1}$. This seems somewhat exaggerated. The variation in eqn. (125) (when corrected for Jahn–Teller distortions, e.g. Cu(II) giving a value of Δ only about 8 kK) may be expressed for the $3d^q$-hexa-aqua ions with oxidation state $+2$ (and $+3$ is given for comparison) in kK:

$$
\begin{aligned}
+2: \quad & 12{\cdot}2 - 0{\cdot}75q + 0{\cdot}4\rho \\
+3: \quad & 16{\cdot}2 - 0{\cdot}4q + 0{\cdot}4\rho
\end{aligned}
\tag{130}
$$

This linear interpolation of the variation with q and ρ is of course a rather rough method, but it suggests an increase of Δ by about 2 kK by the change of the ground sub-shell configuration as above. A corollary to this effect was pointed out by Orgel that when a low-spin and a high-spin ground state cross each other in the Orgel diagram, the values of Δ of the low-spin complexes jump, being all well above the threshold limit of Δ for the high-spin complexes. It is to some extent possible to estimate the variation of Δ with ρ from the fluorescence of manganese (II) complexes, emitting light from the vibrational ground state of the excited electronic level $^4\Gamma_4$, mainly of the configuration $\gamma_5{}^4\gamma_3$, and correspondingly, a smaller value of r_L than the electronic ground state $^6\Gamma_1$ of $\gamma_5{}^3\gamma_3{}^2$. Orgel estimates the accompanying increase of Δ (leading to a smaller wave number of the emitted than of the absorbed light) to be about 20 per cent.

　　Tetrahedral symmetry T_d has the same Bethe quantum numbers Γ_n as O_h (eqn. 43) except that the centre of inversion is absent and accordingly, the wave functions cannot be classified according to parity. The identification of the wave functions from spherical symmetry (eqn. 44) is also the same, except that for odd parities (in spherical symmetry especially the orbitals with odd l), the symbols Γ_4 and Γ_5 (γ_4 and γ_5) are to be interchanged, as also Γ_1 and Γ_2. Thus, in tetrahedral symmetry, the orbitals of the central ion have the γ_n:

$$
\left.
\begin{aligned}
\text{s-electrons} \quad & \gamma_1 \\
\text{p-electrons} \quad & \gamma_5 \\
\text{d-electrons} \quad & \gamma_3 \text{ and } \gamma_5
\end{aligned}
\right\}
\tag{131}
$$

while the orbitals of the ligands (arranged according to λ of the linear microsymmetry of each bond to the central ion) form linear combinations according to:

$$\begin{aligned} \sigma\text{-orbitals:} \quad & \gamma_1 \text{ and } \gamma_5 \\ \pi\text{-orbitals:} \quad & \gamma_3, \gamma_4, \text{ and } \gamma_5 \end{aligned} \Bigg\} \tag{132}$$

If only σ-bonding is considered, γ_5 of the partly filled shell has hence a higher energy than γ_3. This is also expected from the electrostatic model, since the γ_5-electrons of the d-shell are directed against the four ligands (eqn. 47), but also in the four opposite directions, forming together the corners of a cube, while the γ_3-electrons are concentrated in the six directions between the ligands. Ballhausen (1954) calculated that the electrostatic first-order perturbations in a regularly tetrahedral complex with the distance r_L between the nuclei of a ligand and the central ion produces a value of Δ, $-4/9$ times as large as the value of Δ in the corresponding regularly octahedral complex with the same ligands (but now six) and the same r_L.

It is possible to use the Orgel and Tanabe–Sugano diagrams for O_h, extended towards negative values of Δ. (This is equivalent to an exchange of d^n and d^{10-n} in Fig. 6.) The entities in Table 8 are exactly identical, as also the multiplication in Table 5. The ligand field spectra of very many tetrahedral transition group complexes are not known, and most have the electron configurations of the gaseous ion $3d^5$(Mn(II) and Fe(III)) and $3d^7$(Co(II)). This is connected with the Jahn–Teller effect, allowing regularly tetrahedral complexes only for

$$\begin{array}{llll} \text{high-spin} & d^2 & {}^3\Gamma_2 & \gamma_3^2 \\ \text{low-spin} & d^4 & {}^1\Gamma_1 & \gamma_3^4 \\ \text{high-spin} & d^5 & {}^6\Gamma_1 & \gamma_3^2\gamma_5^3 \\ \text{high-spin} & d^7 & {}^4\Gamma_2 & \gamma_3^4\gamma_5^3 \\ & d^{10} & {}^1\Gamma_1 & \gamma_3^4\gamma_5^6 \end{array} \tag{133}$$

(cf. eqn. 126). If the ligand field stabilization is expressed in terms of ρ, it is seen to be absent for d^5 and d^{10} and to have its maximum value for low-spin d^4 and high-spin d^7. There is no certain example known of the former case*, while the strongly coloured, tetrahedral Co(II) complexes represent the latter. Since the Tanabe–Sugano diagram is identical for tetrahedral and octahedral d^5-systems, we cannot conclude directly from the absorption spectrum of a given

---

* Orgel suggested $ReCl_4^-$, and Wilkinson has prepared Re dtc Cl_2 of diethyldithiocarbamate.

Mn(II) or Fe(III) complex, whether it is four- or six-co-ordinated. However, since the Δ-values found for tetrahedral Co(II) are near to the value $-4/9\Delta$ of the corresponding octahedral complexes (agreeing with the spectrochemical series $I^- < Br^- < Cl^- < OH^- < NCS^-$ according to Shimura and Tsuchida), as also in certified $FeCl_4^-$, the high value of Δ found in the Mn(II) aqua ion (slightly smaller than in Mn en_3^{2+}) indicates the constitution $Mn(H_2O)_6^{2+}$, while on the other hand, $MnBr_4^{2-}$ has the expected small value of Δ, $(-)2$ kK. Similar differences occur in the green (octahedral) and pink (tetrahedral) modifications of manganese (II) sulphide.

Dunitz and Orgel (1957) indicated that the Jahn–Teller distortions were to be expected in tetrahedral complexes with other configurations than eqn. (133), and found agreement with the crystallographic data, especially on *spinels*, which are mixed metal oxides, containing metal atoms (of different elements) surrounded by four and six oxygen atoms. Unfortunately, most of these materials are black. However, paramagnetic resonance experiments show that Cr(III) and Ni(II) strongly prefer the octahedral positions, while Co(II) prefer the tetrahedral and most other metals may be distributed between the two types of position. Contrary to the opinion expressed by Pauling, when assuming $Ni(NH_3)_4^{2+}$ to be tetrahedral, the tetrahedral Ni(II) complexes are exceedingly rare. However, $NiCl_4^{2-}$ is found in salt melts (Boston and Smith, 1958; Gruen and McBeth, 1959). Recently, this ion and the two heavier nickel (II) tetrahalides have been studied in organic solvents (Gill and Nyholm, 1959; Cotton and Francis, 1960; Buffagni and Dunn, 1960; and Furlani and Morpurgo, 1961). In the same way as tetrahedral Cu(II) complexes, Ni(II) are Jahn–Teller distorted, and it may be difficult to say whether a given complex is best considered as a distorted square-planar or T_d-complex. An unusual case is the pale yellow Co(III) heteropolywolframate, studied by Shimura and Tsuchida (1957), showing no ligand field bands in the range 10–22 kK. If these authors are correct in assuming tetrahedral microsymmetry of Co(III), it is then a high-spin $\gamma_3^5\gamma_5^3$ $^5\Gamma_3$, having its only spin-allowed ligand field transition to $^5\Gamma_5$ (representing $-\Delta$) well below 10 kK. A low-spin $\gamma_3^4\gamma_5^2$ demands a higher Δ with consequent visible bands. High-spin Fe(II) in ZnS has $\Delta = -3\cdot3$kK according to Low and Weger (1960).

In the tetroxo complexes CrO_4^{2-}, MnO_4^-, OsO_4 (all d^0), and the d^1-complexes CrO_4^{3-}, MnO_4^{2-}, RuO_4^-, ReO_4^{2-}, and the d^2-complexes

MnO_4^{3-}, FeO_4^{2-}, RuO_4^{2-}, it was not certain whether Δ is positive or negative. In the famous calculations of Wolfsberg and Helmholz (1952), the effects of π-bonding, acting on γ_3, were so much stronger than the combined effects of σ- and π-bonding acting on γ_5, that the former orbital turned out to have the highest energy in CrO_4^{2-} and MnO_4^{-}. The present writer has not been able to observe either the narrow bands to be expected from internal transitions in γ_3^2, if Δ is large (the only band of RuO_4^{2-} found at 10·3 kK ($\delta = 0·07$ kK) belongs to OH^- from the solvent, strong sodium hydroxide) nor the broad $\gamma_3 \to \gamma_5$ transitions for smaller Δ. However, the broad $\gamma_3 \to \gamma_5$ transitions seem to occur at 16·0 kK of Cr(V), 14·8 kK of Mn(V), 16·5 kK of Mn(VI), 12·7 kK of Fe(VI), and 21·6 kK of Ru(VI), since they have much smaller wave numbers than expected of the electron transfer bands. It is seen in the section on hexahalide complexes (p. 154) that the electron transfer bands generally shift smoothly towards lower wave number for increasing oxidation number and number of electrons in a given partly filled shell. Hence, the bands ~23, 29 and 33 kK of Mn(VI) are probably related to those at 26·8 and 36·6 kK of Cr(VI), etc. Recently (Carrington, Ingram, Lott, Schonland and Symons, 1960), paramagnetic resonance experiments on Cr(V) and Mn(VI) have been interpreted as γ_3 being lowest, and thus Δ having its expected, negative value. Carrington and the present writer have submitted a paper on this subject to *Molecular Physics*.

Cubic co-ordination of eight ligands with the symmetry O_h is very rare, probably because a monomeric complex would gain energy by turning one of the planes, containing four ligands, 45° parallel to the other plane, and thus forming the *Archimedean anti-prism*, known from several complexes ML_8. However, Low has studied several ions, embedded in fluorite CaF_2, and interpreted the paramagnetic resonance results by the cubic co-ordination of eight fluoride ligands. In one case, Co(II), an absorption spectrum could be measured, corresponding to $\Delta = -3·4$ kK. This does not agree well with the electrostatic model, giving $-8/9$ of the value for an octahedral hexafluoride (Stahl-Brada and Low, 1959), but might be excused by the anomalous large Co-F distance due to the large ionic radius of Ca(II).

As discussed above, it is impossible to predict absolute values of the ground states of complexes. As a special case of this difficulty, we cannot directly compare the stability of tetrahedral and octahedral complexes. It is not surprising that the tetrahedral complexes are formed by d^5 (having $\rho = 0$ in both symmetries) and d^7 (eqn. 133) with bulky anions, such as Br^-. It is also plausible that ligand fields of tetragonal symmetry, changing a four-co-ordinated complex from tetrahedral to square-planar co-ordination, may explain the occurrence of square-planar complexes with partly filled shells while such a development would be impossible with a rigid spherically symmetrical central ion and four ligand atoms which repel each other (rather than be bound together, e.g. in a porphyrine ring). However, we cannot obtain quantitative results, and we cannot easily use M.O. theory for comparing the relative stability of the ground state in different geometrical configurations of a given complex. M.O. theory is much more apt to describe the distribution of the excited states, once we have assumed a given symmetry. Fortunately, in many cases, this symmetry follows from simple considerations on the reciprocal behaviour of the ligands; we would not expect an octahedral complex to distort from O_h without a special cause (such as the Jahn–Teller effect).

We may conclude this chapter with some remarks on the nature of the parameter Δ. We are surprised to see that all the ligand field bands of all octahedral transition group complexes agree with the Orgel diagram with use of this one orbital energy difference, demanded by group theory between the two necessarily equivalent γ_3- and the three necessarily equivalent γ_5-orbitals. Further, we are surprised to see Δ vary rather slightly with the ligands (only by a factor of 1·8, excepting cyanide) and to a similar extent with the central ion, the function g of eqn. (123) and Table 12 varying with a factor 4. Finally, it is surprising to find that Δ is approximately the product of two functions, i.e. the spectrochemical series of ligands eqn. (120) and the spectrochemical series of central ions eqn. (124).

While all these features justify the most juvenile optimism regarding the usefulness of ligand field theory as a *semi-empirical theory* (i.e. a theory which does not begin, a priori, with the properties of the electron and the quantum-mechanical laws, but uses parameters derived from more direct experimental evidence), we must also conclude that the parameter Δ is nearly impossible to predict within a factor 2 or 5 (as it is impossible to predict most other quantities in theoretical chemistry also) except by the very successful interpolation in empirical series. We may consider Δ as consisting of mainly four contributions, given here with their sign:

$$\Delta \simeq + \text{electrostatic first-order perturbation} + \sigma(L \rightarrow M) - \\ - \pi(L \rightarrow M) + \pi(L \leftarrow M) \qquad (134)$$

(cf. the discussion of eqn. (54)). However, we have here neglected the contributions from parameters of interelectronic repulsion (eqn. 84), appearing even for pure M.O. configurations, if l deviates from 2; and we have neglected the perhaps very substantial contributions from intermixing of M.O. configurations. We cannot reasonably consider Δ as a subject for calculation from one or another set of assumptions (such as Hartmann's electrostatic model or Wolfsberg and Helmholz's M.O. model), but it is very useful as an *empirical parameter*. We have thus a continuous theory of all transition group complexes without sharp divisions between electrovalent and covalent complexes; and we have an interesting property to investigate in each new octahedral complex.

K

CHAPTER 8

THE NEPHELAUXETIC SERIES

A GIVEN transition group complex is characterized by two sets of parameters: (1) the orbital energy differences in the partly filled shell (one value, Δ, in octahedral and tetrahedral symmetry, two quantities in linear, three in tetragonal, and four in lower symmetries), and (2) the parameters of interelectronic repulsion. This determines a *two-dimensional classification* of octahedral complexes, according to two variables: Δ in the spectrochemical series and β in the *nephelauxetic series* (see p. 138). β is the ratio between the value of a representative parameter of interelectronic repulsion, e.g. B from eqn. (35), in the complex and in the corresponding gaseous ion. This definition is equivalent to the assumption of the expanded radial function model, conserving the separability of each orbital of the partly filled shell into the product of an arbitrary radial function and a linear combination of angular functions, hydrogenic and corresponding to a well-defined $l = 2$. As expected from the occurrence of nine different parameters in the M.O. theory of $\gamma_5{}^a\gamma_3{}^b$ without the assumption of $l = 2$ (see eqn. 75), the absorption spectra show that the assumption of a single value of β is an over-simplification. Therefore, the comparison of β in different metal ions is somewhat uncertain, also due to the fact that not all the values of B are known from atomic spectroscopy. On the other hand, the comparison of β in different octahedral complexes with varying ligands is much more certain for the same central ion.

We have mainly eight energy differences in the Tanabe–Sugano diagrams (Fig. 6) which are useful for determination of β, since they are nearly independent of Δ, viz.

d^2: narrow, spin-forbidden transitions $^3\Gamma_4 \rightarrow {}^1\Gamma_3, {}^1\Gamma_5$ (see Table 8 and eqn. (67)) and $^1\Gamma_1$, mainly inside the sub-shell configuration $\gamma_5{}^2$. Though this system in principle is Jahn–Teller-unstable, Pryce and Runciman have successfully used V(III) in Al_2O_3 for determining $K(4, 5) \cong 3B + C$. Furthermore OsF_6 has been studied.

d^3: narrow, spin-forbidden transitions $^4\Gamma_2 \rightarrow {}^2\Gamma_3, {}^2\Gamma_4$ (Table 8 and eqn. (68)) and $^2\Gamma_5$, the famous *ruby lines*, mainly inside $\gamma_5{}^3$.

They are only slightly disturbed by sub-shell configuration inter-mixing. The higher $^2\Gamma_5$ is split by an amount proportional to ζ_{nl}^2 by intermediate coupling, Γ_8 being 0·8 kK higher than Γ_7 in Re(IV). The systems studied are numerous complexes of Cr(III), Mn(IV), Mo(III), Re(IV) and Ir(VI). The parameter determined is $K(4,5) \simeq 3B + C$.

d^3: broad, spin-allowed transitions $^4\Gamma_2 \to {}^4\Gamma_5$ (pure $\gamma_5^2\gamma_3$) and $^4\Gamma_4$ (mainly $\gamma_5^2\gamma_3$, according to eqn. (85)). Both Δ and B (identifying $J(3,4) - K(3,4) - J(1,4) + K(1,4)$ with $12B$) can be found easily, if the two first absorption bands are observed. The cases so far studied are V(II), Cr(III), Mn(IV), and Mo(III).

Low-spin d^4: narrow, spin-forbidden transitions $^3\Gamma_4 \to {}^1\Gamma_{3, \, 5}$ and $^1\Gamma_1$. Until now, only the Os(IV) hexahalide complexes and PtF$_6$ have been studied. The energy difference $^3\Gamma_4 \to {}^1\Gamma_1$ (eqn. 67) $5(3B + C)$ is considerably changed by intermediate coupling, stabilizing the ground state component $\Gamma_1 - \frac{1}{2}\zeta_{nl}$ to first order, and this and $^1\Gamma_1$ repulsing each other with a non-diagonal element $\sqrt{(2)}\zeta_{nl}$.

TABLE 15. THE ENERGY DIFFERENCE ^{6}S — ^{4}G (BETWEEN THE BARI-CENTRES) OF SEVERAL GASEOUS IONS AND OF MANGANESE (II) AND IRON (III) COMPLEXES (THE AVERAGE WAVE NUMBER OF THE COMPO-NENTS OF THE FIRST NARROW BAND $^6\Gamma_1 \to {}^4\Gamma_1, \, ^4\Gamma_3$) IN kK.

Cr$^+$ 3d^5	20·516	Fe^{3+} 3d^5	32·8
Mn^0 3d^{5}4s^2	25·279	FeF$_6^{3-}$	25·35
Mn^{2+} 3d^5	26·846	Fe(H$_2$O)$_6^{3+}$	24·45
MnF$_2$	25·3	Fe urea$_6^{3+}$	23·25
Mn(H$_2$O)$_6^{2+}$	25·0	Fe mal$_3^{3-}$	22·8
MnCl$_2$(H$_2$O)$_4$	24·65	Fe formiate$_6^{3-}$	22·75
Mn(II) in 9 M KSCN	24·0	Fe ox$_3^{3-}$	22·2
Mn enta^{2-}	24·0		
MnO	23·8	FeCl$_4^-$	18·8
Mn en$_3^{2+}$	23·8		
Mn py$_2$ Cl$_2$	23·8		
MnBr$_4^{2-}$	22·3		
MnS	21·6		

High-spin d^5: narrow, spin-forbidden transitions $^6\Gamma_1 \to {}^4\Gamma_{1, 3}$ and several other quartet levels (Tables 7 and 8), mainly within $\gamma_5^3\gamma_3^2$. Table 15 gives several wave numbers for the two first transitions, degenerate for $l = 2$ (giving $10B + 5C$, and for the transition to $^4\Gamma_1$ according to general M.O. theory (eqn. (75) and Table 7), $\frac{5}{2}(K(1,4) + K(3,4))$ for Mn(II) and Fe(III) complexes, in addition to some values for the multiplet energy difference $^6S - {}^4G$ in several gaseous ions $3d^5$. This transition is independent of Δ to a very high approximation. On the other hand, Δ is not too well-defined in high-spin d^5-complexes, we have estimated it from the position of the first quartet level $^4\Gamma_4$ relative to $^4\Gamma_1$ (cf. eqn. 94).

Low-spin d^6: broad, spin-allowed transitions $^1\Gamma_1 \to {}^1\Gamma_4$ and $^1\Gamma_5$. The excited levels belong mainly to the sub-shell configuration $\gamma_5^5\gamma_3$, but are subject to some intermixing with other $\gamma_5^a\gamma_3^b$, depressing $^1\Gamma_5$ to the largest extent. For this reason, the limiting value of the energy difference between the two excited singlets for $\Delta \to \infty$ (cf. eqn. (73), which reduces to $16B$ for $l = 2$) is decreased some 10–30 per cent, i.e. roughly $84B^2/\Delta$ in actual complexes. The two bands are known of $Fe(CN)_6^{4-}$, many low-spin Co(III), Rh(III), Ir(III), and PtF_6^{2-}. Δ is found from the first band after a correction with interelectronic repulsion parameters, amounting to about C (see Table 8).

High-spin d^8: narrow, spin-forbidden transitions $^3\Gamma_2 \to {}^1\Gamma_3$ (the wave number representing $2K(1,3)$ of eqn. (71), being $8B + 2C$ for $l = 2$) and $^1\Gamma_1$. The latter transition is not certainly identified. The former transition has been extensively studied in Ni(II) complexes and may probably also be found in CuF_6^{3-} and PdF_2 (the latter compound is paramagnetic, according to Bartlett and Maitland (1958)). It is only slightly perturbed by sub-shell configuration intermixing, but has large intermediate coupling effects with the Γ_3-components of the triplet terms $^3\Gamma_5$ and $^3\Gamma_4$ of $\gamma_5^5\gamma_3^3$ as discussed by Griffith (1961) and the writer (1955–58).

High-spin d^8: broad, spin-allowed transitions $^3\Gamma_2 \to {}^3\Gamma_5$ of $\gamma_5^5\gamma_3^3$ and the two $^3\Gamma_4$, which in Ni(II) complexes are strongly mixed $\gamma_5^5\gamma_3^3$ and $\gamma_5^4\gamma_3^4$ according to eqn. (85). Since the three spin-allowed bands $\sigma_1, \sigma_2, \sigma_3$, are generally all observed, it is possible to infer the quantity $15B$ (or in M.O. language, $\frac{5}{4}$ times the quantity mentioned above for the spin-allowed bands of d^3) from the diagonal sum rule as $\sigma_2 + \sigma_3 - 3\Delta$, where Δ is σ_1, the wave number of the first band, corrected (if necessary) for intermediate coupling effects. Bostrup and the present writer have shown that σ_2 and σ_3 of regularly octahedral nickel (II) complexes agree individually with eqn. (85), while Schäffer has shown this not to be the case for most Cr(III) complexes, of which the third spin-allowed band sometimes is known.

We may also find B in tetrahedral complexes from such transitions. In d^5, the situation is exactly analogous to that described above. In d^7, the Tanabe–Sugano diagrams for octahedral d^3-complexes are used. Especially the third spin-allowed transition $^4\Gamma_2 \to {}^4\Gamma_4$ can be described by eqn. (85) (inserting a positive value of Δ), but it seems rather influenced by first-order L,S-coupling and by intermixing with low doublet terms such as $^2\Gamma_3$ and $^2\Gamma_4$.

TABLE 16. VALUES OF B IN GASEOUS IONS USED FOR THE CALCULATION OF β IN THE NEPHELAUXETIC SERIES, IN kK. THE VALUES WITH THREE DECIMALS ARE DIRECTLY DERIVED FROM ATOMIC SPECTROSCOPY (CF. TABLE 3), WHILE THE OTHERS ARE EXTRAPOLATED FROM OTHER VALUES KNOWN FOR LOWER IONIC CHARGES

$3d^3$	V^{2+}	0·757	$5d^2$	Os^{6+}	0·78
	Cr^{3+}	0·918	$5d^3$	Re^{4+}	0·65
	Mn^{4+}	1·064		Ir^{6+}	0·81
$3d^6$	Co^{3+}	1·1	$5d^4$	Os^{4+}	0·7
$3d^8$	Ni^{2+}	1·041	$5d^6$	Ir^{3+}	0·66
$4d^3$	Mo^{3+}	0·61		Pt^{4+}	0·72
$4d^6$	Rh^{3+}	0·72	$5d^8$	Pt^{2+}	0·6
$4d^8$	Pd^{2+}	0·683			

The discussion above makes B available to us for a given complex, where some of the appropriate absorption bands are known. For determining β, as given in Table 11 together with the values of Δ in each complex, we must by definition also know the value of B in the corresponding gaseous ion. In some cases (Table 16), this can be inferred directly from the baricentre distances between F- and P-terms with maximum value of S (cf. Table 3) in d^2, d^3, d^7 and d^8, when known from atomic spectroscopy (Charlotte E. Moore's tables *Atomic Energy Levels*, National Bureau of Standards Circular No. 467, Vols. I, II, III). In other cases, it is necessary to extrapolate from the regular evolution of the parameters of interelectronic repulsion with ionic charge $Z_0 - 1$ and a number of electrons in the partly filled shell. It is a special problem to compile information about the second and third transition group, where investigations by atomic spectroscopy are rather rare and mainly available for neutral atoms only. It seems fairly well established that B is, in a given $4d^n$-ion, 66 per cent of that in the corresponding $3d^n$-ion with the same charge.

The values of B indicated in Table 16 for $4d^n$ are thus calculated except the directly known value for Pd^{2+}. The similar ratio is much less certain for $5d^n$, due to the strong influence of intermediate coupling in the gaseous ions (stronger than in the complexes), but seems to be about 60 per cent of the corresponding B-value of $3d^n$. In the special case of $3d^5$, we have not calculated B, but compared rather directly the energy difference $^6S - {}^4G$ in gaseous ions with that of the complexes (given in Table 15).

It is seen from Table 11 that the *nephelauxetic series of ligands* is quite different from the spectrochemical series of ligands (eqn. 120), viz. ordered according to increasing values of $(1 - \beta)$:

$$F^- < H_2O < \text{urea} < NH_3 < \text{en} \sim \text{ox}^{2-} < \underline{N}CS^- < \\ < Cl^- \sim CN^- < Br^- < I^- < \text{dtp}^- \tag{135}$$

and it cannot be expressed as a simple series of ligand atoms* such as in eqn. 122. Thus Schäffer has established that oxygen-containing complexes of Cr(III) are distributed widely over eqn. (135), sulphate or hydrogen sulphate complexes (as in concentrated H_2SO_4) occurring before H_2O, most anion complexes and the red ruby in the middle of the nephelauxetic series and the green Cr_2O_3 at a very late place, even after $Cr(CN)_6^{3-}$. For all these oxo complexes, Δ does not vary more than between 14·64 and 18·52 kK.

It is argued above that the corresponding *nephelauxetic series of central ions* is less well-defined, but it seems to be

$$Mn(II) \sim V(II) < Ni(II) \sim Co(II) < Mo(III) < Re(IV) \sim \\ \sim Cr(III) < Fe(III) \sim Os(IV) < Ir(III) \sim Rh(III) < \tag{136} \\ < Co(III) < Pt(IV) \sim Mn(IV) < Ir(VI) < Pt(VI)$$

This series resembles the spectrochemical series of central ions (eqn. 124) in the arrangement according to increasing oxidation number, but it does not show the strong increase in the later transition groups as Δ, but rather a slow decrease $3d^n > 4d^n > 5d^n$ for a given oxidation number.

Analogous to eqn. (123), the nephelauxetic series can be factorized to a good approximation into the product of two functions of one variable only:

$$(1 - \beta) = \text{h (ligands)} \cdot \text{k (central ion)} \tag{137}$$

(we might equally well have chosen the expression $\beta = (1 - h')(1 - k')$) of which the values were given in Table 12.

* Though $F < O < N < Cl < C \sim Br < S \sim I$ would also be the order of Pauling's electronegativities.

For many purposes, the nephelauxetic series of ligands much more resembles the chemist's idea of the tendency towards covalent bonding than does the spectro-chemical series (eqn. 120). Thus, water and fluoride, recognized as forming the least covalent complexes of a given central ion have β near to 1, while the complexes with covalent bonds from anions have highly decreased values of β. In the spectrochemical series, the former complexes are situated in the middle, while most of the covalent complexes have values of Δ either well below or well above the hexa-aqua ions.

The order of ligands in eqn. (135) corresponds well to the idea of the *reducing character of the ligands*, viz. their tendency to lose electrons. It must be emphasized that this chemical idea does not correspond to the ionization potentials, measured in the gaseous state. Thus, the electron affinities of F, Cl, Br, I, do only vary between 3·2 and 3·8 eV (recently, Bailey has even shown that the ionization potential of F^- in gaseous state is between that of Cl^- and Br^-), while the ionization potentials of neutral molecules are between 9 and 14 eV (py 9·23, C_6H_6 9·24, NH_3 10·23, H_2O 12·59, CH_4 12·99, CO 14·01, etc.). These values are obviously not connected directly with the conditions in aqueous solution, where ions are highly stabilized. We may guess that the energy of the most loosely bound electrons in ligands varies from about -8 to -14 eV (as shall be discussed in the chapter on electron transfer spectra) while there are reasons to believe that the d-orbital energy is about -5 to -7 eV. We may here use *Fajans' concept of polarizability*—the iodide and bromide ligands are so much more deformable than fluoride that in the presence of the central ion, the energy of the most loosely bound electron is increased, approaching the d-shell energy. If the dielectric effects of water are sufficient to stabilize the halide ion $e^2/2r_{ion}$, this stabilization would amount to 27 kK for I^-, 32 for Cl^-, and 42 for F^-. These values, when added to the ionization energy of gaseous halide ions, account qualitatively for the wave number of the absorption band of isolated halide ions in aqueous solution. However, the relative difference between the halides is too small by a factor of 2. The iodine atom of the excited state may be more stabilized by the solvent than is the chlorine atom.

It can be implied from electron transfer spectra that the total spreading of energy is about 44 kK, i.e. 6·3 eV (cf. eqn. 97). Surprisingly enough, this spreading is not larger in complexes of platinum group metals, having the oxidation number $+3$ or $+4$, than in the simple process of removing an electron from a halide ion in aqueous solution (or the transfer to some acceptor orbital in alkali halide crystals). We shall see later on that we must accept these quantities as indicating, in a certain sense, an absolute set of energy differences for the most loosely bound electrons of different ligands.

Considering the partly covalent bonding as a second-order perturbation, we expect it to become more pronounced as the energy difference between the orbitals of the partly filled shell and of the ligands' lone pairs decreases. Thus, we can understand the dependence on the reducing character of the ligand, and we may next ask for an influence from the *oxidizing character of the central ion*. Actually, we observe such a variation with the oxidation number $+2$, $+3$, and $+4$ in eqn. (136). When $4d^3$ Mo(III) is distinctly less nephelauxetic than $3d^3$ Cr(III), we may also see there a variation with the oxidizing character. However, we cannot identify this concept with the standard oxidation potentials, as measured in galvanic elements. This is connected with the fact that the electrochemical potentials are measured as equilibria reactions, not conforming to the Franck–Condon principle of optical excitation; the ligand distances r_L have time to establish their preferred values in the reduced and the oxidized form of the metal complex. As will be discussed later, the energy difference between the partly filled shell and the highest filled M.O. is about 15 kK smaller in $+4$ than in $+3$ complexes, having the same ligands and the same electron configuration.

The nephelauxetic series resembles the *hyperchromic series* of increasing intensity among the Laporte-forbidden ligand field bands. Yamada and Tsuchida (1953) wrote this series

$$H_2O < Cl^- < NH_3 < Br^- < SO_4^{2-} < NO_3^- < OH^- <$$
$$< en < ox^{2-} < CO_3^{2-} < NO_2^- < SO_3^{2-} < CrO_4^{2-} < \qquad (138)$$
$$< NCS^-$$

from the behaviour of cobalt (III) pentammine monosubstituted complexes. However, the series is not quite the same for every central ion, and in most cases, Br^- and I^- produce very high intensities. A similar hyperchromic series of central ions may be constructed:

$$V(II) \sim Fe(II) < Ni(II) < Co(II) < Cr(II) < V(III) <$$
$$< Mo(III) < Cr(III) < Cu(II) < Ir(III) \sim Co(III) \sim \qquad (139)$$
$$\sim Pt(II) < Rh(III) < Pd(II) < Pt(IV) \sim Ir(IV) < Pd(IV)$$

The hyperchromic series eqn. (138) and eqn. (139) are connected with the wave number of the first strong Laporte-allowed transition by eqn. (116). Thus, the oxidizing character of the central ion is fairly well related to eqn. (139), though the square-planar complexes seem to have unusually high intensities of their Laporte-forbidden bands (the central ion possibly slightly out of plane).

In the chapter on the interelectronic repulsion in M.O. configurations, the interpretation of the nephelauxetic effect was discussed. By analogy with Δ of eqn. (134), the decrease of β below 1 can be related to several effects, which are not easily experimentally separated. Three proposals have been given in literature:

(1) Van Vleck (1935) discussed symmetry-restricted covalency, the delocalization of the γ_3-sub-shell by σ-antibonding and the γ_5-sub-shell by π-bonding or antibonding. Tanabe and Sugano (1954), Owen (1955) and Orgel (1955) ascribed a part of the nephelauxetic effect to this reason.

(2) Orgel (1955) compared the decreased distance of $^6S - {}^4G$ in $Mn(H_2O)_6^{2+}$ (see Table 15) with the action of the two 4s-electrons in $3d^5 4s^2$ of neutral, gaseous manganese atoms, as compared to $3d^5$ of gaseous Mn^{2+}. In this way, the central-field covalency was mentioned for the first time in literature.

(3) Schläfer (1956) assumed the nephelauxetic effect to consist of differing first-order perturbation of the ligand field on the excited terms of a given configuration. Moffitt and Ballhausen (1956) mentioned that this effect could only be caused by differences in the radial functions of different terms of the same configuration in the gaseous ion. Craig and Magnusson (1958) further studied the variation of the term distances in a spherically symmetrical ligand field.

We may consider these explanations to fall into two large groups: the *Stevens delocalization effects* (1953) and the *variation of the effective charge in the partly filled shell*. The delocalization effects are expressed by a coefficient in the L.C.A.O. description of the orbitals, having a certain density in the ligands. This model has not only been applied to the nephelauxetic effect, but also to the gyromagnetic factor g (eqn. 119) and to hyperfine structure, caused by the nuclear magnetic moment of nuclei of ligands and the central ion, on the paramagnetic resonance curves (Owen and Stevens studied $IrCl_6^{2-}$ and $IrBr_6^{2-}$, Tinkham (1956) $3d^n$-ions embedded in ZnF_2). If the delocalization results in a L.C.A.O. charge density of $(1 - x)$ of its original value in the central ion with subsequent presence of x in the ligands (neglecting effects of overlap integrals), the value of β is expected to be slightly above $(1 - x)^2$. It is one of the arguments for the intermixing of γ_3-orbitals from the d-shell and from the σ-bonding lone-pairs not to have reached the value $x = 0.5$ that none of the β-values found in complexes is so small (Table 11) as slightly above 0.25 (excepting the new complexes Rh dtp$_3$ and Ir dtp$_3$). If x was above 0.5, the original energy of the d-shell before the intermixing (in the simple second-order perturbation case eqn. (18) without overlap) would have been below that of the lone-pairs of the ligands.

On the other hand, we have also an expansion of the partly filled shell due to the entrance of electrons from the ligands to the core of the central ion. Even though the oxidation number of the central ion (at least in a monomeric, octahedral complex) is an integer, derived from the number of electrons in the partly filled shell (and this is possible, because M.O. configurations are applicable for classification), we cannot expect the effective charge to be an integer. Table 15 gives the impression of an effective charge Z_{eff} (it is not exactly Z_* as defined from a proportionality to the interelectronic repulsion parameters) varying from 1.7 to 1.2 in manganese (II) complexes and from 1.7 to 0.7 in iron (III) complexes, if we define Z_{eff} to be 2 in Mn^{2+} and 1 in Cr^+. This cannot be interpreted as only 0.3 to 2.3 electrons being transferred from the ligands to the central ion, since Z_{eff} is as high as 1.75 in $3d^5 4s^2$ of neutral Mn. Rather, we are measuring the effect of other orbitals on the partly filled shell, the 4s-orbital having little influence only, due to its great spatial extension, surrounding the 3d-shell. There seems to occur a *cancellation in* Z_{eff} of the opposite influences of increasing ionic charge and larger covalent bonding, e.g. in $Mn(H_2O)_6^{2+}$ and $Fe(H_2O)_6^{3+}$. This has previously been noted, e.g. by Kröger (1948) in the nearly coincident ruby lines of Cr(III) and Mn(IV) in solid, fluorescent materials. It is interesting that the *principle of electroneutrality*, i.e. that the positive and negative ionic charges tend to cancel by forming covalent bonds, is not sufficiently effective for decreasing Z_{eff} to zero. In metals with high oxidation numbers, such as MnF_6^{2-} and PtF_6^{2-}, there is a tendency towards large nephelauxetic effects even in the fluoride complexes, decreasing Z_{eff} to below $+1$, and it can be concluded that the evolution of the nephelauxetic effect in the ligands is not as pronounced as indicated by the function h in Table 12.

Actually, since the neutral atoms Os, Ir, and Pt are extrapolated to have larger values of $3B + C$ than observed in OsF_6, IrF_6 and PtF_6, the latter compounds would have negative values of Z_{eff}. It can hardly be assumed in these hexafluorides that negative charge is accumulated on the central atom; rather a proportion of symmetry-restricted covalency has expanded the partly filled shell and diluted its probability density in the central atom.

There is no doubt that both delocalization (symmetry-restricted covalency) and change of Z_{eff} (central field covalency) occur in actual complexes, and the distribution of the nephelauxetic effect on the two reasons is not easier to estimate than Δ on effects of electrostatic first-order perturbations and covalent second-order perturbations. The delocalization can be estimated from Owen's study of the hyperfine structure of paramagnetic resonance, and it can also be inferred from the average radius $\bar{r}$ of Hartree-Fock self-consistent radial 3d-functions, which is 0·615 Å in gaseous V^{2+}, 0·57 Å in Mn^{2+}, and 0·45 Å in Zn^{2+}. Since these functions tail off very slowly for large r, and since the ionic radii of divalent transition group ions are about 0·7–0·8 Å, there must be some overlapping with the ligands with consequent delocalization.

Table 15 gives a clear idea that the wave function of the partly filled shell $3d^5$ in Mn^{2+} changes smoothly in the complexes, progressing in the nephelauxetic series. Some arguments can be given that the central-field covalency part of the nephelauxetic effect is even more fundamental, regarding the influence of the ligands on a partly filled shell, than are the ligand field splittings of the energy levels known from atomic spectroscopy. Thus, in the lanthanides, the ligand field splittings are very small (amounting to about 0·2 kK), while nephelauxetic effects can clearly be observed in anion complexes. Thus, Ephraim studied (1926–29) the reflection spectra of anhydrous and hydrated lanthanide compounds and found a shift of the wave numbers of the narrow $4f^n$-bands

$$RF_3 > R(H_2O)_9{}^{3+} > RCl_3 > RBr_3 \sim R_2O_3 \qquad (140)$$

which he explained partly by an expansion of the 4f-shell by covalent bonding (our explanation) and partly by a hypothetical contraction by electrostatic forces. Actually, even before Bethe's paper, Brunetti (1928) attempted to apply a sort of ligand field theory to eqn. (140). Boulanger (1952) studied the reflection spectra of several solid compounds, and the present writer (1956) the complexes with tartrate, citrate, ethylenediaminetetra-acetate, etc., in solution. In the latter case, the shift towards smaller wave numbers, compared to the aqua ion bands, is 0·8–1·2 per cent in Pr(III), 0·4–0·7 per cent in Nd(III), 0·6–1·0 per cent in Sm(III), and 0·3 per cent in Gd(III). Unfortunately, the term distances are not known in the gaseous ions, but if we assume the function h of Table 12 to be about 1·5 to 1·8 in these complexes, we arrive at the values of the central ion function k, given in Table 18. It is seen that they are not very small, compared to the value k = 0·08 of the least nephelauxetic $3d^n$-central ion. They are about twice as large in the actinides as in the lanthanides. The largest nephelauxetic effect known in any lanthanide compound is $\beta = 0\cdot91$ in one of the modifications of Pr_2O_3 (assuming $\beta = 0\cdot97$ in the aqua ion). This shows the large effect of oxide as ligand which, however, cannot be expressed by a constant value of the function h.

The appearance of substantial nephelauxetic effects in lanthanides shows that we cannot assume that the ratio between the covalent second-order perturbation effects and the electrostatic first-order effects necessarily decreases, as usually assumed among physicists, when both quantities are small. The present writer feels that this ratio increases in the direction 5d < 4d < 3d < 5f < 4f of smaller ligand field splittings. This may be due to several reasons: quantitatively, the ligand field of low-symmetry (often nine ligands in trigonal symmetry) f^n-complexes approaches that of spherical symmetry; and the addition of this spherically symmetric part of the ligand field to the central field inevitably changes the radial function of the partly filled shell, even when it is deeply placed inside the core of the central ion. Actually, the average radius of the f-shell cannot be as low as assumed by many authors, applying a hydrogenic radial function for interpreting ζ_{nl}. Though a direct calculation from F^k is slightly uncertain (eqn. 38), the value of $\bar{r}$ is probably 0·5 Å in Pr(III), 0·35 Å in Tm(III), and 0·65 Å in U(IV), somewhat below the ionic radii around 1 Å.

It was shown in eqns. (79)–(82) that the Orgel explanation of second-order perturbation of the radial function from the spherical part of the ligand field and the Schläfer–Craig–Magnusson explanation of different first-order perturbation on different terms of the same configuration, actually are two parts of the same effect, the change of the central field by the ligands.

The *variation of the nephelauxetic effect with internuclear distances* produces some of the effects discussed in the preceding chapter on the variation of Δ. Thus, the band widths δ of the bands in Table 15 are undoubtedly as large as 0·3 kK, caused by the decrease of the interelectronic repulsion parameters for decreasing r_L. Bostrup and the present writer (1957) studied $Ni(H_2O)_6{}^{2+}$ in different crystalline salts and in solution, and found an evolution in terms of Δ exactly as expected from the Tanabe–Sugano diagram, superposed the intermediate coupling effects on the double band in the red. Thus, Δ is exceptionally large in the double salt $(NH_4)_2[Ni(H_2O_6)](SO_4)_2$. It is well known that decreasing *temperature* has a hypsochromic effect on the ligand field bands, especially of the aqua complexes. Thus, the green salts of $Ni(H_2O)_6{}^{2+}$ turn sky-blue, and the red $Co(H_2O)_6{}^{2+}$ turns yellow at the temperature of liquid air. In addition to the broadening of absorption bands with increasing temperature, as discussed on p. 99, this shift of the maxima towards lower wave numbers produces the effect that the high-wave number side of a band does not vary much, while the decline on the low-wave number side shifts considerably by warming. (Cf. Holmes and McClure, 1957.) These effects can be connected with the anharmonic part of the vibrations of the ligands, producing the thermal expansion of the materials. Since Δ decreases so quickly with increasing r_L, these effects may amount to about 1 per cent per 100°K. Parsons and Drickamer (1958) studied the influence of an increase in *pressure* from 0 to 120,000 atm on several ligand field bands and reported an increase of Δ by about 10–15 per cent and rather irregular changes of B. Keating and Drickamer (1961) found a generally increasing nephelauxetic effect in lanthanide compounds under high pressure (β decreasing about 0·003 at 150,000 atm). Recent measurements (Drickamer, Minomura, Stephens and Zahner, private communication) have indicated a systematic decrease of β (about 0·03) in 3d-compounds. In particular, the narrow bands of $MnCl_2$ and $MnBr_2$ are shifted 700 to 1000K to the red at 100,000 atm.

Finally, we have a negative contribution from the nephelauxetic effect to the ligand field stabilization. Thus, the sextet ground state of high-spin d^5 has a particularly low energy, due to favourable high average values of the interelectronic distances (cf. eqn. 149). If this stabilization is decreased in covalent Mn(II) and Fe(III) complexes, they will be destabilized, compared to the less covalent aqua and fluoro complexes.

CHAPTER 9

ELECTRON TRANSFER SPECTRA

WHILE the weak, Laporte-forbidden bands of transition group complexes have been successfully described by the ligand field theory, with the interest concentrated on the partly filled shell, it has only recently been demonstrated that M.O. theory is equally well adapted to the strong, Laporte-allowed bands, at least in the octahedral complexes of a central ion with six halide ligands.

Qualitatively, a satisfactory description of these strong bands may be given by saying that their wave number is the lower, the more oxidizing the central ion and the more reducing the ligands. This is exactly as expected for *electron transfer* from the ligand to the central ion during the optical excitation. It is a *redox process*, decreasing the oxidation number of the central ion by one, while the ligand field bands correspond to the same oxidation number in the excited and the ground state. Consequently, the d^n-systems with $n < 10$, i.e. a partly filled shell which can take up an electron, have the much lower wave numbers of these bands than the corresponding d^{10}-systems.

Two review papers have been written by Rabinowitch (1942) and Orgel (1954). However, these papers were not very much concerned with complexes of metals, but rather with the absorption spectra of crystalline, dissolved and gaseous alkali halides and with some molecular addition compounds, occurring in organic chemistry, etc. There has always been a close connexion between the study of electron transfer spectra and of *photochemical reactions*: the change of oxidation number in the excited state may make it chemically a very reactive entity; other reactions may be started, or ligands dissociated off, often with changed oxidation number.

Fajans, Fromherz, and their assistants made a thorough study on the halide complexes of d^{10}-systems such as Cu(I), Ag(I), Hg(II), and $d^{10}s^2$-systems such as Sn(II), Tl(I), Pb(II). In this way, some confusion arose between the genuine electron transfer bands of the former systems and the particular $s^2 \rightarrow sp$ transitions to be discussed in the next chapter in the latter case. However, it was realized that the bands of Cl⁻, Br⁻ and I⁻ (as known in aqueous solution or in

146

alkali halide crystals) are shifted towards lower wave numbers by oxidizing central ions, and the development was described by the idea of mutual polarizability. If we except the s^2-systems, the shift towards lower wave numbers of the first strong band of completely co-ordinated chloride (or bromide) complexes is approximately:

$$
\left.
\begin{array}{l}
\text{Ir(III)} \sim \text{Mo(III)} \sim \text{Ti(IV)} \sim \text{Sn(IV)} > \text{Pt(II)} \sim \text{Hg(II)} > \\
> \text{Tl(III)} > \text{Rh(III)} > \text{Pt(IV)} > \text{Sb(V)} > \text{Pd(II)} \sim \\
\sim \text{Os(III)} \sim \text{Re(IV)} \sim \text{U(IV)} > \text{Pb(IV)} \sim \text{Au(III)} > \\
> \text{Pd(IV)} > \text{Ru(III)} > \text{Fe(III)} > \text{Os(IV)} > \text{Cu(II)} > \\
> \text{W(VI)} > \text{Ir(IV)} \sim \text{Ru(IV)} > \text{Rh(IV)}
\end{array}
\right\} \quad (141)
$$

while the similar series is for peroxo complexes:

$$
\begin{array}{l}
\text{Zr(IV)} > \text{Ta(V)} > \text{W(VI)} > \text{Nb(V)} > \text{Re(VII)} > \text{Mo(VI)} > \\
\qquad\qquad\qquad\qquad\qquad\qquad > \text{Ti(IV)} > \text{V(V)}
\end{array} \quad (142)
$$

and for tetroxo complexes:

$$
\left.
\begin{array}{l}
\text{Ta(V)} > \text{W(VI)} > \text{Nb(V)} > \text{Mo(VI)} \sim \text{Re(VII)} > \\
> \text{V(V)} > \text{Tc(VII)} \sim \text{Os(VIII)} > \text{Cr(VI)} \sim \text{Ru(VIII)} \sim \\
\sim \text{Ru(VII)} > \text{Mn(VII)} > \text{Fe(VI)}
\end{array}
\right\} \quad (143)
$$

The order in series (141) is undoubtedly determined by the oxidizing character of the central ion and the presence or absence of a partly filled shell. Exceptions may be the low-spin d^8-systems, where the Laporte-allowed bands may be rather different from usual electron transfer spectra, due to the presence of a rather low, empty orbital of odd parity (composed of the p-orbital z of the central ion, etc.).

It was assumed for many years that an increasing number of atoms of a given type, e.g. chloride, always shifted the wave number of the electron transfer spectrum towards lower wave numbers. This is undoubtedly the case with the low-spin d^6-complexes, as will be explained below, and with Cu(II) (which may be somewhat exceptional, due to its low symmetry) and the $s^2 \rightarrow sp$ transitions, discussed in the next chapter. However, the measurements of Ru(III) complexes and the theory given below do not imply a general variation of the orbital energy differences between the partly filled shell and the lone-pairs of the ligands. This is contrary to any primitive electro-static model, where the electron transfer from a negatively charged ligand to a positively charged central ion has an energy highly dependent on the presence of other negative charges at the same distance from the central ion. The quantities of energy involved in

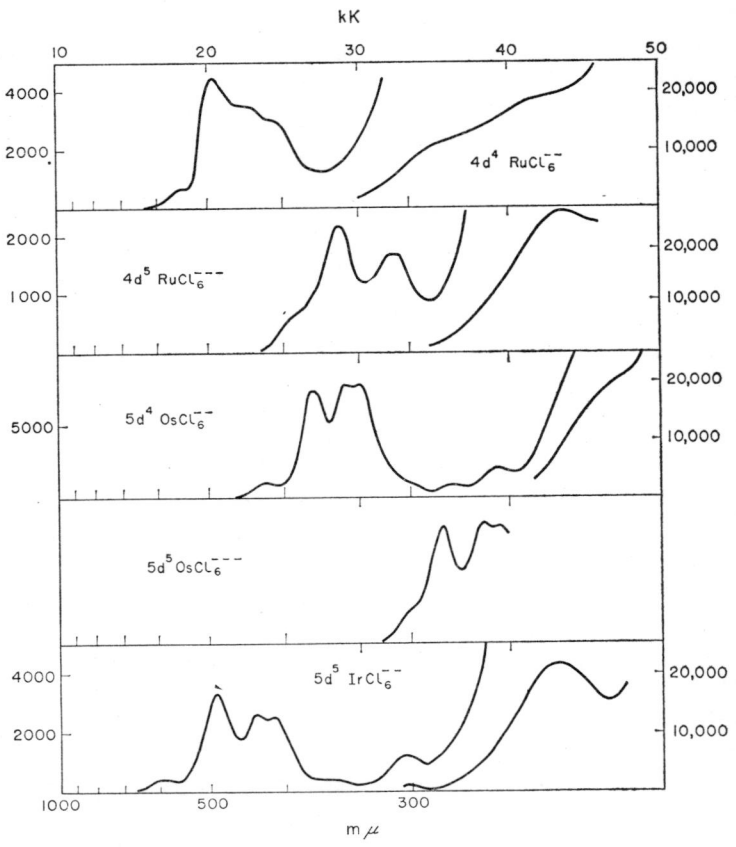

FIG. 13. Electron transfer spectra of d^4- and d^5-hexachloro complexes. The figure is reproduced with permission from *Molecular Physics* (ref. 299).

such a model are always very large, of the order of magnitude $\frac{1}{2}$ry $\sim$ 50 kK. Rather, we must recognize that the electron transfer transitions simply occur by an electron jumping from one of the highest filled M.O. to the partly filled shell or an empty orbital slightly above (cf. eqn. 54). Thus, the difference between electron transfer bands and ligand field bands can only be maintained if the energy levels involved have fairly pure M.O. configurations. Since these orbitals are not essentially different, we have a combined theory of both types of transitions, based on group theory. The most conspicuous contrast between the two types of bands is the intensity, owing to the odd parity of most of the observed, excited levels of the electron transfer bands, while the excited levels of the ligand field bands have nearly pure even parity, only very slightly perturbed by vibronic interactions.

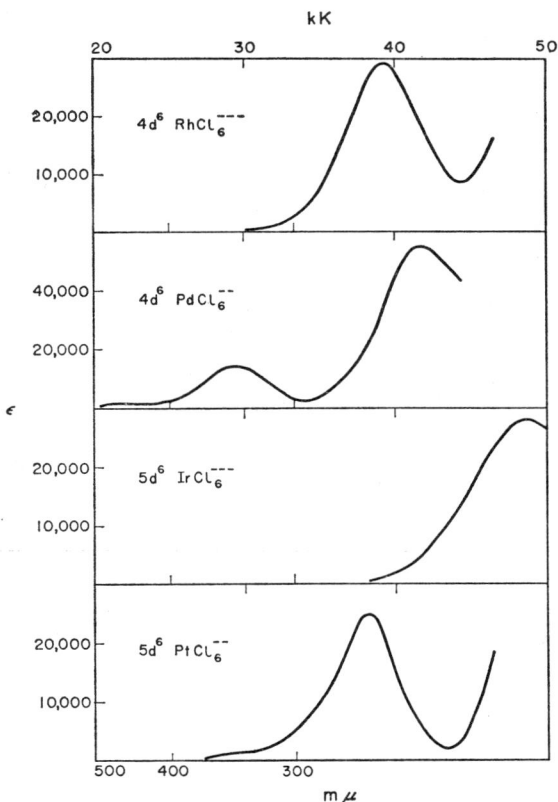

Fig. 14. Electron transfer spectra of d^6-hexachloro complexes. The figure is reproduced with permission from *Molecular Physics* (ref. 299).

L

In M.O. theory, we must revise some of our preliminary ideas of electron transfer to some extent. Since both the orbitals of slightly higher and slightly lower energy than the partly filled shells are delocalized, we cannot consider the excited electron to be entirely transferred from the ligand to the central ion. In most cases, it has a considerable possibility of being in both places in the excited and the ground state. Further, the filled M.O. is not concentrated

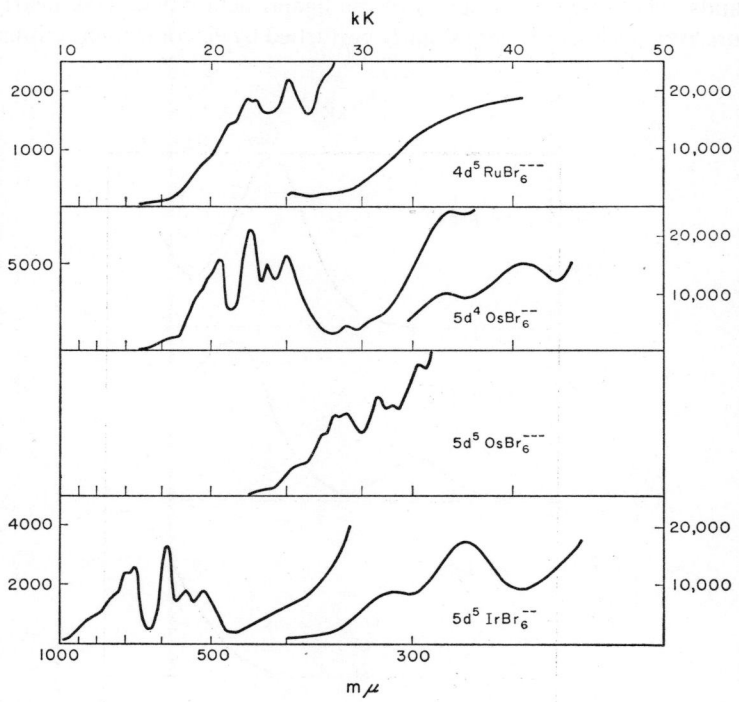

Fig. 15. Electron transfer spectra of d^4- and d^5-hexabromo complexes. The figure is reproduced with permission from *Molecular Physics* (ref. 299).

on one ligand, but is delocalized on several ligand atoms. If this delocalization is without consequences for the energy levels, we refer to the *one-ligand excitation model*. Finally, it is not always true, as in atomic spectroscopy, that increasing average distance $\bar{r}$ from the nucleus corresponds to increased orbital energy. It is obvious that the nuclei of the ligands create areas around themselves of such low potential energy that orbitals are available there with much lower energy than, for example, the partly filled shell of the central ion. Thus, an optical excitation in a complex may as well produce a decreased value of $\bar{r}$ as an increased value. We may even consider the possibility of an excited orbital, having, for example, so many nodes that its energy is much higher than that of another orbital with the same $\bar{r}$.

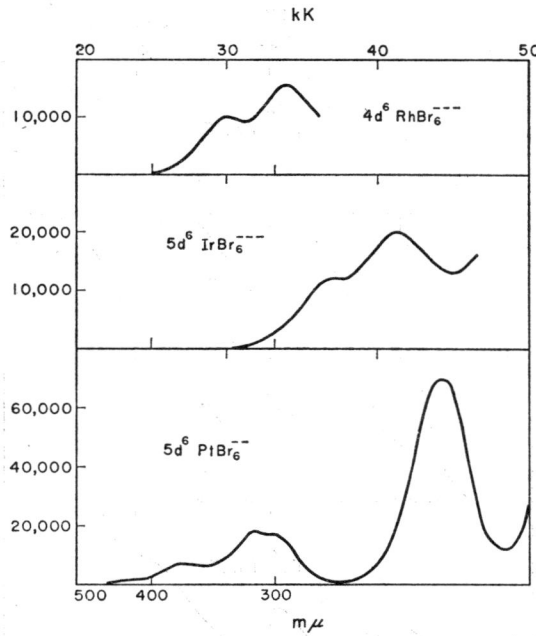

FIG. 16. Electron transfer spectra of d^6-hexabromo complexes. The figure is reproduced with permission from *Molecular Physics* (ref. 299).

In an octahedral complex of six halide ligands, the order of orbitals is that indicated in eqn. (54). Figs. 13–17 give the absorption spectra of a series of such complexes, where the corresponding gaseous ions have the configurations $4d^4$, $4d^5$, $4d^6$, $5d^3$, $5d^4$, $5d^5$, and $5d^6$. We note several characteristic features:

(1) The d^3, d^4, d^5, but not the d^6, have a group of narrow bands with low intensity at relatively low wave numbers.

(2) These narrow band groups have the same overall distribution, only shifted on the wave number scale, for the same set of ligands, independent of whether the electron configuration is $4d^4$, $4d^5$, $5d^4$ or $5d^5$.

(3) At higher wave numbers, all complexes seem to have a set of broad, strong bands. There is one such band in hexachloro, two (the first with lower intensity) in hexabromo, and three bands in hexaiodo complexes.

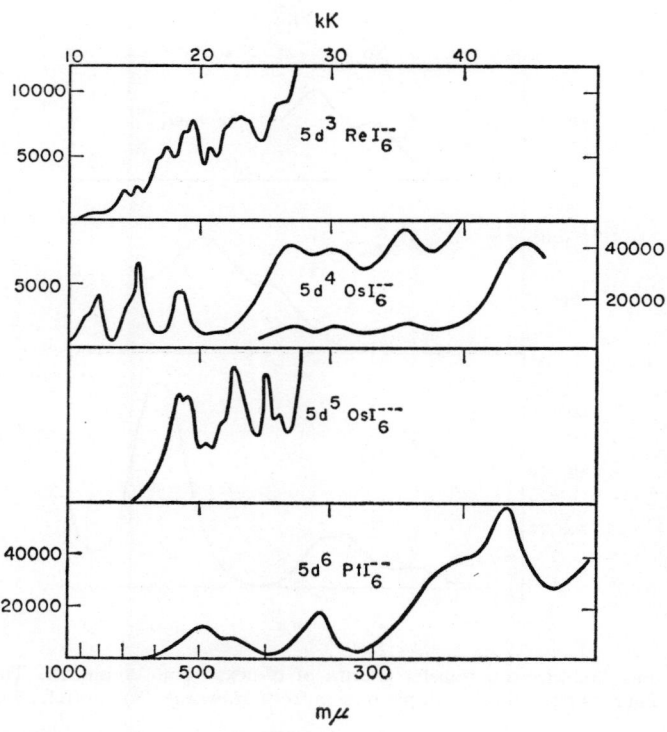

FIG. 17. Electron transfer spectra of d^3-, d^4-, d^5-, and d^6-hexaiodo complexes. The figure is reproduced with permission from *Molecular Physics* (ref. 299).

(4) In the cases where the range of measured wave numbers permit the observation, a much stronger band than those mentioned under point (3) occurs. So far, it has been found in OsI_6^{2-}, $PtBr_6^{2-}$, $PdCl_6^{2-}$ and PtI_6^{2-}.

If we consider the linear microsymmetry of the system of one ligand and the central ion, we can classify the lone-pair orbitals of the ligands into σ- and π-orbitals. Since the first-order perturbation from the central ion certainly produce a lower diagonal element of energy for the σ- than for the π-orbitals, we expect four types of electron transfer, assuming relatively pure M.O. configurations:

$$\left. \begin{array}{ll} \text{(a) } \pi \to \text{even } \gamma_5. & \text{(c) } \sigma \to \text{even } \gamma_5. \\ \text{(b) } \pi \to \text{even } \gamma_3. & \text{(d) } \sigma \to \text{even } \gamma_3. \end{array} \right\} \quad (144)$$

Of these transitions, (a) certainly corresponds to the lowest wave numbers and (d) to the highest, while the relative order of (b) and (c) is dependent on whether the energy difference between π- and σ-electrons is smaller or larger than Δ, the energy difference between even γ_5 and even γ_3.

The group of narrow bands (1) is most certainly to be identified with the transitions (a), since they are absent in low-spin d^6-complexes, having a filled γ_5-sub-shell without room for more electrons. On the other hand, the strong bands (3) and (4) must, at least in low-spin d^6-complexes, be identified with the transitions (b) and (d) to the upper sub-shell. The conclusions of the following arguments were exposed in eqn. (54), p. 61.

For understanding the structures of these groups, it is necessary to take several effects into account:

(5) The deviations from linear microsymmetry produce a mixing of σ- and π-orbitals into the octahedral orbital odd γ_4, while the other orbitals have either $\lambda = 0$ or 1 (cf. eqn. 53), if higher values of λ are not considered.

(6) The presence of the central ion produces intermixing with our octahedral orbitals from orbitals, belonging to the spherical microsymmetry of the central ion. Thus, $(n + 1)p$ interacts with odd γ_4 and nd with even γ_3 and even γ_5 according to eqn. (53).

(7) The introduction of intermediate coupling in the halogen ligands produces a redistribution of the energy levels, belonging to the total set of σ- and π-wave functions to be mixed. This development as function of increasing Landé parameter ζ_{nl} in the halogen can be described by orbitals having a well-defined value of $\omega = \frac{1}{2}$ or $\frac{3}{2}$ in the linear microsymmetry (p. 57).

Before studying the three effects more closely, we may seek rules for the average wave number of the band groups (1) and (3), giving strong evidence for their identification as the transitions (a) and (b) of equ. (144).

If we consider the variation of the narrow band group (1) with different central ions and ligands, we observe a wave number increase from 6 to 9 kK by going from $4d^n$ to the analogous $5d^n$. This is known as the hypsochromy by going to later transition groups, as exemplified by the electron transfer (eqn. 141 and 143) and ligand field (eqn. 124) bands. If we compare two isoelectronic d^5-complexes with the same ligands, the oxidation number $+3$ corresponds to 15 kK higher wave number than $+4$.

The broad bands (3) must be due to transitions to γ_3 in the low-spin d^6 complexes. They are subject to similar regularities as the narrow bands (1), e.g. the oxidation number $+3$ has 10 kK higher wave number than $+4$. Arguments will be given below to show that the broad bands of d^4- and d^5-systems are due to the same reasons as in d^6, and hence are not the transitions (c) of eqn. (144). With this assumption, we can give a linear interpolation formula for the wave number in kK, valid with an average accuracy of some 2 kK for all the hexahalide complexes measured: (cf. Table 32, p. 298).

$$\begin{Bmatrix} 4d^q\ 51-7q \\ \\ 5d^q\ 58-7q \end{Bmatrix} + \begin{Bmatrix} F & +28 \\ Cl & 0 \\ Br & -6 \\ I & -16 \end{Bmatrix} + \begin{Bmatrix} +3\ (\pi \to \gamma_5)\ +15 \\ +4\ (\pi \to \gamma_5)\ \ \ \ 0 \\ +3\ (\pi \to \gamma_3)\ +30 \\ +4\ (\pi \to \gamma_3)\ +20 \end{Bmatrix} \quad (145)$$

The energy levels cannot be considered as the sum of the M.O. energies, it is necessary to take the interelectronic repulsion energy into account. As a heuristic attempt, let us do this for the central ion only, assuming it to be dependent on the total spin S for the configuration d^q:

$$q(q-1)A/2 - DS(S+1) \quad (146)$$

The quantity A is approximately equal to A of eqn. (35), it may be written $A = A - \frac{8}{5}B + \frac{7}{9}C$ in terms of the latter equation, while D is $\frac{35}{12}B + \frac{7}{6}C$ for obtaining the best agreement with specific calculations. Eqn. (146) is a rather good approximation. If $C = 4B$ and the lowest term of each sub-shell configuration $\gamma_5{}^a\gamma_3{}^b$ is considered in Tanabe and Sugano's determinants (except for $\gamma_5{}^6$ and $\gamma_5{}^5\gamma_3$ which are to be increased $9A - 14B + 7C$ in addition to the reversal of Δ from d^4) the value of D according to eqn. (146) is found to vary between 5 and $7B$, while our average value is $91B/12$. If only terms with maximum seniority number are considered, D equals $\frac{5}{2}B + C$.

We assume next that the orbital energy of the γ_5-electrons varies linearly with the number q of electrons in the partly filled shell, and hence we write it $W - qE$, while the orbital energy of γ_3-electrons then is $W - qE + \Delta$. If the energy of the π-electrons is assumed to be constant (or its variation being linear in q and, thus, it can be incorporated in E), the variation of the wave number with q in a series of low-spin $4d^q$ or $5d^q$-complexes is

$$
\begin{array}{ll}
\pi \to \gamma_5 & V + k_D D + qA - qE \\
\pi \to \gamma_3 & V + k_D' D + qA - qE + \Delta
\end{array}
\tag{147}
$$

where V is a wave number, characteristic of the variables of eqn. (145) (ligands, oxidation number, 4d vs. 5d) and the constants are:

q =	0	1	2	3	4	5	6	7	8	9	
$k_D =$	$-\frac{3}{4}$	$-\frac{5}{4}$	$-\frac{7}{4}$	$+\frac{7}{4}$	$+\frac{5}{4}$	$+\frac{3}{4}$	—	—	—	—	(148)
$k_D' =$	$-\frac{3}{4}$	$-\frac{5}{4}$	$-\frac{7}{4}$	$-\frac{9}{4}$	$-\frac{7}{4}$	$-\frac{5}{4}$	$-\frac{3}{4}$	$-\frac{5}{4}$	$+\frac{5}{4}$	$+\frac{3}{4}$	

The development in the series $ReCl_6^{2-}$, $OsCl_6^{2-}$, $IrCl_6^{2-}$, $PtCl_6^{2-}$ can be accounted for by the set of parameters $D = 3$ kK and $(E - A) = 6$ kK, while Δ is known from ligand field bands to be some 29 kK.

The broad bands of d^5-systems at slightly higher wave numbers than the corresponding bands of the d^6-systems (cf. the pairs $RuCl_6^{3-}$ (43·6 kK), $RhCl_6^{3-}$ (39·2) or $IrCl_6^{2-}$ (43·1), $PtCl_6^{2-}$ (38·2)) are presumably due to transitions to the γ_3-sub-shell. According to eqn. (148), they should not have exactly the wave number Δ above the transitions to γ_5, but only $\Delta - 2D$ in d^5, $\Delta - 3D$ in d^4, and $\Delta - 4D$ in d^3. Actually, the observed distances between the narrow band groups and the broad bands are $\Delta - 6$ kK in d^5, $\Delta - 9$ kK in d^4, and $\Delta - 12$ kK in d^3. This is a surprisingly good agreement with our simplification eqn. (146), since we have not considered the interelectronic repulsion between the partly filled shell and the filled M.O., from which one electron is excited.

Eqn. (146) may also be applied to the ionization $d^q \to d^{q-1}$ of gaseous ions with high-spin behaviour, i.e. maximum value of S, if the somewhat dubious assumption is made that A does not vary with q, and that the orbital energy is a linear function $- I - Z_0 Q - qE$ of the ionic charge $(Z_0 - 1)$ and of q, giving the ionization energy:

$$q \leq 5 \quad I + (Z_0 - q + 1)Q + (2q - 1)E - (q - 1)A +$$
$$+ (2q + 1)D/4$$
$$q > 5 \quad I + (Z_0 - q + 1)Q + (2q - 1)E - (q - 1)A + \qquad (149)$$
$$+ (2q - 23)D/4$$

representing two straight lines with the same slope in a graph vs. q, but with the hump 6D at the half-filled shell d^5. It has previously been shown (1956) that the ionization potential $M^{2+} \to M^{3+}$ for the gaseous $3d^n$-ions has the slope 2·2 eV and a d^5-hump 4·8 eV, corresponding to D = 6·4 kK. The standard oxidation potentials for the M(III): M(II) hexa-aqua ions of the first transition group, when corrected for the ligand field stabilization $0·2\rho\Delta$ (equ. 128), have the slope 1·5 V and the d^5-hump 4 V (D = 5·3 kK). (Contrary to equ. (149), the Franck–Condon principle is assumed to prevent the orbital energy to drop from $W - qE$ to $W - (q + 1)E$ in equ. (147).)

The *half widths* $\delta(-)$ *of the electron transfer* spectra support the identifications of the narrow bands as $\pi \to \gamma_5$ and the broad bands as $\pi \to \gamma_3$ and $\sigma \to \gamma_3$. It was discussed above (eqn. 112) how δ is a measure of the slope of the potential curve of the excited electronic state at the value of r_L for the ground state, i.e. it is a measure of the antibonding character of the excited state. It is surprising that the half widths $\delta(-)$ of the narrow bands of d^4- and d^5-hexachloro complexes between 1 and 1·5 kK have the same magnitude as the ligand field bands, and are much smaller, 0·5–0·8 kK in hexabromo and 0·3–0·5 kK in hexaiodo complexes, approaching the values for bands caused by transitions within the same sub-shell configuration. We could hardly assume that $\sigma \to \gamma_5$ transitions would not produce increased internuclear equilibrium distances, while $\pi \to \gamma_5$ does not necessarily change the electron density in the complex very much*. The development as function of the halogen mass m might be interpreted as the decreasing amplitude of the vibrational ground state (equ. 107). However, this argument is only valid in the not too probable case that $\frac{1}{2}hc\sigma_c$ is large, compared to kT at room temperature. Rather, the smaller δ of iodo than of chloro complexes is caused by an intrinsic change of the π-electron and γ_5-delocalization. The much larger half widths (2–3 kK) of the $\pi \to \gamma_3$ bands can be explained easily, since we may assume the half width to be the sum

---

* Actually, the high-pressure experiments (Stephens and Drickamer, 1959, and private communication) on Re(IV), Os(IV), and Ir(IV) demonstate that the equilibrium r_L is even *smaller* of the excited state than of the ground state. Similar red-shifts are observed in organic solvents, compared to the aqueous solution.

of the half widths of $\gamma_5 \to \gamma_3$ ligand field and $\pi \to \gamma_5$ electron transfer transitions.

If we interpret the *intensities of electron transfer bands* as the squared dipole moment between two different orbitals ψ_1 and ψ_2, assuming pure M.O. configurations (eqn. 118), we see that all the transitions in eqn. (144) are group-theoretically allowed, if we make a suitable choice (i.e. the odd γ_4 and γ_5) of the orbitals in Table 4. However, if we make qualitative drawings of the overlap regions of such orbitals with their accompanying sign of $\psi_1\psi_2$ and the resulting dipole moments, we see that the dipole moment is very small for $\pi \to$ even γ_3 and $\sigma \to$ even γ_5, while it has a reasonable size (perhaps producing the oscillator strength $P \sim 0.1$) for $\pi \to$ even γ_5 and is very large (probably $P \sim 1$) for $\sigma \to$ even γ_3.

Thus it is not surprising that $\sigma \to \gamma_5$ transitions in d^3, d^4, d^5 are not observed at somewhat higher wave number than the narrow $\pi \to \gamma_5$ bands. It is more peculiar that the broad $\pi \to \gamma_3$ bands have oscillator strengths P some five to twenty times larger than the narrow $\pi \to \gamma_5$ bands, though they should have much smaller intensity as one-electron transitions. However, we have here neglected the inter-mixing of λ due to breakdown of the linear microsymmetry at the ligands. When d^6-hexahalide complexes such as $PdCl_6^{2-}$ and $PtBr_6^{2-}$ have a super-intense band ($P = 1.4$) at 12 kK higher wave number than the ordinary broad band ($P = 0.4$), it can hardly be assumed that both bands are caused by $\pi \to \gamma_3$ transitions (especially since the spreading of the group of $\pi \to \gamma_5$ bands in d^4 and d^5 is only some 7 kK). Neither can it be assumed that both bands are caused by $\sigma \to \gamma_3$ transitions (since only one odd set of σ-orbitals exist, and since the configuration $\gamma_4^{-1}\gamma_3$ has only one level, $^1\Gamma_4$, to which transitions from $^1\Gamma_1$ in d^6 are symmetry-allowed). Thus, we must conclude that the two strong bands are caused by the transition from two different odd γ_4-orbitals to the even γ_3. The two odd γ_4-orbitals are mixtures of some 25 per cent σ- and 75 per cent π-characters and vice versa. This intermixing can be described by a secular determinant, having as diagonal elements the energies of the wave functions, marked with the same letter, e.g. a, in Table 4, being (1) the x-orbital of the central ion, (2) the σ-orbital odd γ_4 on the x-axis, and (3) the π-orbital odd γ_4 perpendicular on the yz-plane (Fig. 2). All these orbitals have mutual overlap integrals, resulting in a mixture according to eqn. (57). As shall be discussed below, the intermediate coupling in the halogen (e.g. in $PtBr_6^{2-}$)

further redistributes the property odd γ_4 on the other state, which is purely odd γ_5 (π) for $\zeta_{np} = 0$, producing one further, but weak band at lower wave number than the first of the two strong bands.

If the interaction between the π-orbitals and the orbitals of the central ion was the only one considered, the expected order* of decreasing energy would be:

$$\text{even } \gamma_4 > \text{odd } \gamma_5 > \text{even } \gamma_5 \sim \text{odd } \gamma_4 \text{ perturbed by} \atop \geq \text{g} \qquad\quad \text{f} \qquad\quad \text{d} \qquad\quad \text{p and f -orbitals} \qquad (150)$$

since p- and d-orbitals presumably interact more strongly with the low-energy π-orbitals than do the highly excited orbitals with higher l. Correspondingly, the electron transfer transitions were expected to increase in wave numbers according to eqn. (150). The low-spin d^5-systems are especially interesting, because the excited state contains the closed shell $\gamma_5{}^6$ of the central ion, forming configurations each consisting of only one doublet term $^2\Gamma_k$ as an image of the M.O. energies of γ_k. These doublet terms are not split by intermediate coupling in the d-shell for any value of ζ_{nd}. No transition would be spin-forbidden.

Actually, the d^5-hexachloro complexes exhibit a weak band, followed by two stronger, narrow bands (Fig. 13). The latter of these two bands is often double. If we do not accept the rather improbable opinion that one of these bands is caused by $\sigma \rightarrow \gamma_5$, the first, weak band must be Laporte-forbidden, since we can only procure two Laporte-allowed bands without use of large ζ_{np} (of the halogen or of the central ion), the transitions from odd γ_5 and odd γ_4 in eqn. (150). The doublet structure of the second, Laporte-allowed $\pi \rightarrow \gamma_5$ band may be caused by the near degeneracy of even γ_5 and odd γ_4 in eqn. (150), or by the intermixing of some 20 per cent of the central ion p-orbital in odd γ_4. Unfortunately, ζ_{6p} is not known in Ir(IV), but has probably the order of magnitude 5 kK. If the wave function contains some 20 per cent of this contribution, $^2\Gamma_4$ would be split some 1·5 kK with Γ_8 at lowest energy (the same direction as in Fig. 18), which might explain the doublet structure at 23·1 and 24·4 kK of $IrCl_6{}^{2-}$. However, we do not know definitely whether the order of odd γ_5 and odd γ_4 in eqn. (150) cannot be reversed, due to the intermixing of λ in the latter orbital, as discussed above.

---

* McClure (1959) pointed out that the halogen–halogen interactions also would produce eqn. (150), the number of node planes being 4, 3, 2 and 1 respectively.

The development observed in the bromo and iodo complexes can be connected with the increasing ζ_{np} in the gaseous atoms and ions, in kK:

$$F \quad 0{\cdot}269 \quad Cl \quad 0{\cdot}587 \quad Br \quad 2{\cdot}457 \quad I \quad 5{\cdot}069$$
$$Ne^+ \; 0{\cdot}521 \quad A^+ \; 0{\cdot}955 \quad Kr^+ \; 3{\cdot}581 \quad Xe^+ \; 7{\cdot}025 \tag{151}$$

In the same way as the positive Z_{eff} of central ions is decreased by covalent bonding, we might expect Z_{eff} of the halide ligands to be less negative than -1, producing increasing values of ζ_{np}. The hexabromo d^4- and d^5-complexes have seven narrow $\pi \rightarrow \gamma_5$ bands, of which three are particularly strong (Fig. 15). Qualitatively similar features prevail in the hexaiodo complexes. The two intervals between the three band groups is remarkably identical in all cases, from $2{\cdot}5$ to 3 kK.

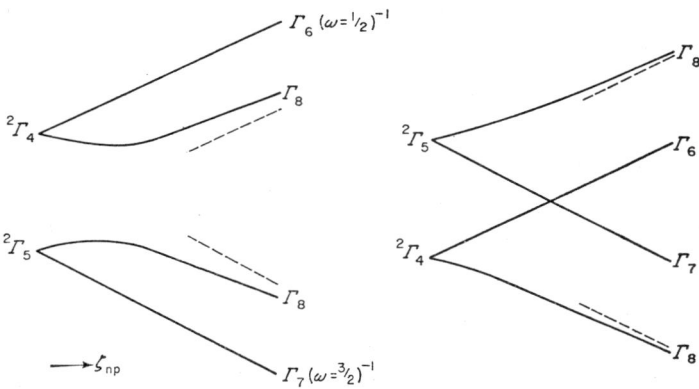

FIG. 18. The effects of intermediate coupling in the halogen on the energy levels, formed by electron transfer from the π orbitals of the ligands to one effective state of the central atom (e.g. $\gamma_5{}^6$ formed from low-spin $\gamma_5{}^5$ complexes). As seen in eqn. (150) the left-hand side is expected to correspond to the excited states of odd parity, $^2\Gamma_5$ being the lowest energy level of π^{-1}, while the right-hand side corresponds to the even parity with $^2\Gamma_4$ lowest. The figure is reproduced with permission from *Molecular Physics* (ref. 299).

Fig. 18 shows the splitting of $^2\Gamma_4$ and $^2\Gamma_5$ from $\pi^{-1}\gamma_5{}^6$ as function of increasing ζ_{np}. The left-hand side illustrates the case expected from eqn. (150) that the odd $^2\Gamma_5$ has the lower energy; but we also give the opposite case for comparison and for treatment of the even levels according to eqn. (150). The splitting is introduced via the splitting of each π-orbital into states with $\omega = \frac{3}{2}$ (with the L,S coupling energy $+\frac{1}{2}\zeta_{np}$) and $\omega = \frac{1}{2}$ $(-\frac{1}{2}\zeta_{np})$. By consideration of the Γ_J-values of the σ-electrons (cf. Tables 4 and 17 and eqn. (154)), which also have $\omega = \frac{1}{2}$, we learn that both the odd and even Γ_6-levels belong to the pure states $\omega = \frac{1}{2}$ (having increasing energy among the transitions in Fig. 18) while correspondingly, the Γ_7-levels belong to states with $\omega = \frac{3}{2}$. The two sets of Γ_8-levels are mixed and form a hyperbola (Fig. 1) in the diagram Fig. 18 as eigenvalues of

$$\begin{vmatrix} {}^2\Gamma_5 + \frac{1}{4}\zeta_{np} - E & \sqrt{3}\,\zeta_{np}/4 \\ \sqrt{3}\,\zeta_{np}/4 & {}^2\Gamma_4 - \frac{1}{4}\zeta_{np} - E \end{vmatrix} = 0 \qquad (152)$$

as can be found by application of eqn. (88) and the baricentre rule for the splitting of ${}^2\Gamma_5$ into the Γ_J-components Γ_7 and Γ_8 and of ${}^2\Gamma_4$ into Γ_6 and Γ_8.

We may include some remarks on the effect of L,S coupling in $\gamma_5{}^n$, which is exactly analogous to the effect in p^{6-n} in spherical symmetry, if the γ_5-orbitals correspond to $l = 2$ (cf. eqn. 77) as pointed out by J. S. Griffith. Thus, the lowest levels are, in units of ζ_{nd}: (cf. also Fig. 24, p. 179).

$$\gamma_5{}^5 \quad {}^2\Gamma_5\text{:} \quad \Gamma_7 \text{ at } -1, \Gamma_8 \text{ at } +\tfrac{1}{2}$$
$$\gamma_5{}^4 \quad {}^3\Gamma_4\text{:} \quad \Gamma_1 \text{ below } -\tfrac{1}{2}, \Gamma_4 \text{ at } +\tfrac{1}{2}, \Gamma_3 \text{ and } \Gamma_5 \text{ below } +1 \quad (153)$$
$$\gamma_5{}^3 \quad {}^4\Gamma_2\text{:} \quad \Gamma_8 \text{ slightly below } 0$$

Hence, for large values of ζ_{nd}, $\gamma_5{}^5$ and $\gamma_5{}^4$ are Jahn–Teller stable, because Γ_7 is a Kramers doublet and Γ_1 has $e = 1$, while $\gamma_5{}^3$ is not Jahn–Teller stable, because Γ_8 has $e = 4$ (cf. the multiplication table for the double-group representations Γ_6, Γ_7 and Γ_8 in O_a, Table 17). The values of Γ_J for a half-integral value of J or S were indicated by Bethe in analogy to eqn. (44):

$$\begin{array}{ll} \tfrac{1}{2} : \Gamma_6 & \tfrac{7}{2} : \Gamma_6 + \Gamma_7 + \Gamma_8 \\ \tfrac{3}{2} : \Gamma_8 & \tfrac{9}{2} : \Gamma_6 + 2\Gamma_8 \\ \tfrac{5}{2} : \Gamma_7 + \Gamma_8 & \tfrac{11}{2} : \Gamma_6 + \Gamma_7 + 2\Gamma_8 \end{array} \qquad (154)$$

While none of the Laporte-allowed transitions $\gamma_4 \to \gamma_5$ and $\gamma_5 \to \gamma_5$ is symmetry-forbidden, one of the four $\pi \to \gamma_5$ transitions expected in intermediate coupling of low-spin d^5-complexes with large ζ_{nd}, is forbidden. It can be seen from eqn. (114) and Table 17 that the transition even $\Gamma_7 \leftrightarrow$ odd Γ_6 has no electrical dipole moment, and we can only expect three Laporte-allowed transitions, from Γ_7 of γ_5^5 (eqn. 153) to Γ_7 and the two Γ_8 of $\pi^{-1}\gamma_5^6$ in Fig. 18.

Actually, the mutual distances of the three strong band groups of bromo and iodo complexes do not agree perfectly with any of the two cases in Fig. 18. One of the reasons can be the interaction between the $\omega = \frac{1}{2}$ components of the σ- and π-orbitals having the non-diagonal element $\sqrt{2}\,\zeta_{np}/2$. If the energy difference between σ and π is small, compared to ζ_{np}, we get an asymptotical re-validization of j ($= \frac{3}{2}$ or $\frac{1}{2}$) for large ζ_{np}, as shown in Fig. 19. The depression of the middle Γ_8-level in the case of eqn. (150) produces a fair agreement with the observed three band groups.

TABLE 17. THE MULTIPLICATION FOR THE DOUBLE-GROUP ("HALF-VALUED") Γ_J IN CUBIC- OCTAHEDRAL SYMMETRY O_h

	Γ_6	Γ_7	Γ_8
Γ_1	Γ_6	Γ_7	Γ_8
Γ_2	Γ_7	Γ_6	Γ_8
Γ_3	Γ_8	Γ_8	$\Gamma_6 + \Gamma_7 + \Gamma_8$
Γ_4	$\Gamma_6 + \Gamma_8$	$\Gamma_7 + \Gamma_8$	$\Gamma_6 + \Gamma_7 + 2\Gamma_8$
Γ_5	$\Gamma_7 + \Gamma_8$	$\Gamma_6 + \Gamma_8$	$\Gamma_6 + \Gamma_7 + 2\Gamma_8$
Γ_6	$\Gamma_1 + \Gamma_4$	$\Gamma_2 + \Gamma_5$	$\Gamma_3 + \Gamma_4 + \Gamma_5$
Γ_7	$\Gamma_2 + \Gamma_5$	$\Gamma_1 + \Gamma_4$	$\Gamma_3 + \Gamma_4 + \Gamma_5$
Γ_8	$\Gamma_3 + \Gamma_4 + \Gamma_5$	$\Gamma_3 + \Gamma_4 + \Gamma_5$	$\Gamma_1 + \Gamma_2 + \Gamma_3 + 2\Gamma_4 + 2\Gamma_5$

It is rather unexpected that d^4-systems such as Os(IV) present nearly the same distribution of electron transfer bands as the d^5-systems such as Ir(IV). The former have two accessible excited states of central ion plus one electron, Γ_7 and Γ_8 of γ_5^5 (see eqn. 153), separated by $\frac{3}{2}\zeta_{5d}$ which is at least 4 kK in $OsCl_6^{2-}$, as can be implied from ligand field transitions within γ_5^4. However, we do not observe two sets of electron transfer bands, because Γ_1 of γ_5^4 has a nearly pure (some 88 per cent) γ_j-configuration γ_8^4, allowing the uptake of a γ_7-electron as a one-electron jump (p. 105), but not the formation of Γ_8, being $\gamma_8^3\gamma_7^2$. On the other hand, d^3-systems with four Γ_J-components of $^3\Gamma_4$ of γ_5^4 (eqn. 153) as possible excited states are expected to show a much more complicated structure, as actually observed (Fig. 17).

Even when we have rationalized the fact that d^4-systems have only one effective excited state of the central ion, we still have reasons for being surprised by the close analogy to the d^5-systems. Group theory tells us that each of the $^2\Gamma_4$ or $^2\Gamma_5$-levels of the π-holes will be a parent term (multiplied by Γ_7 of the central ion) of five Γ_j-levels of the complex. Thus, according to Table 17

$$^2\Gamma_4 \times \Gamma_7 = \Gamma_6 \times \Gamma_4 \times \Gamma_7 = (\Gamma_6 + \Gamma_8)\Gamma_7 = \Gamma_2 + \Gamma_3 + \\ + \Gamma_4 + 2\Gamma_5 \quad (155)$$

Even though only electric dipole transitions to Γ_4 are allowed from the ground state Γ_1, we would expect considerable vibronically created intensity of the symmetry-forbidden transitions.

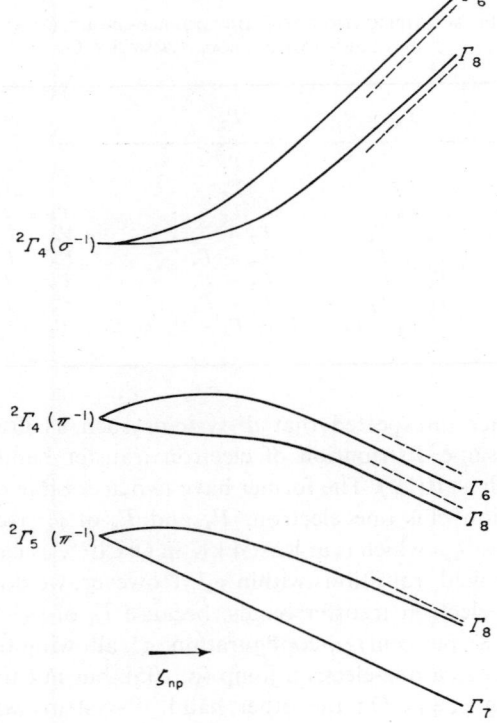

FIG. 19. The intermediate coupling effects as the left-hand side of Fig. 18, taking into account the presence of σ^{-1} somewhat above π^{-1}, having a non-diagonal element of energy between the two $\omega = \frac{1}{2}$. The figure is reproduced with permission from *Molecular Physics* (ref. 299).

We must conclude that the type of interelectronic repulsion parameters, separating the energy levels of a given configuration (in contrast to the common contribution) is very small in the $\pi^{-1}\gamma_5{}^n$ configurations. This is not a general rule for complexes; we shall see in the next chapter (eqn. 163) that the exchange integral $K(6s, 6p)$ in s^2-complexes is some 3 to 4 kK, and the K-integrals in the γ_5- and γ_3-sub-shells (eqn. 75) have the order of magnitude 4 kK. With our present, very restricted knowledge of the wave functions of the electron transfer states, it is not possible to calculate such K-integrals *a priori*, and due to the intermixing of M.O. configurations, it is far from certain that they would correspond to the observed energy differences. We may expect configurations other than $\pi^{-1}\gamma_5{}^n$ to show larger spreading of the energy levels. Thus, the terms of $(\sigma,\ odd\ \gamma_4)^{-1}\gamma_3$ are expected to have the order of energy $^3\Gamma_4 <$ $^1\Gamma_4 < {}^3\Gamma_5 \sim {}^1\Gamma_5$, since the J-integrals involved in Γ_4 will be between the closely adjacent orbitals $(\sigma_5 - \sigma_6)$ and $z^2 - \frac{1}{2}(x^2 + y^2)$ of Table 4, while the J-integrals in Γ_5 are between, e.g. $(\sigma_5 - \sigma_6)$ and $(x^2 - y^2)$. These terms are not split to first order by L,S-coupling, neither by contributions from ζ_{np} of the halogen nor from ζ_{nd} of the central ion.

An interesting case is the monohalide pentammine d^6-systems such as $Co(NH_3)_5Cl^{2+}$, $Co(NH_3)_5Br^{2+}$, $Co(NH_3)_5I^{2+}$, $Rh(NH_3)_5Br^{2+}$, and $Rh(NH_3)_5I^{2+}$, having two bands in the ultra-violet with the nearly constant energy difference 8 kK. The first band has a small intensity, proportional to $\zeta_{np}{}^2$ in the halogen, while the second has $\epsilon_0 \sim 20{,}000$ and $P \sim 0.5$. Orgel (1956) and Yamatera (1957) assume the weak band to be caused by the transfer of a π- and the strong band a σ-electron to the same acceptor orbital $(z^2 - \frac{1}{2}(x^2 + y^2)$ pointing towards the halogen and highly attracted by a hole in the halogen electron cloud). In the notation of linear symmetry, the configuration $\pi^{-1}\gamma_{t_1}$ corresponds to the terms $^3\Pi$ and $^1\Pi$, while the σ-transfer corresponds to $^3\Sigma$ and $^1\Sigma$. In this case, there is no strong intermixing of λ as in a hexahalide complex, and the intensity of the π-transfer may be very small and the actual band intensity mainly caused by intermediate coupling.

In the d^6-complexes, we observe a strong decrease of the wave number of the electron transfer bands by introducing a larger number of halide ligands. However, this seems to be connected with the fact that Δ, the energy difference between γ_5 and γ_3, is some 14 kK smaller in $RhCl_6{}^{3-}$ than in $Rh(NH_3)_6{}^{3+}$. In d^5-systems, the transitions

$\pi \rightarrow \gamma_5$ have wave numbers remarkably independent of the number of halide ligands. Thus, Hartmann and Buschbeck (1957) found bands of $Ru(NH_3)_5Cl^{2+}$ at 30·5 kK, cis-$Ru(NH_3)_4Cl_2^+$ at 28·4 and 32·3 kK, and Cady and Connick (1958) bands of $Ru(H_2O)_5Cl^{2+}$ at 31·8 kK and of $Ru(H_2O)_4Cl_2^+$ at 33 kK, while the narrow band group of $RuCl_6^{3-}$ is situated at 28·6 and 32·4 kK. If the band at 44·5 kK of $Ru(H_2O)_6^{3+}$, reported by the latter authors, is identified as a transfer of π-electrons from the water molecules, this would imply a value of $+15$ for H_2O of the second parenthesis of eqn. (145).

Most d^{10}-systems with low oxidation numbers have linear co-ordination such as Cu(I), Ag(I), Au(I), and often Hg(II) or tetrahedral co-ordination such as Hg(II). However, some systems with higher oxidation number, Sn(IV), Sb(V), and Pb(IV), form octahedral hexachloride complexes. The wave numbers of their electron transfer bands are very high, relative to the chemical oxidizing character of the central ions. The spectra seem closely analogous to those of low-spin d^6-hexahalides; thus, $PbCl_6^{2-}$ has two strong, very broad bands at 32·6 kK ($\epsilon_0 = 9700$) and 48·1 kK ($\epsilon_0 = 24,000$). They are probably caused by transitions from the two odd γ_4-orbitals, the first mainly of π- and the second mainly of σ-character, as discussed above, to the empty, even γ_1-orbital which is filled in, for example, the Pb(II) complexes. It might be argued that an electron might also jump from the filled even γ_3-orbital to the first empty odd γ_4-orbital. However, this is not so probable when the high wave numbers of the transition even $\gamma_1 \rightarrow$ odd γ_4 in Tl(I) and Pb(II) are considered. It is generally expected that the genuine electron transfer bands shift towards lower wave number with increasing oxidation number. Hence, it is surprising that Cu(I) and Ag(I) chloro complexes have Laporte-allowed bands at distinctly lower σ than Zn(II) and Cd(II). This effect may be connected with the difference in co-ordination number. However, in tetrahedral complexes, $\gamma_5(d) \rightarrow \gamma_1$ (s) may be rather intense.

The d^0-systems (except the tetroxo complexes) have generally very broad absorption bands and often only an increasing slope in the ultra-violet. It may either be caused by a highly antibonding character of the excited states, or by the superposition of many adjacent bands. A special case occurs in the dioxo ions of actinides with the oxidation number $+6$ (while the bands are not certainly identified for the similar dioxo ions $+5$) such as UO_2^{2+}, NpO_2^{2+}, PuO_2^{2+} and AmO_2^{2+}. The increasing absorption at higher wave numbers begins with one or more, very weak bands with pronounced vibrational structure. In UO_2^{2+}, these bands are centred around 24 and 30 kK and have P as low as 10^{-4}. Since the lowest empty orbital is probably derived mainly from the 5f-shell, the low intensity may be explained

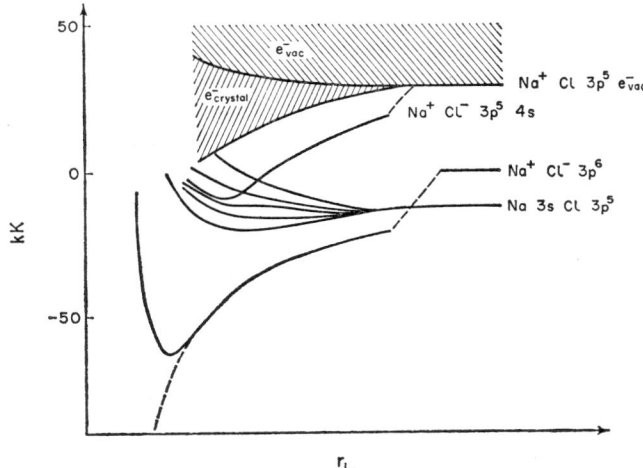

FIG. 20. The energy levels (potential curves) of sodium chloride in a regular, cubic lattice as function of the distance r_L between adjacent Na and Cl nuclei. With r_L infinitely large, the neutral atoms Na and Cl have 11·4 kK lower energy than the ions Na^+ and Cl^-, but the Madelung energy (eqn. 185) stabilizes the latter state considerably at lower values of r_L. It is a very important difference between this ionic ground state and the "electron transfer" states formed by neutral Na and Cl atoms in the lattice that the degeneracy number e is 1 for Na^+Cl^-, but it is a multiple of the number N of chlorine nuclei in the crystal for the Na^0Cl^0-excited states. A further complication arises from the possible intermixing between the Na^0Cl^0 states, and states where Cl^- is excited to $3p^54s$. The crystal with finite r_L has two different ionization energies, one marked e_{vac}^- for the expulsion of electrons from the crystal to the surrounding space, and one marked $e_{crystal}^-$ for the formation of quasi-continuous M.O. for the electron, extending over the whole crystal. It might be expected that the "external" and "internal" ionization would be accompanied by photo-emission (Einstein effect) and photoconductivity, respectively. It is not yet known with certainty to what extent these continuum states come down below the discrete electron transfer states in actual crystals of different salts.

by a very small dipole moment (due to small spatial overlap with the filled M.O.) or by Laporte-forbiddenness from an odd M.O.

The aqueous solution of iodide ions has two strong bands (44·2 kK, $\epsilon_0 = 13,100$ and 52 kK), while bromide ions also show a double

M

band at 54 and 58 kK, and chloride ions a single band at 62 kK. Franck and Scheibe (1928) explained these bands as caused by electron transfer to one effective state of the solvent, leaving the neutral halogen atom in one of the two states $^2P_{\frac{3}{2}}$ or $^2P_{\frac{1}{2}}$ (at a wave number $\frac{3}{2}\zeta_{np}$ higher, see eqn. (151)) known from atomic spectroscopy. This approximation of spherical symmetry is fairly valid in aqueous solution, where the halide ion probably does not form strongly directed chemical bonds. Katzin (1955) considers all electron transfer bands as being these transitions of the free halide ions, shifted towards smaller wave numbers by the increasing electron affinity of the central ion, as also qualitatively expressed by Fromherz. However, as seen above, this approximation to the complexes (for explanation of the double bands of iodo and bromo complexes, etc.) is valid only so far as ζ_{np} is larger than the splitting into π- and σ-orbitals introduced by the central ion.

The *crystals of alkali metal halides* have absorption bands at nearly the same places as the aqueous solution. However, the bands often show some fine structure, thus CsI has at least four bands at the place of the excited state $^2P_{\frac{3}{2}}$, the most prominent at 46·4 and 48·3 kK. Overhauser (1956) applied group theory to show that among the thirty levels of the first excited configuration of a halide ion in the NaCl-structure (and the forty of CsCl), four (five) levels are accessible for symmetry-allowed electric dipole radiation (see eqn. 115) and of these transitions, only two are spin-allowed for small ζ_{np}. Since the alkali halides have some five strong bands more in the range up to 90 kK, several acceptor orbitals must be available. These acceptor orbitals may be rather widely distributed over a number of halide and metal ions. Petersen (1957) made an analogy between the alkali halide bands and the excited configuration p^5s of the inert gases. Since I$^-$ is isoelectronic with xenon, the excitation of a 5p- to a 6s-electron represents to some extent the Laporte-allowed transitions in the crystal though not having perfect spherical microsymmetry. It is highly interesting that the wave number shifts 8 kK between I$^-$ and Br$^-$ and 6 kK between Br$^-$ and Cl$^-$ are nearly the same as derived in eqn. (145) for all electron transfer spectra of hexahalide complexes of transition group metals, showing a quite invariant orbital energy difference between different halogens in any type of compound.

Solid-state physicists often apply descriptions of crystals, which classify energy levels in a way that would be impossible for separate ions in dilute solution. Bloch proposed to write wave functions, extended over the whole crystal, but repeating themselves periodically in each unit cell (cf. p. 233). In this way, the "energy bands" of the available M.O. are considered as continuous groups of highly delocalized M.O. with nearly the same energy. It is now recognized (cf. Seitz, *Modern Theory of Solids*) that the transitions from the highest filled band in alkali halides to these continuum-like states probably does not correspond to the usual absorption bands somewhat above 45 kK, since the photoconductivity does not occur to the extent expected by a Bloch-metallic state (even though the arguments about photoconductivity are very uncertain, due to the strong influence of impurities on this type of experiment). In an electrostatic model the potential at the position of a negative ion in a lattice is $\alpha e^2/r_L$, where α is the Madelung constant (1·748 for the NaCl-lattice) and r_L the cation–anion distance. The potential at the position of a positive ion has the opposite value. It is thus possible to calculate the energy needed for transfer of an electron from Cl^- in the lattice to a very distant Na^+ in the lattice, forming the neutral atoms with $3p^5$ and 3s-configurations, respectively. This energy, which is negative, -11 kK, for gaseous Na^+ and Cl^- ($r_L = \infty$) is 133 kK for r_L equal to the value in crystalline salt. However, if the transfer of the electron takes place to a nearby sodium ion with the distance r_{Na}, less energy (e^2/r_{Na}) is required. Hence, a set of transitions should occur, the lowest of which at 91 kK. Von Hippel (1936) attempted to refine this model somewhat and arrived at a value of 67 kK, close to the observed wave number 63 kK of the first absorption band.

This type of excitation is then believed to form "excitons" in the crystal. It is obvious that if the excited state of one halide ion is q times degenerate, the N halides in a crystal produce a level, containing Nq states. These states are approximately degenerate, since the interactions between different halide ions are fairly weak. Quantum mechanics opens the interesting possibility that the excitation may wander between many unit cells in the crystal, all being identical in an ideal, pure crystal. However, it is doubtful whether such a phenomenon actually occurs, when alkali halides and similar crystals absorb ultra-violet radiation. A more likely possibility is that the energy simply is degraded into heat and infra-red radiation after excitation of the first halide ion and its immediate environment of cations. This is a question of the lifetime of the excited state. It is known from crystals of anthracene that very small traces of other molecules (10^{-6}) may take over all the energy irradiated to the crystal, and fluoresce with their own excited states rather than the anthracene itself (hence, the emission spectrum of anthracene fluorescence was very difficult to obtain, because a very pure material was needed). However, it is not certain that similar energy transfer between molecules occurs in ordinary salts. In the literature, it has become common, without much further proof, to talk about exciton energy levels as the excited states of materials like CdS (narrow bands at 20·6 kK), Cu_2O (three series of bands, at 16·4, 17·4 and 18·3 kK), CuBr (23·9 and 25·2 kK), CuI (24·6 and 29·8), red HgI_2 (18·9) and PbI_2 (20·2). The spectra often show an interesting fine structure (the lines being some 0·01 kK wide) when the compounds are studied in thin layers at very low temperature. This structure has been ascribed to the hydrogen-like levels produced by an exciton, separated at finite distance, as discussed above. Obviously, the two first wave numbers are used to estimate the two parameters (the series limit and the effective value of the "Rydberg constant" $\sim$1 kK) in the exciton theory; but in some cases, such as Cu_2O and PbI_2, five members of the series are observed. However, the agreement may perhaps frequently be fortuitous; the description of the solid as a homogeneous material with high dielectric constant seems rather crude for the occurrence of excited states of this nature.

The electrostatic model does not give satisfactory results, when applied to other halides, oxides and sulphides. However, an important qualitative difference exists between the halides on one hand, and O^{2-} and S^{2-} on the other: the latter ions are not stable in the gaseous state, but require approximately 56 and 40 kK for their formation from O,S, and electrons. Hence, oxide ions need a close packing of cations to be stabilized; in a certain sense, oxides are very reducing for large r_L, while they are very stable for small r_L. It is well known that the alkali metal oxides do not have very large heats of formation, and are liable to further oxidation to M_2O_2 and M_2O_4. This can also be seen from the absorption spectra: Cs_2O is orange-red (probably having a band 20–25 kK), BaO has, according to Zollweg (1958), a band at 32·8 kK, while La_2O_3 is much less coloured. This is the opposite trend to that expected in an electron transfer band for increasing oxidation number of the metal ion, and is probably connected qualitatively with the influence of ionic size on O^{2-}. Hence, the evolution from red Cu_2O to white ZnO (having a strong band at 27·2 kK, broadening into the visible at higher temperature), or from brown CdO to yellow In_2O_3, and pale yellow SnO_2 and Sb_2O_5 is no proof of "inverse" electron transfer, even though impurities (e.g. of other oxidation numbers) and crystalline defects may be very important to the colour of bulk material, and even though $3d^{10} \rightarrow 3d^9 4s$ transitions might be expected to occur in the Cu(I) compounds. In aqueous solution, Cs^+, Ba^{2+}, Zn^{2+} and Cd^{2+} do not show the slightest sign of light absorption below 50 kK.

It is not quite unambiguous from the absorption spectra, whether O^{2-} or S^{2-} is the most reducing. Zollweg (1958) reports bands of BaS at 31·3 and 32·6 kK, of BaSe at 28·9 and 31·7 kK, while the bands of SrS at 38·4 and 39·6, and of SrSe at 35·7 and 37·9 kK indicate larger distances to the acceptor orbitals in the smaller cation Sr(II), which is not more oxidizing than Ba(II) by ordinary, chemical criteria. The doublet structure in the sulphides and selenides is presumably connected with the fairly large values of ζ_{3p} (0·4 kK in neutral S) and ζ_{4p} (1·8 kK in Se) though it cannot be completely explained by a splitting of ^2P of p^5 to its two J-levels. Rather, the Overhauser-splitting discussed above, is important where the behaviour of six σ-orbitals in octahedral symmetry (formed by linear combinations of six s-orbitals from the cations and possibly excited s- and p-orbitals of the anion), determines the lowest group of energy levels. The number and $^{2S+1}\Gamma_n$ of the cluster $NaCl_6$, considering s-orbitals of Na and p-orbitals of Cl, are the same as of the cluster $ClNa_6$, though the latter description is probably physically more significant. Differences in the number of states of the clusters occur in lattices with a different charge on the cation and the anion. For example, OCu_4 in Cu_2O (see p. 233) and FCa_4 in CaF_2 have forty-eight states of this type ($^{1,\,3}\Gamma_{13455}$), while CaF_8 has ninety-six states.

In the transition groups, S^{2-} undoubtedly plays the role of a more reducing ligand than O^{2-}, and it probably has an even larger nephelauxetic effect than I^-. Thus, the sulphur-containing ligands mentioned on p. 109 suggest values of h (Table 12) $\sim 2·8$. This partly explains the high tendency towards formation of low-spin sulphides, though co-operative effects in the solids also are important. MnS is green or pink, ZnS is nearly colourless (the first band at 31·5 kK), while FeS, CoS, NiS, Cu_2S and CuS are all black, and thus have strong absorption in the whole visible range. The study of sulphur-containing ligands will probably be quite interesting in the future.

Solid-state physicists distinguish between the Heitler–London description of less excited states of the insulators, using the wave functions of the individual atoms as starting points for the calculations, while the higher excited states, and all states of the electronic conductors, are given in the Bloch scheme with orbitals delocalized over the whole crystal. It may seem strange to them that the present writer criticizes the Heitler–London scheme so strongly in isolated molecules, preferring the Hund–Mulliken M.O. (cf. the discussion of the sulphate ion p. 216) and on the other hand prefers a localized excitation of well-defined chemical entities (e.g. the $ClNa_6$ or SBa_6 clusters mentioned above) for the more logical extension of the M.O. to the whole crystal. But here we have an example of the wisdom of choosing the appropriate microsymmetry, neither too high (in an unrealistic way), nor to low (paying too much respect to the group theory). The present writer freely admits that completely delocalized M.O. are necessary for the metallic bonding, but he recognizes so many features of the spectra of chemical entities, known from solution, and also in the spectra of crystals, and believes that these entities very often have their distribution of energy levels only slightly perturbed in the solid. In the same way as we must use configurations as an approximation in atoms and molecules, we must also sometimes neglect the somewhat horrifying consequences of the presence of N identical units in a crystal. The units must really be treated as identical, i.e. having the excited levels known from one unit only, in order to conserve an intelligible picture which, fortunately, very often agrees well with experience.

The energy band description assumes that metallic (electronic) conductance occurs when an energy band is partly filled with electrons. There is no doubt that this condition is not sufficient, e.g. the lanthanide compounds containing partly filled 4f-shells, or the paramagnetic d^n-compounds (such as NiO) must be said to contain partly filled energy bands. As explained (eqn. 188) the error has been to neglect the interelectronic repulsion energy depending on the average distance of the electrons with equal orbital energy. Thus, in NiO or in CeF_3, the number of electrons in each metal atom in the partly filled shell is very well determined, and we do not obtain low excited states redistributing this number as predicted by the energy band theory.

The *gaseous molecules of alkali halides* also have Laporte-allowed bands, but at lower wave numbers than the crystals (30·8 and 38·8 kK of the iodides, 36·4 and 39·2 kK of the bromides, and ∼40·6 kK of the chlorides). While the ground state of the diatomic molecules such as CsI can be fairly well described by the atomic states ^1S of Cs^+ and ^1S of I^- (and thus, the molecule is rather an ion pair than a covalent structure), the excited levels have to some extent the states ^2S of Cs and ^2P of I, and bands are even known at higher wave number, arising from ^2D and ^2P of Cs combined with ^2P of I. The decreased energy differences in the gaseous state are explained by the large mutual stabilization of ions in crystals, as compared to the much less symmetric diatomic molecule. However, the wave numbers, corresponding to ionization of halide ions in gaseous state are even smaller, between 26 and 30 kK. The electron transfer spectra of polyatomic molecules may also be found in the gaseous state. Thus, the tetrahedral $TiCl_4$ has a band ($\epsilon_0 = 7300$) at 35·6 kK. The octahedral molecules MoF_6 and WF_6 have their first strong band at 54 and 57 kK, respectively. It is from this evidence that the value $+28$ for F in the second term of eqn. (145) has been implied. The present writer pointed out (1960) that the $\pi \rightarrow \gamma_5$ bands (one weak and two strong, cf. Fig. 13) of gaseous OsF_6, IrF_6 and PtF_6 agree with eqn. (147). When UF_6 has its first band at so low a wave number as 24 kK, it shows the much higher electron affinity of the empty 5f-shell in U(VI) than of the 5d-shell in W(VI). Langseth and Quiller (1934) demonstrated that solutions of OsO_4 in water or hexane have nearly the same band positions as gaseous OsO_4, only with much of the line structure blurred out. In the same way, the present writer has shown that the complicated band system (Fig. 17) of OsI_6^{2-} in aqueous solution is conserved, and shifted some 3 per cent towards lower σ in the reflection spectrum of $[N(C_2H_5)_4]_2$ $[OsI_6]$.

If only the results of group theory were compared with the experimental data available on electron transfer spectra, it would be very difficult to identify the energy levels, and it would seem especially strange that the splitting of the excited configurations into numerous energy levels (such as eqn. 155) is not generally observed. We would in most cases predict a crowding of hundreds of excited levels, of which neither the order nor the exact position could be calculated. However, if we combine the purely mathematical aspects of M.O. theory with experimental evidence of the orbital energies and with the principle of the effective microsymmetry, arranging the perturbations of lower symmetry according to decreasing importance, we obtain an intelligible description. In the first application of M.O. theory to transition group complexes, the ligand field theory, the orbital energy differences within the partly filled shell were found to have the same order of magnitude as the interelectronic repulsion parameters, separating the terms of a given configuration. Consequently, it was necessary to study the whole theory of two-electron operator quantities to obtain agreement with experiment, and it was found extremely practical to use the electrostatic model or the expanded radial function model for obtaining a set of reasonable numerical parameters of interelectronic repulsion for comparison. When the M.O. theory was next applied to the electron transfer spectra of transition group complexes, it was necessary to find a series of orbital energy differences in eqn. (54) by comparing all the measurements of hexahalide complexes (of which many were to be made for this purpose) with the expectations, as derived from the one-ligand excitation model, comparison with ligand field theory, and other somewhat intuitive sources. On the other hand, at least for the $\pi \to \gamma_5$ and $\pi \to \gamma_3$ transitions, it was shown that the interelectronic repulsion parameters between the π- and the partly filled shell-orbitals (not those within the partly filled shell) could be neglected, returning to the primitive M.O. description without diagonal elements of interelectronic repulsion. This result could never have been predicted; but it may be accepted now, as we realize the possibility of small "squared overlap" between the π-orbitals and the partly filled shell. The consideration of intermediate coupling in bromo and iodo complexes is, however, a necessary refinement for the satisfactory description of the more complicated spectra of these species, as compared to the hexachloro complexes.

CHAPTER 10

ENERGY LEVELS IN COMPLEXES
WITH ALMOST SPHERICAL SYMMETRY

THERE is no reason for a M.O. in all cases to resemble the orbitals adapted to spherical symmetry, as is known from atomic spectroscopy. In the next chapter we shall consider transitions which certainly cannot be thus described. The three ordinary transition groups present a partly filled shell which to some extent is intermediate between the subject of this and the following chapter. On the one hand, the assumption of $l = 2$ in the electrostatic model or in the expanded radial function model explains almost all the pertinent features of the ligand field bands; on the other hand, the splitting into two (octahedral and tetrahedral complexes) or more sub-shells produces so large an orbital energy difference that the multiplet term system of the corresponding gaseous ion is not easily recognized.

If we except the recent study by Herzfeld of the energy levels of nitrogen atoms, trapped in solid nitrogen lattice, we have no other case of approximate spherical symmetry in chemistry as conspicuous as the $4f^n$-transitions in the lanthanides. These absorption bands have been known since about 1880 for their narrowness (δ between 0·016 and 0·3 kK, since the Franck–Condon principle enforces no vibrational excitation of the nearly parallel potential curves), their low intensity (ϵ_0 between 0·01 and 15, they are all Laporte-forbidden), and their relative small shifts (the nephelauxetic effect) and band splittings (the ligand field effects) as functions of the chemical compound of a given central ion. Many experimental studies have been made of these lanthanide compounds, whose bands can be seen sharply in a spectroscope and are easy to photograph. Thus, Auer von Welsbach and Georges Urbain used them for analytical purposes, H. Becquerel was interested in them as a physicist, H. C. Jones studied the influence of solvents, of temperature etc., the Leyden laboratory the influence of magnetic fields (the *Zeeman effect*), Gorter *et al.* measured oscillator strengths, and many authors such as

Yntema, Prandtl, Stewart, Holleck, etc., continued the investigation to find all the weak bands. For the identification of the J-values of the energy levels of the corresponding gaseous ions (which have not been studied by atomic spectroscopists, because the spectra,

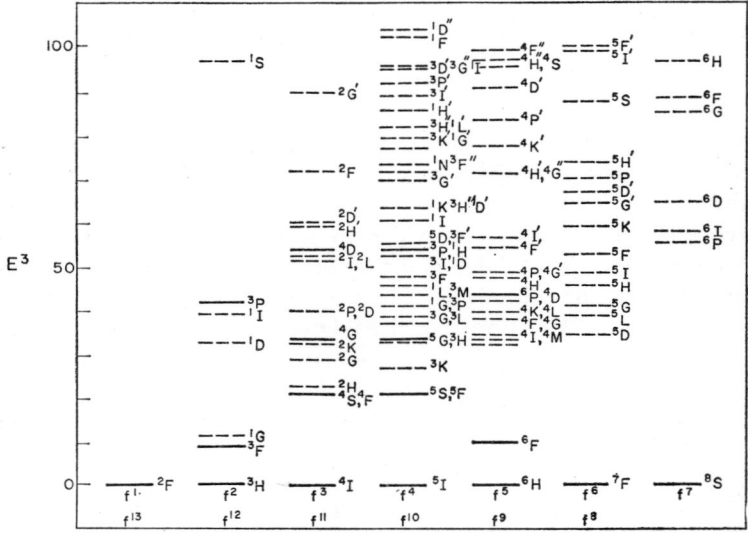

Fig. 21. The relative position of the baricenters of multiplet terms of f^n-systems in units of E^3 according to the assumptions of Elliott, Judd and Runciman (ref. 133). The dotted lines indicate terms with a lower value of S than the maximum value.

also involving excited states, consist of hundreds of thousands of lines, in contrast to the simple ions with only one or two electrons in partly filled shells) it has been proposed to measure the number of *sub-levels* at very low temperature (suppressing the population of excited sublevels of the ground state) which at most can be $2J + 1$ for an even number of f-electrons and $J + \frac{1}{2}$ (Kramer's doublets) for an odd number. This is not an infallible method of determining J of the excited state, since vibrational levels may imitate electronic sub-levels. Hellwege used it with great success (1951) for levels of Pr(III) and Eu(III) and Satten (1953) for Nd(III). Many authors have tried to explain the distribution of the sub-levels of each such level by first-order electrostatic perturbations and often obtained good results, if the ligand field parameters were chosen as empirical parameters in the same way as Δ in the octahedral complexes. This is probably a translation of relatively large second-order perturbations contributing to the small ligand field splitting.

However, the method of identification of J of each excited level from the distribution of the sub-levels has not been the most fruitful way of increasing our knowledge of the $4f^n$-energy levels. Rather, it has been useful to compare the overall distribution of spin-allowed and spin-forbidden bands with the Slater–Condon–Shortley theory of energy levels, which for the special case of f^n-configurations has been highly simplified by Racah (1949). Ellis (1936) was the first to identify the three bands of Pr(III) at 20·7, 21·4 and 22·6 kK with transitions to $^3P_0$, $^3P_1$, $^3P_2$ of $4f^2$. In the following years, Bethe and Spedding (1937) studied $4f^{12}$ Tm(III), Spedding (1940) Pr(III) and Satten (1953) $4f^3$ Nd(III). Actually, these authors only identified levels with maximum value of S correctly, while the excited levels of the spin-forbidden bands were erroneously assigned, due to the assumption of the unrealistic relation $F^2 \sim F^4 \sim 10F^6$. This does not affect the relative distances between the spin-allowed bands, which are multiples of the linear combination E^3 only (eqn. 36). Fig. 21 gives the order of multiplet term baricentres of $4f^n$ with a plausible ratio $F^4 = 0·75F^2$ and $F^6 = 0·60F^2$. Table 18 gives the values of E^3 in the trivalent lanthanides, increasing smoothly with the atomic number.

Gobrecht (1938) identified the J-levels expected in each lanthanide from the lowest multiplet term:

$$
\begin{array}{ccccccc}
f^1, f^{13} & f^2, f^{12} & f^3, f^{11} & f^4, f^{10} & f^5, f^9 & f^6, f^8 & f^7 \\
{}^2F & {}^3H & {}^4I & {}^5I & {}^6H & {}^7F & {}^8S
\end{array}
\tag{156}
$$

and determined thus the values of ζ_{4f} from the total spreading $(L + \tfrac{1}{2})\zeta_{4f}$ of this multiplet. Table 18 gives the somewhat more accurate values, derived by Judd from a partial correction for intermediate coupling effects. Elliott, Judd and Runciman (1957) have generally discussed how to predict the f^n-energy levels as function of F^k and ζ_{4f}. However, the exact identification of the definite levels is at present only certain for all the eleven known excited levels of Pr(III), for a certain number in Nd(III) (among those the extremely narrow band at 23·4 kK, having as excited level $^2P_{\frac{1}{2}}$), for the ten first levels of Er(III) (Wybourne (1960), and Kahle (1961)), all the eleven levels of Tm(III) (Jørgensen (1955), Johnsen (1958), and

Gruber and Conway (1960)) again excepting the high-lying ^1S; and a few others. Fluorescent transitions, which usually go from the lowest excited state with a given S to the ground state (cf. Cr(III) and Mn(IV) ruby lines, Gd(III), the first electron transfer state of UO_2^{2+}, etc.), may also go to less excited states. This has been of great importance for the study of Sm(III), Eu(III), Tb(III), Dy(III), Pu(III) and Am(III), where the corresponding absorption lines, going from the ground state, are situated in the infra-red.

TABLE 18. THE PARAMETER OF INTERELECTRONIC REPULSION E^3 AND THE LANDÉ PARAMETER ζ_{nf} IN COMPLEXES (ROMAN NUMERALS INDICATING THE OXIDATION NUMBER) AND GASEOUS IONS OF THE LANTHANIDES AND ACTINIDES, IN kK. THE FUNCTION k FOR THE NEPHELAUXETIC EFFECT OF EQU. (137) IS ESTIMATED FROM A COMPARISON OF AQUO AND VARIOUS ANION COMPLEXES

	E^3	ζ_{4f}	k		E^3	ζ_{5f}	k
4f² Pr(III)	0·46	0·8	0·03	5f Ac²⁺	—	0·750	—
4f³ Nd(III)	0·48	0·86	0·02	Th³⁺	—	1·236	—
4f⁵ Sm(III)	0·48	1·18	0·02	5f² Th²⁺	0·298	1·035	—
4f⁶ Sm(II)	0·42	1·1	—	U(IV)	0·36	1·6	0·04
Eu(III)	—	1·36	—	5f³ U(III)	0·3	1·7	0·05
4f⁷ Gd(III)	0·52	—	0·01	5f⁶ Am(III)	—	3·5	—
4f⁸ Tb(III)	—	1·72	—	5f⁷ Cm(III)	0·3	—	0·02
4f⁹ Dy(III)	0·57	1·92	—				
4f¹⁰ Ho(III)	—	2·08	—				
4f¹¹ Er(III)	—	2·3	0·00				
4f¹² Tm(III)	0·63	2·6	—				
4f¹³ Yb(III)	—	2·95	—				

It is the main concept of *Seaborg's actinide hypothesis* that the actinide compounds have only 5f-electrons in addition to the closed emanation shell of eighty-six electrons. This hypothesis can be proved with certainty either by extrapolation from atomic spectroscopy or from the absorption spectra of the complexes. The atomic number Z, at which the 5f orbital energy falls below that of 6d is highly dependent on the ionic charge $Z_0 - 1$ (eqn. 158). Thus the neutral gaseous atoms contain, in most cases, two 7s- and one 6d-electron in addition to the 5f-shell. These electrons are removed at first by ionization, leaving Th³⁺ with the configuration [Em]5f some 10 kK below that of [Em]6d. It can be safely extrapolated from the behaviour of other transition groups that Pa⁴⁺, U⁵⁺ and all following systems with eighty-seven electrons are most stable as [Em]5f. Further, it can be extrapolated that all actinide ions with the ionic charge at least $+3$ have the configuration [Em]5f^n. Thus, while [Em]5f^2 is an excited configuration, extended from 15·96 to 33·68 kK (with one further level $^1S_0$ at 51·48 kK) above the ground state of Th^{2+}, it can be extrapolated that [Em]5f^2 occurs at least 50 kK below [Em]5f6d of gaseous U^{4+}.

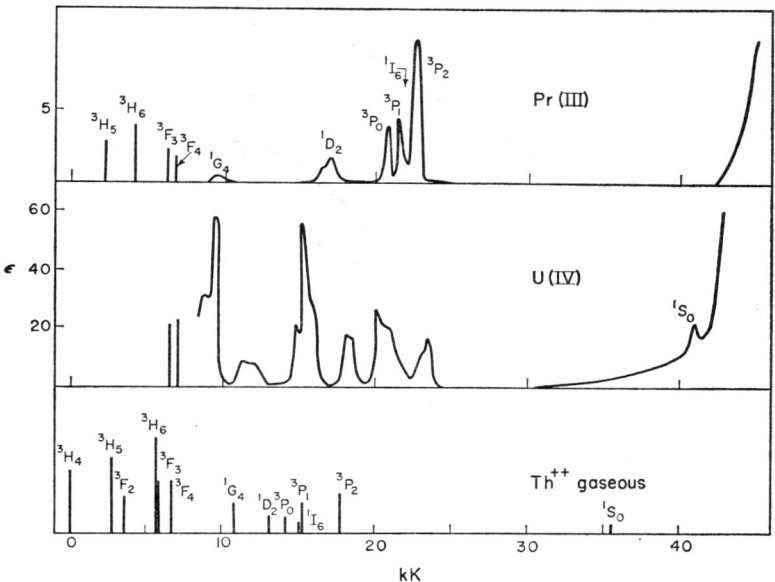

FIG. 22. Energy levels of 4f^2- and 5f^2-systems. Pr(III) aquo ions from refs. 119, 223, and 560. U(IV) aquo ions from refs. 122, 268, and 336 (cf. also ref. 95). Gaseous Th^{2+} from refs. 325 and 489. For a general discussion of 5f^2, see ref. 298.

This behaviour is also found in the chemical compounds. Fried and Hindman (1954) report the same type of transitions in Pa(IV) as described below (eqn. 160) for Ce(III). The actinides with at least two electrons outside the emanation shell have the narrow bands with only small chemical shifts known from the lanthanides. There are only observed from two to five times the ligand field splittings in 5f^n, compared to 4f^n, and the nephelauxetic effect (see Table 18) is also only about twice as large. The intermediate coupling effects, removing the spin-forbiddenness, are pronounced in the actinide complexes, which have ζ_{5f} equal to about twice the value of the corresponding ζ_{4f}. The most conspicuous difference between the Laporte-forbidden bands in $5f^n$ and $4f^n$ is the intensity, which is some thirty times larger (incorporated in this factor are the band widths, which are on average twice as large). This can be explained by the smaller distance to the levels with opposite parity (cf. eqn. 116), $5f^{n-1}6d$ discussed below, and by a larger interaction with the environment.

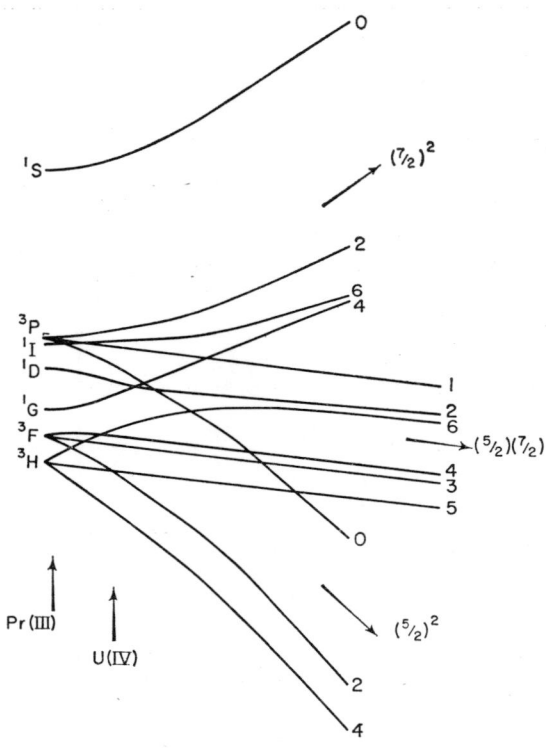

FIG. 23. The effects of intermediate coupling in f^2. The determinants were given by Spedding (ref. 550) and applied by Gruen (ref. 188). For large ζ_{nf}, the j values $\frac{5}{2}$ and $\frac{7}{2}$ of the individual f-electrons begin to form pure configurations. The actual cases Pr(III) and U(IV) are shown in Fig. 22.

The absorption spectrum of uranium (IV) aquo ions (see Fig. 22) has a very characteristic distribution of some nine narrow absorption bands in the range 6 to 23 kK, and after that only one narrow band at 40·8 kK. This is exactly as expected from the Slater theory for $5f^2$, having the twelve first energy levels (^3H, ^3F, ^1G, ^1D, ^1I, ^3P)

within a range slightly above $42E^3 + \frac{7}{2}\zeta_{5f}$, while the thirteenth level $^1S_0$ follows some 55 E^3 higher. This indicates slightly higher values of E^3 and ζ_{5f} in U(IV) (Table 18) than known in Th^{2+}. The present writer does not believe that the ligand field effects are so small that each observed sub-level of U(IV) corresponds to exactly one J-level of spherical symmetry; J may very well be intermixed between levels having a distance of, say, 2 kK. On the other hand, the overall distribution of multiplets is closely analogous to the case of 5f^2 in spherical symmetry. Conway (1959) has drawn similar conclusions for U(IV) in CaF$_2$, reporting the values $E^3 = 300$ K and $\zeta_{5f} = 1870$ K.

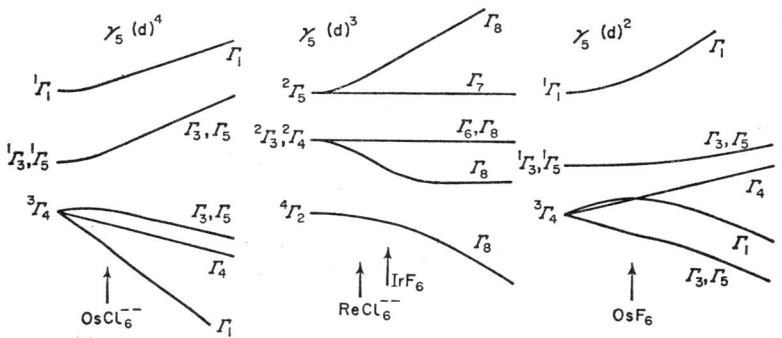

FIG. 24. The effects of intermediate coupling in p^2, p^3, and p^4 in spherical symmetry. As discussed by Kotani (refs. 334 and 559) and Griffith (ref. 181), the determinants are the same as those for γ_5^4, γ_5^3, and γ_5^2, when these electrons have the angular dependence characteristic for $l = 2$. The position of 5d^3 ReCl$_6^{2-}$ and 5d^4 OsCl$_6^{2-}$ in the diagram has previously been discussed by Griffith and the present writer (ref. 295), while Moffitt, Goodman, Fred and Weinstock (ref. 409) investigated 5d^2 OsF$_6$ and 5d^3 IrF$_6$.

Unfortunately, most lanthanide and actinide compounds have environments of very low symmetry. The only octahedral complexes well known are 5f^1 NpF$_6$, 5f^2 PuF$_6$ and UCl$_6^{2-}$. It would be very interesting to interpret the absorption spectra of these complexes in terms of the two orbital energy differences* demanded by group theory between the orbital odd γ_2 at lowest energy (see Table 1) and three equivalent, odd γ_5 (being π-antibonding) and the three equivalent, σ-antibonding odd γ_4 (L,S-coupling in f^1 produces the levels Γ_7, Γ_8, Γ_7, Γ_8, and Γ_6). These orbital energy differences are

*Axe (1960) demonstrated that these two M.O. energy differences in Pa(IV) in Cs$_2$ZrCl$_6$ are 1·50 kK and 2·16 kK, while ζ_{5f} is 1·49 kK.

only some 2 kK in NpF_6, while ζ_{5f} is some 2·2 kK. PuF_6 may indicate $E^3 \sim 0·4$ kK. It is remembered that near to the weak field limit (p. 81), the total splitting of a given level of atomic spectroscopy is in most cases considerably smaller than the ligand field parameters, since the M.O. configurations are highly mixed. Goodman and Fred (1959) assume that the energy difference $\gamma_2 - \gamma_4$ is as large as 21 kK. However, they consider the bands at 23 and 28 kK of NpF_6 as ligand field transitions. These bands are closely analogous to the electron transfer bands of UF_6, which also have vibrational structure, at 27 kK. The relative invariability of all internal 5f-transitions with the environment makes such large values of the ligand field parameters highly improbable.

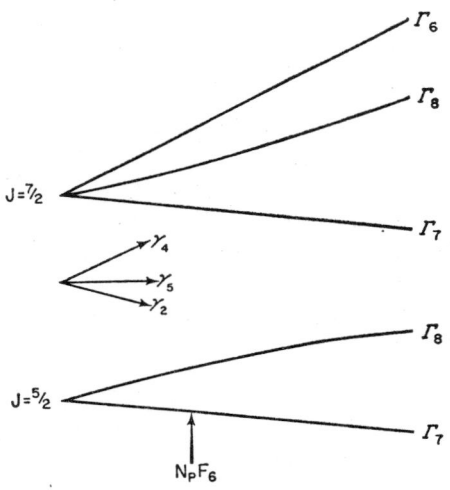

FIG. 25. The effect of intermediate coupling and octahedral ligand fields on a single f-electron. The variable is the orbital energy differences between γ_4, γ_5 and γ_2. The position of NpF_6 is indicated, as discussed on this page.

An unsolved problem is the splitting of energy of orbitals with different λ (=3, 2, 1, 0) from the 5f-shell in actinide dioxo ions MO_2^{2+} and MO_2^+. If the two oxygen atoms have strongly π- and σ-antibonding effects on the 5f-shell, the orbitals with $\lambda = 1$ and 0, respectively, would have higher energy. Actually, the complexes with at least one 5f-electron exhibit extremely narrow, rather intense bands, but it is not certain whether the energy levels involved have mainly retained J from atomic spectroscopy, or if the components with different Ω are strongly separated.

Since the 5f-electrons are so much more easily removed in actinide compounds (producing higher and more variable oxidation numbers) than are 4f in lanthanides, it is often felt by chemists that the 5f- and 4f-orbitals have quite differing properties, though they both may have $l = 3$. If we except the more uncertain case of the dioxo ions and concentrate our interest on the usual compounds of $+3$ and $+4$, we must conclude that the absorption spectra by no means present evidence of a much larger perturbation from the ligands on 5f- than on 4f-electrons. If we apply the argument of inverse proportionality between the parameters of interelectronic repulsion and the average radius of the partly filled shell, $\bar{r}$ in the actinides is some 60 per cent larger, while the crystallographic ionic radii are nearly the same. This situation in the actinides also occurs in the case of Mn^{2+}, where $\bar{r}$ is known to be $0\cdot57$ Å and the ionic radius in Mn(II) compounds is $0\cdot80$ Å. Actually, the nephelauxetic effect is not conspicuously large in the actinides (Table 18), though it can clearly be observed in the recently published absorption spectra of curium (III), where eqn. (140) is valid for CmF_3, the aquo ion, and Cm(III) in anhydrous $LaCl_3$. The discovery of narrow bands, belonging to internal f⁷-transitions as in Gd(III), is very valuable, since the interelectronic repulsion parameters are seen to be decreased some 40 per cent, compared to the lanthanides.

The intermediate coupling effects (which in a half-filled shell are always second-order perturbations and thus proportional to ζ_{nf}^2) are rather strong. Previously, only the broad bands of radiation-introduced impurities (H_2O_2, Cl_2, etc.) were reported of Cm(III). Asprey and Keenan (1958) measured reflection spectra of AmF_3, AmF_4, CmF_3 and CmF_4. CmF_4 is an f⁶-system, having the sharp peak at $19\cdot9$ kK of AmF_3 shifted to $22\cdot1$ kK, showing the influence of increasing ionic charge, as previously discussed by the writer.

Magnetochemists have often argued for the presence of 6d- rather than 5f-electrons in the actinides. This was in some cases (e.g. U(IV)) supported by a difference of 40 per cent of the paramagnetic deviation from 1 of the macroscopic, magnetic susceptibility. In these cases, the experimental values are compared to the expressions valid, when the ligand field splittings of fⁿ-energy levels are negligible, or when the ligand field allows a dⁿ-complex to retain a

N

paramagnetic contribution to the magnetic susceptibility, proportional to $S(S + 1)$, divided by the absolute temperature (Curie's law). The *magnetic moment* μ in Bohr units is in that case $g\sqrt{\{S(S + 1)\}}$ (see eqn. 119) being derived from *Curie–Weiss' law* for the *molar magnetic susceptibility* χ_m

$$\chi = 1 + C_m\chi_m \text{ with } \chi_m = \frac{N\mu^2}{3k(T - \vartheta)} + \chi_{corr} \qquad (157)$$

where N is Avogadro's number, ϑ is the Weiss correction (of dimension as temperature) and C_m the molar concentration of the paramagnetic substance. The *diamagnetic correction* χ_{corr} is composed of: (a) a small, negative contribution, common to all materials and approximately proportional to the number of electrons; and (b) a positive contribution, the *temperature-independent paramagnetism*, characteristic of complexes with low-lying electronic states, with which the ground state has non-diagonal elements of magnetic moment. The expression of this quantity, inversely proportional to the energy of excitation to the low-lying states, is often called Van Vleck's high-frequency formula.

Actually, the assumption of negligible ligand field splittings in $5f^n$-complexes is not valid, because the splittings are to be compared to the Boltzmann energy kT, which is only 0·21 kK at room temperature. When the splittings are some 0·5 to 1 kK, they may produce almost any type of magnetic behaviour, though the effects on the visible absorption bands are not very serious. Thus, the $5f^2$-systems generally exhibit too small a paramagnetism (compared to g, as calculated in spherical symmetry) and do not follow eqn. (157) (due to the varying population of sub-levels of the lowest level $^3H_4$), and PuF_6 is nearly diamagnetic, showing only a feeble temperature-independent paramagnetism. This only proves the existence of M.O. energy differences considerably larger than 0·21 kK, but the differences are not necessarily large compared to those between levels in spherical symmetry. The *condition of diamagnetism* (or temperature-independent paramagnetism) in systems with an even number of electrons is actually that the ground state is not degenerate ($e = 1$) and is fairly well removed from other, excited states (producing temperature-independent paramagnetism if the energy distance is some 1 to 50 kK, and strong paramagnetism for lower distances). Ordinarily, the ground states of systems with an odd number of electrons, having always at least $e = 2$, correspond to usual paramagnetism (eqn. 157). However, the gyromagnetic factor g may accidentally assume the value zero for some ground states ($^2\Pi_{\frac{1}{2}}$ of NO at low temperature; the component Γ_8 of $^2\Gamma_5$ of octahedral d^1-systems) leading to only χ_{corr}. Further, the interaction between two ions, each with an odd number of electrons, may be sufficiently large to produce a diamagnetic ground state (eqn. 168).

Laporte-allowed $f^n \rightarrow f^{n-1}d$ *transitions* are observed in Ce(III), Sm(II), Eu(II), Tb(III), Yb(II), Pa(IV), U(III), Np(III) and Pu(III). It is evident that the distance between the orbital energies of 4f and 5d, or of 5f and 6d, increases with the oxidation number. The (strongly reducing) divalent lanthanides present these bands at rather low wave numbers (17·9–30·2 kK in Sm(II), 31·2 and 40·4 kK in Eu(II)) while in the trivalent ions they are only known for Ce(III) and Tb(III). This is due to the steady increase of their wave number with Z for a given ionic charge, bringing the bands outside the measured range already in Pr(III). The reappearance of a band at 45·6 kK of Tb(III) is a half-filled shell effect (eqn. 149), the transition being $4f^8 \rightarrow 4f^75d$. Actually, the hump at f^7 of the ground state energy, assuming eqn. (146), is 8D, where $D = \frac{9}{8}E^1$ is approximately $13E^3$ (eqn. 36) in f^n-systems. Consequently, 8D is about 50 kK in the lanthanides and 30 kK in the actinides (Table 18). The effect on the position of the lowest level of $f^{n-1}d$ is perhaps somewhat smaller, owing to the interelectronic repulsion between the f-shell and the d-electron. The energy differences in kK between the lowest level of $5f^n$ and of $5f^{n-1}6d$, as derived from gaseous actinide ions and their complexes, is a function of Z_0 (the ionic charge plus one) and the atomic number Z for, at most, seven f-electrons:

$$7(Z - 96) + 18(Z_0 - 1) \tag{158}$$

Thus, it would be expected that the $5f^8$-system Bk(III) exhibits the first Laporte-allowed band at 31 rather than 61 kK derived from eqn. (158).

The energy levels of gaseous Ce^{3+} are situated at:

$$\left.\begin{array}{llll} ^2F_{\frac{5}{2}} & 0 & ^2D_{\frac{3}{2}} & 49\cdot74 \text{ kK} \\ ^2F_{\frac{7}{2}} & 2\cdot25 & ^2D_{\frac{5}{2}} & 52\cdot23 \end{array}\right\} \tag{159}$$

The aquo ion is probably nine-co-ordinated, having trigonal

microsymmetry, which can be approximated by the linear symmetry, identifying the excited orbitals:

ω	$Ce(III)\sigma_0$	ϵ_0	$Pa(IV)\sigma_0$	
$\sigma\frac{1}{2}$	33·7	16	—	
$\delta\frac{3}{2}$	39·6	710	36·3	(160)
$\frac{5}{2}$	41·7	600	39·2	
$\pi\frac{1}{2}$	45·1	380	44·8	
$\frac{3}{2}$	47·4	270		

ζ_{5d} of Ce^{3+} is seen from eqn. (159) to be 1·00 kK, explaining a reasonable splitting into ω-levels of the λ-levels in eqn. (160). The

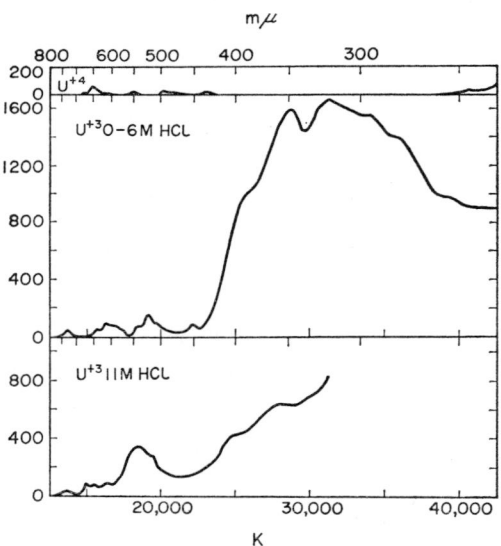

FIG. 26. The absorption spectra of uranium (IV) aquo ions (cf. Fig. 22), grey uranium (III) aquo ions in 1 M $HClO_4$ or 2 M HCl, and red uranium (III) chloro complexes in 11 M HCl. The Laporte-allowed transitions from $5f^3$ to $5f^2 6d$ are seen of U(III). The figure is reproduced with permission from *Acta Chem. Scand.* (ref. 281).

low intensity of the first band must be due to some selection rule of symmetry-forbiddenness. However, the present writer (1956) has previously suggested as an alternative possibility that Ce(III) aquo ions might exist with a lower co-ordination number, producing the weak band. The intensity of the weak band varies strongly with temperature, with the presence of other ions, and even by the substitution of heavy water as solvent. By comparison of eqn. (159) and (160), it is seen that the energy-difference 4f — 5d is some 19 per cent smaller in the Ce(III) aquo ion than in the gaseous Ce^{3+}. The decrease is some 28 per cent in complexes of ethylenediaminetetraacetate, citrate, chloride, etc., representing the same analogy to the nephelauxetic effect as the $s^2 \to sp$ transitions discussed below.

In the systems with more f-electrons in addition to an excited d-electron, it is not possible at the present to assign the broad, Laporte-allowed bands to specific energy levels from among the many possible ones. However, the strongest bands of U(III) aquo ions (see Fig. 26) with maxima at 28·6 and 31·2 kK and a shoulder at 25·5 kK might suggest parent terms of λ,ω-classification of the 6d-electron (eqn. 160) with a definite level of $5f^2$, e.g. the ground state $^3H_4$.

Solid-state physicists have been much interested in the absorption spectra and fluorescence properties of thallium (I) ions, embedded in alkali halide crystals. Seitz (1938) proposed that the weaker of two bands was due to the transition $^1S_0 - {}^3P_1$ and the stronger to $^1S_0 - {}^1P_1$ of the configurations $6s^2$ and $6s6p$, respectively. Actually, Fromherz had studied for many years similar bands of lead (II) complexes; and Macbeth and Maxwell (1923) found the rather narrow band at 30·5 kK of bismuth (III) in hydrochloric acid. These $s^2 \to sp$ *transitions* can be observed of the whole series of metals, having oxidation numbers two lower than the group-number in the periodical system: In(I), Sn(II), Sb(III), Tl(I), Pb(II), and Bi(III). If considered as problems of atomic spectroscopy, the energy of the levels with $J = l$ of a given configuration ls are the eigen values of a *Houston determinant*

$$\begin{vmatrix} E(ls) + K(ls) - E & \sqrt{[l(l+1)]}\zeta_{nl}/2 \\ \sqrt{[l(l+1)]}\zeta_{nl}/2 & E(ls) - K(ls) - \frac{1}{2}\zeta_{nl} - E \end{vmatrix} = 0 \quad (161)$$

where in Condon and Shortley's book, the exchange integral $K(ls)$ is denoted by $G_l = G^l/(2l + 1)$. The two other levels of a configuration ls with the values of $J = l - 1$ and $l + 1$ belong both to the triplet term only and have the energies $E(ls) - K(ls) - (l + 1)\zeta_{nl}/2$ and $E(ls) - K(ls) + l\zeta_{nl}/2$, respectively.

In the case of sp, the transitions to the two levels $^3P_0$ and $^3P_2$ are completely spin-forbidden, if no other configurations are taken into account. On the other hand, the two levels $^3P_1$ and $^1P_1$ with the original distance $2K(ls)$ are mixed by the non-diagonal element $\zeta_{nl}/\sqrt{2}$ according to eqn. (161). Table 19 gives the wave numbers of the transitions in the gaseous ions and in the complexes. While the second, spin-allowed transition corresponds to bands with oscillator

TABLE 19. THE WAVE NUMBERS IN kK OF THE TRANSITIONS TO THE TRIPLET AND SINGLET STATES OF 5s5p OR 6s6p IN 5s²- AND 6s²-COMPLEXES

	gaseous	H_2O	OH^-	enta⁴⁻	Cl^-	Br^-	I^-
In(I)	43·35	—	—	—	33, 36	34	32
	63·03	—	—	—	43	39	41
Sn(II)	55·20		—	—	34·7, 38·2	32	28
	79·91		—	—	45·0	41	34·5
Sb(III)	66·70	—	—	—	34·8	—	—
	95·95	—	—	—	43·9	—	—
Te(IV)	78·02	—	—	—	26·8?	—	—
	111·71	—	—	—	33·2?	—	—
Tl(I)	52·39	46·7	—	—	40·5	38·1	35·5
	75·66	—	—	—	51·0	47·2	43·0
Pb(II)	64·39	48·0	41·65	41·4	36·8	33·2	27·6
	95·34	—	—	—	51·0	44·9	—
Bi(III)	75·93	45·0	—	37·9	30·5	26·8	(22·2)
	114·60	—	—	—	45·0	38·5	29·8

strength $P \sim 0.8$, the first bands are much weaker ($P \sim 0.02$) for 5s²- than for 6s²-complexes ($P \sim 0.15$). This agrees well with the higher values of ζ_{6p} than of ζ_{5p}. In the gaseous atoms and ions, the values are (in kK):

$$
\left.
\begin{array}{lccccccc}
 & \text{Cd} & \text{In}^+ & \text{Sn}^{2+} & \text{Hg} & \text{Tl}^+ & \text{Pb}^{2+} & \text{Bi}^{3+} \\
\zeta_{np} & 1.14 & 2.37 & 3.79 & 4.26 & 8.18 & 12.39 & 16.99 \\
K(np, ns) & 5.68 & 9.11 & 11.25 & 5.59 & 8.16 & 9.66 & 10.86 \\
E(sp) - E(s^2) & 36.67 & 53.19 & 67.56 & 46.74 & 64.02 & 79.86 & 95.26
\end{array}
\right\} \quad (162)
$$

It is seen from Table 19 that the increase with ionic charge of the gaseous ions of the three parameters of importance: the distance $E(sp) - E(s^2)$ between the average (not the baricentre) of $^3P_1$ and $^1P_0$ and the ground state; ζ_{np} and $K(np, ns)$ of eqn. (162) do not occur in isoelectronic complexes with different oxidation number.

Actually, a certain principle of electroneutrality is acting; in the chloro complexes the parameters are (in kK):

	In(I)	Sn(II)	Sb(III)	Tl(I)	Pb(II)	Bi(III)
ζ_{np}	—	—	1·9	4·2(3·6)	5·7 (4·8)	5·8(4·9)
$K(np, ns)$	—	—	3·8	2·1(3·6)	2·85(4·8)	2·9(4·9)
$E(sp) - E(s^2)$	39	41	39·4	44·8	43·9	37·8

$$(163)$$

The relative size of ζ_{np} and $K(np, ns)$ in complexes can in principle be inferred from the ratio between the oscillator strengths of the spin-forbidden and spin-allowed bands. The values given for 6s6p in eqn. (163) are calculated for $\zeta_{6p} = 2K(6s, 6p)$ (giving φ of eqn. (23) 21·6° and $\sin^2\varphi = 0·136$), while the values in parenthesis are calculated from the observed band distances in Table 19 with $\zeta_{6p} = K(6s, 6p)$. The truth is probably somewhere between these limits. The values for Sb(III) are calculated by the analogous method, assuming $\zeta_{5p} = \frac{1}{2}K(5s, 5p)$.

Though the values of ζ_{np} and the K-integral in eqn. (163), taken separately, are rather uncertain, they give an impression of analogy to neutral Cd and Hg in eqn. (162), as also does the orbital energy difference between s and p. The development of the latter quantity follows closely the nephelauxetic series, decreasing in the order (see Table 19):

$$\text{gaseous ion} > H_2O > OH^- \sim \text{enta}^{4-} > Cl^- > Br^- > I^- \quad (164)$$

The decrease of the three quantities in complexes (eqn. 163 and 164), as compared to gaseous ions (eqn. 162) may be interpreted in two different ways, analogous to the discussion of the nephelauxetic effect. We might ascribe it solely to central-field covalency effects, having the spherical symmetrical part V_0 of the ligand field added to the central field with consequent changes in the orbital energies of ns and np. This would be a genuine Schläfer–Craig–Magnusson effect (p. 141), as may also act on the orbital energy differences 4f — 5d in Ce(III) and 5f — 6d in U(III) complexes. Notably, the exchange integrals K are very sensitive to a relative change of the two radial functions.

We may also consider symmetry-restricted covalency. The mere existence of s²-complexes is rather surprising, since the s-orbital in any symmetry will be σ-antibonding. In octahedral complexes (eqn. 54), the p-orbital is both σ- and π-antibonding. The energy difference might either increase or decrease by this effect, and we may very well ascribe a part of the decrease in eqn. (163) to the formation of covalent bonds. The σ-antibonding character of the ground state is shown by the difficulty in forming amine and cyanide complexes, and by the great tendency to form complexes with the more polarizable halides, Cl^-, Br^- and I^-. The anomalously large formation constants of the halide complexes may partly be due to relative instability of the aquo ions. The behaviour with OH^- is variable; Tl(I) does not form complexes at all, or at most, weakly bound ion pairs; Sn(II), Sb(III) and Pb(II) form precipitates, easily soluble in an excess of OH^-; and Bi(III) form precipitates at very low hydroxyl concentration (highly acidic solution), perhaps containing oxygen bridges.

However, the value of $K(ns, np)$ in eqn. (163) must not be taken too seriously. Already in the gaseous ion, the different J-levels of the configuration sp might have slightly different radial functions (cf. eqn. 40), and this might evolve further in the complexes, making eqn. 161 an inappropriate expression of the energy difference between the mainly triplet and mainly singlet level. Actually, the band widths are some 80 per cent larger in the singlet band of Pb(II) and Bi(III) chloro complexes than in the triplet bands. Williams (1951) suggests that the internuclear equilibrium distance r_L is smaller in the excited state, at least the triplet, than in the ground state. The triplet and singlet levels of Tl(I) in KI have different fine-structures (Yuster and Delbecq, 1953). This is somewhat surprising, since low microsymmetries (e.g. rhombic) should produce the same number of sublevels of the two levels with $J = 1$. However, it can be shown that the splitting of $^3P_1$ is only $-\frac{1}{2}$ as large as the splitting of $^1P_1$.

Finally, quite another deviation from the case of atomic spectroscopy may occur: the highest filled M.O. of the ground state may be rather far removed from having the symmetry s. Orgel (1958) studied the stereochemistry of this family and concludes that the lone pair, at least in Pb(II) compounds, has a strong tendency to concentrate at one side of the metal ion, corresponding to strong intermixing with higher values of l. Nevertheless, crystalline $PbCl_2$, which has this stereochemistry with irregularly varying distances between the ligands, shows the usual spectrum with a triplet and a singlet band. Once more, we cannot estimate the exact form of the M.O. involved from the absorption spectra.

Tl(I) ions in solid alkali halides have been much discussed by Williams and Johnson (1951, 1952) and In(I), as given in Table 19, is only known in such crystals (Williams, Segall, and Johnson 1957; Dunina, Morgenstern, and Shamosky, 1958; Katz and Nikolsky, 1958). In solution, Merritt, Hershenson and Rogers (1953) studied Tl(I), Pb(II) and Bi(III) halide complexes for analytical purposes (and also Hg(II)), and Newman and Hume (1957) made a profound study of Bi(III) chloro and bromo complexes.

A further possible case, where the excited state may involve a configuration from spherical symmetry, is the $3d^n \to 3d^{n-1}4s$ *transitions* producing Laporte-forbidden, broad absorption bands in strongly reducing complexes of the first transition group. Dainton (1952) correlated the wave numbers of these bands (the hexa-aquo ions of V(II) at 33, Cr(II) at 40, and Fe(II) at 41 kK) with the standard oxidation potentials, suggesting an "inverted" electron transfer process, as discussed in the next chapter. The transitions might be Laporte-allowed with accidentally small dipole moment, or spin-forbidden (the K-integrals between 3d and 4s in gaseous atoms are invariably from 1·2 to 1·3 kK). However, it seems probable that the excited orbital has $l = 0$ to some extent and is the even γ_1 (eqn. 54) to be filled in a (somewhat hypothetical) octahedral complex of Ga(I).

The *X-ray absorption edges* of chemical compounds, e.g. the K-edges, representing the excitation of a 1s-electron of a given atom, have been much studied. Stelling (1928–1934) observed a variation between 19,955 kK in S^{2-} and 20,050 kK in SO_4^{2-} among sulphur compounds and found a regular increase of the wave number with the oxidation number of many elements. Later, Mitchell and Beeman (1952) measured several transition group complexes and found chemical shifts and fine structure of the absorption edge of the metal ion. These transitions were now postulated to go to 4p, 5p, . . . shells, and the Pauling hybridization theory was applied to explain that diamagnetic square-planar nickel (II) complexes have a narrow band at lower energies than the paramagnetic, octahedral Ni(II). This description in terms of an empty or a filled 4p-shell was continued by Kauer (1956), Böke (1957) and Collet (1959).

However, it is obvious that the K-absorption spectrum simply indicates the available range of empty, preferably odd γ_4 (eqn. 115), orbitals, if M.O. configurations can be used. The energy and wave functions of these empty M.O. depend as well on the ligands as on the central ion. For instance, aromatic ligands with empty orbitals of relatively low energy show the transitions at lower wave numbers (cf. the Ir-py band to be discussed in the next chapter). Stelling (1932) found one of the two K-edges of covalent chloro-complexes at a lower wave number than of ionic Cl^-. This is not so well explained by Cl having an oxidation number below -1 as by the available empty orbitals in metal complexes.

A particular fact is that $Ni(NH_3)_6^{2+}$ and $Cu(NH_3)_4^{2+}$ both have two bands with an energy difference 30 kK, while $Znen_3^{2+}$ only has one (Cotton and Hansen, 1958). This influence of the partly filled d-shell might be connected with interelectronic repulsion energies, but is probably too large to be explained in that way. It is probably not possible to use ligand field theory on the 4p-shell (Cotton and Ballhausen, 1956) since the splitting is also observed in the octahedral $Ni(NH_3)_6^{2+}$ and $Co(NH_3)_6^{3+}$.

It would be useful to have more experimental facts available, especially on the more loosely bound electrons (2p, 3d, etc.) of complexes of heavy metals. However, the resolution limit is not good, some 20 kK, for bands in the K-region, though the relative accuracy of the wave number is very high. Collet (1959) measured the chemical shifts of several 2p absorption bands of rhenium (IV), platinum (II) and platinum (IV) complexes.

CHAPTER 11

MOLECULAR ORBITALS DISTINCTLY LACKING SPHERICAL SYMMETRY

WHILE the transitions $f^n \to f^n$, $d^n \to d^n$, $f^n \to f^{n-1}d$, $s^2 \to sp$, and perhaps $d^n \to d^{n-1}s$ to a certain approximation describe some of the absorption bands of transition and side group complexes, and while other transitions are electron transfers of the type filled M.O. $\to$ d, f, or s, there certainly also exist absorption bands of complexes, which it would not be appropriate to classify in that way. We may divide these transitions into two main groups: the *internal transitions in ligands* and the *electron transfer from partly filled shells of the central ion to the ligands*. There is no sharp distinction possible from the other transitions mentioned above; the internal transitions in ligands are influenced by the central ion and usually shifted towards lower wave number for increasing oxidation number of the metal ion. We saw above that the electron transfer spectra of halide complexes can to some extent be described as the transitions known from halide ions, shifted towards lower energy by decreasing energy of the acceptor orbital. On the other hand, the $f^n \to f^{n-1}d$ transitions and several other cases can be considered to some extent as electron transfers to the ligands. In addition to all these absorption bands, belonging to the complex as an entity (e.g. obeying Beer's law, eqn. (98), for dilution of the complex in solvents, which do not dissociate the complex measurably), there is a third group, caused by the *co-operative effects of more metal atoms*. Though the bands of polynuclear complexes in solution may occur, they are most common in solid substances, which have particularly strong colours.

One of the internal transitions in ligands, which has been studied by the present writer, is the band at 39 kK of pyridine C_5H_5N, having a vibrational structure of five components with the consecutive distance 0·92 kK, the highest having $\epsilon = 3130$. The presence of a proton in the pyridinium ion $C_5H_5NH^+$ increases the intensity by some 70 per cent and shifts the bands some 0·1 kK towards higher wave number. In rhodium (III) complexes such as cis- and trans-Rhpy$_2$Cl$_4^-$ the band maxima have ϵ about 5000 per pyridine molecule, and the vibrational structure is shifted 1 kK to lower wave numbers than of py. In iridium (III) complexes, containing many anions, such as IrpyCl$_5^{2-}$, the pyridine band is broadened

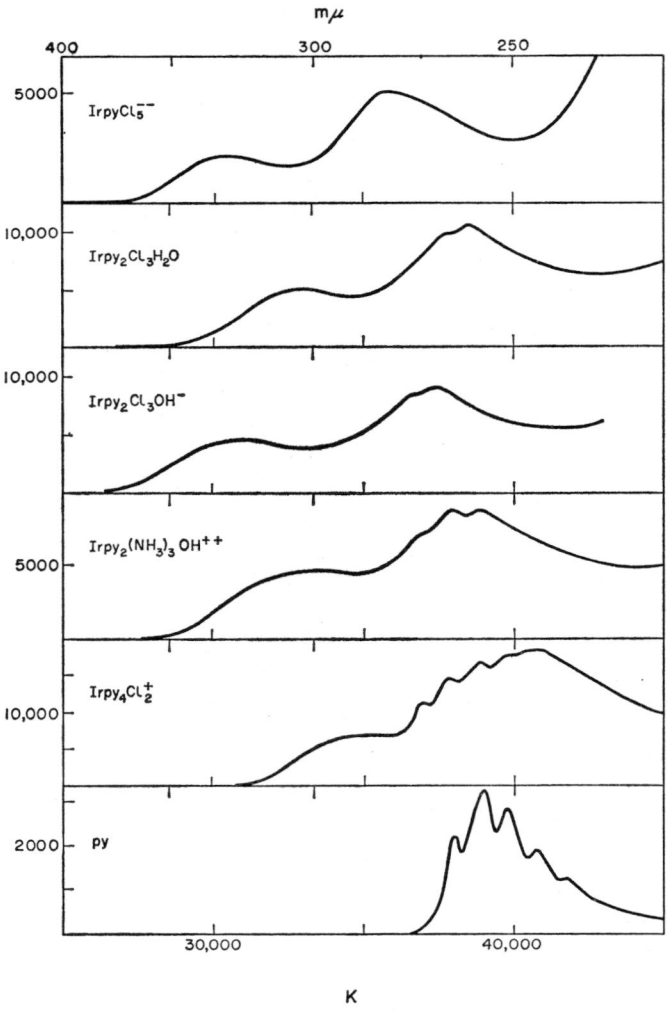

FIG. 27. The Ir–py bands and pyridine bands. The figure is reproduced with permission from *Acta Chem. Scand.* (ref. 285).

somewhat and occurs at 35·8 kK, while in $Irpy_4Cl_2{}^+$, the components of the vibrational structure can still be observed at 36·9, 37·8 and 38·9 kK (Fig. 27). Thus, it is seen that the pyridine spectrum is very little affected by the complex formation. However, in addition to the py bands, the iridium (III) complexes (but neither Rh(III) nor Pt(II)) exhibit a particular band at lower wave numbers, broad ($\delta = 1\cdot5$ to $2\cdot2$ kK) without structure and having ϵ some 2000 per pyridine molecule. The only plausible explanation of this *Ir–py band* is an electron transfer from the filled γ_5-sub-shell of iridium to the empty π-orbitals of pyridine.* The wave number of the Ir–py band varies with the number of anions present in the complex, as seen in Table 20. Kosower (1956) has studied the similar charge transfer band of pyridinium iodide ion-pairs in different solvents. Martin and Waind (1958) report the Ir-dip band in the dipyridyl complex Ir $dip_3{}^{3+}$, which, as all other dip-complexes, also shows the internal dip-transitions. König and Schläfer (1960) investigated a long series of pyridine complexes, such as $Crpy_3Cl_3$ and $Mopy_3Cl_3$.

TABLE 20. THE WAVE NUMBER IN kK OF THE Ir–py BAND OF IRIDIUM (III) COMPLEXES

$IrpyCl_5{}^{2-}$	30·4	$Irpy(NH_3)_4Cl^{2+}$	34·5
$IrpyCl_4(H_2O)^-$	31·4	$Irpy_2(NH_3)_3(H_2O)^{3+}$	(35·4)
$IrpyCl_4(OH)^{2-}$	30·0	$Irpy_2(NH_3)_3OH^{2+}$	32·9
cis-$Irpy_2Cl_4{}^-$	32·0	$Irpy_2(NH_3)_3Cl^{2+}$	35·0
trans-$Irpy_2Cl_4{}^-$	31·6	$Irpy_2(NH_3)_2Cl_2{}^+$	33·9
cis-$Irpy_3Cl_3$	30·6	$Irpy_2Br_4{}^-$	31·4
trans-$Irpy_3Cl_3$	31·4	cis-$Irpy_3Br_3$	29·8
$Irpy_4Cl_2{}^+$	35·2	trans-$Irpy_3Br_3$	31·2

After the presentation of these typical examples of internal ligand transitions and electron transfer from the central ion to empty orbitals of the ligands, we may summarize other cases known. Sone (1953) studied the high ultra-violet bands of acetylacetonate (aca^-),

* The usual classification of σ- and π-orbitals in aromatic compounds has another significance than in linear symmetry. In the former case, we are considering mainly planar molecules, where the plane introduces an approximate microsymmetry $D_{\infty h}$. However, we distinguish here between σ orbitals having the same sign above and below the plane, and π orbitals, where the wave function has opposite signs. In this way, an s-orbital is classified as σ together with x and y of the p-shell, while z alone represents π of the p-shell. This is the opposite to naming $\lambda = 0$ and $\lambda = 1$ with respect to the linear symmetry along the z axis. When text-books indicate that the bond C=C (being a part of a planar molecule) has $\sigma\pi$ and C≡C $\sigma\pi\pi$ characters, they are using the symbols σ and π in a highly different way in the two cases.

8-hydroxyquinolinate, and other ligands, where a shift with increasing oxidation number of the central ion was observed. The variation of the wave numbers with the divalent ions is generally

$$Mg(II) > Mn(II) \sim Zn(II) \sim Co(II) > Ni(II) \sim Cu(II) \\ > Pd(II) \qquad (165)$$

suggesting a variation with the tendency of covalent bonding. Kiss and Szabo (1943) draw similar conclusions from the absorption spectra of Schiff's base anions such as bis(salicylaldehyde)ethylenediimine and salicylaldoxime complexes. Belford and Calvin (1957) studied organic fluoro-derivatives of aca$^-$ as free ligands and in Cu(II) complexes. Holm and Cotton (1958) investigated aca$^-$ complexes. In analogy to pyridine, the heterocyclic di-imines dip and phen (Table 10) have two kinds of bands in the complexes: strong bands (ϵ some 10,000 to 40,000 per molecule) with a complicated structure in the ultra-violet, and bands in the visible with reducing central ions such as Fe(II), Cu(I), Ru(II), Os(II), V($-$I), V(0), V(I), and Cr(0). The latter bands are most certainly identified as electron transfer transitions from the central ion[*] to the empty π-orbitals, as discussed by Williams (1955). In addition to these bands, we may have electron transfer in the usual direction, as found in iron (III) complexes of aca$^-$ and the heterocyclic di-imines (the latter being low-spin d^5-systems).

A complete theory of the internal transitions in ligands involves a description of the energy levels of nearly any type of coloured molecule. Empirically, the study of *organic dyes* produced a set of rules for the bathochromic or hypsochromic effects of other atoms or radicals, the auxochrome groups, on the band positions of the intrinsically coloured chromophoric groups, usually involving carbon atoms with bonds of fractional bond number. These rules resemble the rules in the spectrochemical series for the band positions of transition group complexes. When quantum mechanics was applied to cyclic compounds (at first to benzene) it was realized that the presence of empty π-orbitals at rather low energy above the filled π-orbitals is the reason of the absorption bands at relatively low wave number of most aromatic compounds (especially the *triphenylmethyl cations*, derivatives of $(C_6H_5)_3C^+$, or free radicals such as $(C_6H_5)_3C$, the latter missing one electron in a M.O.), while

[*] Similar transitions have been found in the low-spin d^6-molecule Cr(CO)$_5$py by Strohmeier and Gerlach (1961).

the σ-orbitals produce absorption bands in the far ultra-violet; this phenomenon is also found in the aliphatic compounds. In a way, the effect of loosely bound π-orbitals on the colour is comparable to the presence of partly filled shells in transition group ions. Special cases, important in the cases of pigments occurring in living organisms, are the *porphyrine ring systems*, e.g. chlorophyll, haemoglobin, vitamin B12, and the synthetic dye Monastral (phthalocyanine) blue. The latter compounds are complexes of Mg(II), Fe(III), Co(III) and Cu(II), respectively, co-ordinated by the ring system with four nitrogen atoms in a plane. They are characterized by a set of very intense, narrow absorption bands (as reviewed by Williams, 1956). The complexes involving the transition group elements are often more stable, and the catalytic effects on redox reactions in the organism may be enhanced by the variation of the oxidation number and the capacity of co-ordinating other ligands perpendicular to the porphyrin ring. However, the colours are not connected with the presence of a partly filled d-shell. The small shifting effects of central ions was already used by Formánek (1905) for qualitative analysis of metal ions by addition of alkanna tincture, a plant extract, forming complexes with slightly different wave number of their absorption maxima. Similar effects are known from the coloured, amorphous hydroxide precipitates, formed by metals and alizarin (red with Al(III), purple with Sn(IV), etc.). Recently, similar, but soluble substances have been used as *metal indicators*, such as aurinetricarboxylic acid. The different anions of this acid are formed according to pH, the solution in acetate buffer being yellow, and in alkaline solution red. If Al(III) is added to the acetate buffer, a red complex is formed. The ligand is in nearly the same state as in the alkaline solutions. Pyrocatechine violet and Eriochromschwarz T function according to the same principle, the latter being red as trivalent anion and in complexes with Mg(II) and Ca(II), while it is blue in the divalent anion, containing one further proton.

Fe(III) (and other oxidizing metal ions, such as Ce(IV)) react with polydentate ligands, containing at least one phenolate group. Thus, sulphosalicylic acid, having the visibly colourless anion $H(ssal)^{2-}$ ($^-O_3S)C_6H_3(OH)(COO^-)$), may split off a further proton from the phenolic OH-group, forming the violet Fe(ssal), the red $Fe(ssal)_2^{3-}$, and the yellow $Fe(ssal)_3^{6-}$. The corresponding maxima at 18·4, 21·5 and 22·6 kK (Vareille, 1955) are most probably electron transfer bands, and not shifted internal transitions of the ligands.

Similar remarks apply to the Fe(III) complexes of "Tiron", pyro-catechinedisulphonate, being blue-green (14·1 kK), violet (18·2) and red (20·8) with one, two and three of the bidentate ligands, respectively. In studying metal complexes of complicated organic molecules, it is always wise to be wary of the formation of strongly coloured oxidation products, which are not necessarily complexes. It is known from Perkin's discovery of mauveine, that aniline forms such dark-coloured, multicyclic oxidation products, but even aliphatic amines form, by standing in air, yellow oxidation products with high band maxima in the near ultra-violet. It is a well-known fact from organic synthesis that yellow, brown or black tarry, impurities often result from inevitable side reactions.

TABLE 21. THE WAVE NUMBER IN kK OF THE THIOCYANATE BAND

$Cr(NH_3)_2(NCS)_4^-$	33·0	$Rh(SCN)_6^{3-}$	34·5
$Cr(NCS)_6^{3-}$	32·4	$Pd(SCN)_4^{2-}$	32·5
$Coen_2(NCS)_2^+$	32·5	$Os(SCN)_6^{2-}$	30·5
$Co(NCS)(H_2O)_5^+$	36·7	$Pt(SCN)_6^{2-}$	34·8
$Co(NCS)_4^{2-}$	32·8	$Pt(SCN)_4^{2-}$	37·4
Ni(II) 9 M KSCN	(33·9)	$Hg(SCN)_4^{2-}$	(35·7)

In the following discussion, we shall concentrate on the spectral behaviour of simpler ligands. *Oxalate* ions, as most other carboxylates, have a band above 50 kK. This band is slightly shifted towards smaller wave numbers in many complexes. However, rhodium (III) and iridium (III) oxalate complexes (such as *cis*- and *trans*-$Rhox_2Cl_2^{3-}$ and $Irox_2Cl_2^{3-}$, $Irox_3^{3-}$ and $IroxCl_4^{3-}$) all have a broad band at from 33 to 35 kK with ϵ about 1000 per oxalate group. It cannot be considered as an ordinary electron transfer band, since it would make ox^{2-} more bathochromic even than Br^-, and it would not be easy to understand the comparable wave number in Rh(III) and Ir(III). It is more probably quite a strongly shifted, internal transition in the oxalate group.

The *thiocyanate* complexes present a band between 30 and 38 kK (Table 21) with ϵ some 8000 per thiocyanate in most of the complexes. SCN^- in aqueous solution has a band well above 50 kK (55 kK of solid KSCN), giving the impression of a similar reducing character as Br^-. However, the bands given in Table 21 can hardly be considered as electron transfer bands, since the wave numbers are so independent of the oxidizing character of the central ion. They are also independent of whether the nitrogen or the sulphur end is bound to the central ion. The estimate of this direction has been made from the position of the ion in the spectrochemical series (eqn. 120) and from the evidence from X-ray crystallography, now appearing (Zdanov and Zvonkova, 1953; Rasmussen, 1958). The division between N- and S-bound complexes is not a classification $3d^n$ vs. $4d^n$ and $5d^n$, as Table 21 might suggest. The strongly oxidizing ions such as Fe(III), and probably also Cu(I), Cu(II) and Ag(I), are bound to the S-end. Genuine electron transfer spectra of thiocyanate complexes are found in the visible bands of red Fe(III), Mo(V), Tc(V) and yellow W(V) complexes.

Cyanide complexes also present many of the problems mentioned for thiocyanate. The carbon end is nearly always bound to the central ion, except in the case $Ni(CN)_2$, NH_3, C_6H_6, where half the nickel atoms are surrounded by four carbon atoms in a plane, and half by six nitrogen atoms (two from CN^- and four from NH_3) in an octahedron. Consequently, half the nickel (II) atoms are paramagnetic with $S = 1$, and the rest diamagnetic. Dr. Ole Bostrup has kindly informed me that the reflection spectrum of this pale blue compound is a superposition of the usual spectra of high-spin and low-spin d^8-complexes.

It is generally indicated in literature that the triple bond between C and N corresponds to the presence of a set of two low-lying, empty π-orbitals. It is argued that these orbitals interact with the γ_5-sub-shell of lower energy in the central ion producing the high values of Δ observed in octahedral cyanide complexes. Actually, crystals of KCN do not have band maxima before 59 kK, and ϵ of the aqueous solution is only 1·5 at 45 kK. Hence, the first excited level of the free cyanide ion is very high, as it is also in the isoelectronic CO and N_2. This is no absolute argument against the importance of π-bonding in the complexes, since the change of diagonal elements of energy by polarization may be considerable, and since the non-diagonal elements may be unusually large.

Some high bands of hexacyanide complexes, such as those at 38 kK of Cr(III), 30 kK of Mn(III), and 24 kK of Fe(III) are probably simply $\pi \to \gamma_5$ electron transfer spectra, of these d^3, low-spin d^4 and low-spin d^5 systems. Owing to the addition of $\Delta \sim 34$ kK, the $\pi \to \gamma_3$ transition in the hexacyanides of Fe(II) and Co(III) is expected to occur at some 50 kK (cf. eqn. 147), in agreement with experiments.

o

Actually, it is rather difficult to know the parameters of eqn. (145) for the first transition group, since the hexachloro complexes are not known as monomeric entities in solution. The tetroxo complexes (eqn. 143) might suggest a shift 17 kK between $3d^n$ and $4d^n$, making the first term of eqn. (145) $34 - 7q$, assuming the dependence on q to be the same. With this rather uncertain assumption, the $\pi \to \gamma_3$ band of a hypothetical low-spin $CoCl_6^{3-}$ should occur at 22 kK. This seems rather low, compared to the weak (36·5 kK) and strong band (44·0 kK) of $Co(NH_3)_5Cl^{2+}$, since the Δ-shift discussed on p. 163 was only expected to shift the former band some 10 kK. It can only be concluded* that CN^- probably has a value for the second term of eqn. (145) between 0 and $+5$, i.e. intermediate between the reducing characteristics of H_2O and Cl^-.

TABLE 22. THE ABSORPTION BANDS OF SQUARE-PLANAR d^8- AND OF d^{10}-CYANIDE COMPLEXES. σ_0 AND ϵ_0 ARE THE WAVE NUMBER IN kK AND THE MOLAR EXTINCTION COEFFICIENT OF THE MAXIMA (SHOULDERS IN PARENTHESIS) AND $\delta(-)$ THE HALF WIDTH TOWARDS SMALLER (OR $(+)$, LARGER) WAVE NUMBERS

	σ_0	ϵ_0	$\delta(-)$		σ_0	ϵ_0	$\delta(-)$
$3d^8$ Ni(CN)$_4^{2-}$	(27·8)	100		$3d^{10}$ Cu(CN)$_2^-$	36·5	650	
	32·1	1000			42·7	11,000	
	35·0	5000		Cu(CN)$_3^{2-}$	32·5	160	
	37·5	11,700			35·1	260	
$4d^8$ Pd(CN)$_4^{2-}$	41·6	1200	0·5		42·0	11,600	
	45·4	7200		Cu(CN)$_4^{3-}$	33·9	170	
	47·2	9000			42·5	11,000	
$5d^8$ Pt(CN)$_4^{2-}$	35·72	1590	0·48	$5d^{10}$ Au(CN)$_2^-$	41·8	3300	0·38
	(38·68)	26,000	1·13		43·4	3450	0·8
	39·18	29,500	$(+)0·93$		47·4	9400	0·9
	(41·32)	1850			48·8	11,100	0·6

The linear dicyanide complexes of Cu(I) and Au(I), and the higher complexes $Cu(CN)_3^{2-}$ and $Cu(CN)_4^{3-}$ were studied by Brigando (1957) and Simpson and Waind (1958), while the similar complexes of Ag(I) and Hg(II) do not have such bands below 50 kK. Table 22 gives the bands of Cu(I) and Au(I), as also of the square-planar tetracyano complexes of the diamagnetic d^8-systems Ni(II), Pd(II) and Pt(II). The first band of $Au(CN)_2^-$, $Pd(CN)_4^{2-}$ and $Pt(CN)_4^{2-}$ are each characterized by an unusually small half width $\delta(-)$. The

* Actually, the first band of $Ru(CN)_6^{3-}$ at 22 kK (DeFord and Davidson, 1951) might suggest that CN^- is nearly as reducing as Br^-. However, the partly filled subshell γ_5 may be relatively depressed in the cyanide complex, as is presumably also the case in isonitrile (e.g. dark violet $Mn(CNR)_6^{++}$) and arsine complexes.

evolution of the wave numbers as function of the atomic number is not monotonic:

$$Ni(II) < Pd(II) > Pt(II) \text{ and } Cu(I) \ll Ag(I) > Au(I) \quad (166)$$

We may think of three main reasons for the bands of d^8-tetracyanide complexes: transitions from the filled (mainly) d-orbitals to low, empty orbitals of the central ion, especially the z-orbital of the p-shell (symmetry, odd γ_{t2}), or to the empty π-orbitals of the cyanide (being "inverted" electron transfer), or finally transitions from the filled π-orbitals of cyanide to the empty (mainly) d-orbital γ_{t3} or z. The set of π-orbitals in tetragonal symmetry are t2, t4 and t5, combined with even and odd parity. Among the transitions cited, only a small number is symmetry-allowed, viz. those for which the product of the two γ_{tn} (see Table 6) contains either odd t2 or odd t5 (eqn. 114).

For comparison, Table 23 gives the two groups of strong bands of *diamagnetic d^8-tetrahalide* complexes, differing by a factor ~4 in the intensities, but having the same half widths. The wave number difference varies from 9 to 12 kK between the two groups, and the differences in the pairs Cl–Br and Br–I are some 5 and 10 kK, respectively (cf. eqn. 145). However, we saw in the case of Bi(III) complexes (Table 19) that such a shift is no certain indication of ordinary electron transfer, the distances between other M.O. may vary in the same way.

The most probable processes are $\pi \rightarrow d$ (even t3) and $\sigma \rightarrow d$ (even t3), in analogy to the hexahalide systems, and d (the filled orbitals) $\rightarrow p_z$. According to eqn. (145), the $\pi \rightarrow \gamma_3$ transitions in an octahedral, hypothetical, $AuCl_6^{3-}$ would occur at 32 kK, and if the shift by changing the oxidation number from $+3$ to $+2$ is 10 kK, the bands of $PdCl_6^{4-}$ and $PtCl_6^{4-}$ should occur at 35 and 42 kK. Actually, these values agree extremely well with the first band group of the tetrachloro complexes (Table 23), while the second group might then be identified as transitions of a σ-electron, as discussed on p. 153 for the d^6-hexahalides. However, we meet one difficulty in this interpretation: since the spreading in orbital energies of the partly filled shell is some 50 per cent larger in the divalent d^8-tetrahalides, compared to the trivalent d^6-hexahalides, the four filled orbitals of the d^8-complexes must have some 10 to 20 kK lower energy than the corresponding γ_5-orbitals of a hexahalide complex of a divalent d^6-ion. Perhaps this increase in the oxidizing character of Pd(II) and Pt(II), compared to central ions in octahedral complexes (cf. eqn. 141) may be admitted, implying a close analogy between low-spin d^8- and d^6-electron transfer spectra. According to eqn. (188), the contribution of interelectronic repulsion $J(d, \gamma_{t3}) - J(\pi, \gamma_{t3})$ may be smaller in the square-planar complexes than $J(d, d) - J(\pi, d)$ for an octahedral complex, if γ_{t3} is much more delocalized than the other orbitals, marked d. However, since the excited z-orbital of 5p in Pd(II) and 6p in Pt(II) and Au(III) may perhaps show the same variation of energy with oxidation number and ligands, we cannot exclude the excitation of a (mainly) d-electron to this orbital. In this case, it would have an energy 20 kK above the empty even t3-orbital in $PtCl_4^{2-}$, 15 kK in $PdCl_4^{2-}$, and 10 kK in $PdBr_4^{2-}$.

TABLE 23. THE ABSORPTION BANDS OF SQUARE-PLANAR d^8-TETRAHALIDE COMPLEXES. THE NOTATION AS IN TABLE 22

	First group			Second group		
	σ_0	ϵ_0	$\delta(-)$	σ_0	ϵ_0	$\delta(-)$
$4d^8$ $PdCl_4^{2-}$	35·8	10,500	2·1 (+) 2·0	44·9	30,000	2·4 (+) 2·6
$PdBr_4^{2-}$	30·2	11,000	1·9 (+) 1·8	37·0	25,000	
				40·9	30,000	
PdI_4^{2-}	20·5	4000				
	24·5	8500	1·8 (+) 1·8			
$5d^8$ $PtCl_4^{2-}$	(43·1)	6800	1·4			
	46·1	8600				
$AuCl_4^-$	31·8	5100	2·0 (+) 2·2	44·1	34,400	2·0 (+) 2·4
$PtBr_4^{2-}$	37·3	6700	(+)1·8			
$AuBr_4^-$	26·3	4600	(+)2·0	39·4	24,300	2·0 (+) 1·7
PtI_4^{2-}	25·7	4400	2·2	35·8	13,000	
	30·8	4100				

After this discourse on the tetrahalide complexes, which cannot be presumed to have easily accessible, empty π-orbitals to the same extent as the tetracyanides, we may return to Table 22. It is seen that $Ni(CN)_4^{2-}$ and $Pd(CN)_4^{2-}$ have the wave numbers expected for electron transfer from filled orbitals to the empty even t3-orbital of the central ion, if CN^- is some 10 kK less reducing than Cl^-, as discussed above for the $3d^n$-hexacyanides, and assuming the same energy of the empty t3-orbital in the tetracyano as in the tetrachloro complexes. This is a rather dubious assumption, moving the filled d-orbitals further down in the energy scale. Actually, the first weak shoulder of $Ni(CN)_4^{2-}$ at 27·8 kK is probably a ligand field band, but the order of the orbitals in the cyanide complexes is not known.

On the other hand, the reversal in eqn. (166) suggests that at least some of the bands of $Pt(CN)_4^{2-}$ are due to a reason other than the bands of Ni(II) and Pd(II), and the qualitative appearance is also different.

The situation is even less certain for the d^{10}-cyanide complexes. From group theory, the orbitals formed in linear symmetry are known:

$$
\begin{array}{lll}
\text{central ion:} & & \text{ligands:} \\
(n+1)\text{p} & \text{odd } \sigma & \text{empty } \pi \text{: odd } \pi \\
& \text{odd } \pi & \quad\quad\quad\text{even } \pi \\
(n+1)\text{s} & \text{even } \sigma & \\
& & \text{filled } \pi \text{: even } \pi \\
n\text{d} & \text{even } \sigma & \quad\quad\quad\text{odd } \pi \\
& \text{even } \pi & \text{filled } \sigma \text{: odd } \sigma \\
& \text{even } \delta & \quad\quad\quad\text{even } \sigma
\end{array}
\tag{167}
$$

but their relative energy and their degree of intermixing (for the same λ and parity) is only known very incompletely. A special problem, of interest in d^{10}-chemistry, is the relation between the empty and the filled σ-orbital of the central ion, as to whether $(n+1)$s and nd are strongly intermixed, as suggested by Orgel (1958). The weak bands at 44·4, 47·5 and 51·9 kK with ϵ below 1500 of silver (I) aquo ions (Fromherz and Menschick, 1929; Volbert, 1930) are probably 4d → 5s transitions. On the other hand, the bands of halide complexes in solutions (Ag(I) chloride at 46·4, bromide at 44, Cu(I) chloride at 36·8 and bromide at 36·2 kK) are so strong (ϵ 3400–30,000) that some kind of electron transfer is suggested. It is not known whether the intensity of Cu(I) and Ag(I) in alkali halides, which seems to be only some 1 per cent of the solution, is explained by low solubility in the crystal or by an effective prohibition of a Laporte-forbidden transition (the γ_5-electrons of the filled shell in tetrahedral complexes in solution being mixed with p-orbitals). It is not known either, whether Cu(I) and Ag(I) halide complexes, in addition to these bands, actually have weak bands at lower wave number $\sim$ 25 kK. Recently, Okamoto (1956) has found that AgCl and AgBr at low temperatures (below 90°K) have rather sharp absorption bands at 41·4 and 34·5 kK, respectively. It has not yet been elucidated whether perfect crystals of the silver halides would also show the weak tail of absorption, reaching down to 25 kK, which is responsible for the photographic effect.

Finally, it may be mentioned that the trihalide ions have bands with ϵ exceeding 10,000, Cl_3^- at 45 kK, Br_3^- at 37·6 kK and I_3^- at 28·0 and 34·5 kK. These bands often disturb the measurement of absorption spectra of halide complexes, since it may be necessary to stabilize the complexes of oxidizing central ions ($PdCl_6^{2-}$ with Cl_2, $IrBr_6^{2-}$ with Br_2) with the free halogens, reacting with their anions to form the trihalide complexes, and since acidic iodide solutions oxidize in air to form strongly coloured, brown tri-iodide ions. The much weaker absorption of the halogen molecules will be discussed in the chapter on chemical bonding, as also the transitions in O_2. Several ions, such as NO_3^-, NO_2^- and SO_3^{2-} have weak bands at rather low wave numbers (33·2, 28·2 and 33·3 kK) in addition to the high bands around 50 kK. The former band makes it impossible to use nitrate and nitric acid solutions of moderate or high concentration as solvents for spectroscopic measurements, even in the near ultra-violet. The nitrite and sulphite ions, containing N(III) and S(IV), belong to the same class as Sn(II) and Pb(II) discussed above though the lone pair can no longer be described to any reasonable extent as an s-orbital. However, stereochemistry exhibits interesting analogies, as shall be discussed in the next chapter.

The *co-operative effects of more metal atoms* produce some of the most spectacular colours known in inorganic chemistry. Several elements, when present with *more than one oxidation number*, may show much stronger absorption bands in the visible than the pure compounds with one oxidation number. This phenomenon has mainly been studied in solid compounds, such as the mixtures of CeO_2 and UO_2 (prepared by ignition of the co-precipitated hydroxides or by decomposition of the nitrates) which are dark blue, while CeO_2 is pale yellow (traces of Pr, probably forming Pr(IV) or (V), gives it a red-brown colour) and UO_2 is red-brown. Also the intermediate oxidation numbers of uranium (as the dark olive-green U_3O_8) have much darker oxides than those interpolated between UO_2 and the yellow UO_3. A well-known case is the almost white ferrous hydroxide, oxidized by traces of air to dark green intermediates, while the complete oxidation leads to the much less coloured red-brown $Fe(OH)_3$. Weyl (1951) pointed out that the dark colour of many minerals is caused by the simultaneous presence of Fe(II) and Fe(III). While pure $Fe(CN)_6^{4-}$ is colourless (it decomposes slowly in aqueous solution to the yellow $Fe(CN)_5(H_2O)^{3-}$) and $Fe(CN)_6^{3-}$ is tomato-red, the mixed ferroferricyanides are the dark blue pigments such as Prussian and Turnbull's blue. Many ferrocyanides of the heavy metals are insoluble precipitates, having strong colours not agreeing with the principle of *additivity of ionic colours*, as found in most salts. Thus, there seems to exist binuclear complexes* such as the brown $CuFe(CN)_6^{2-}$ or even the dark red $UO_2Fe(CN)_6^{2-}$. Another deviation from the addition of ionic colours is the sky-blue precipitate of nickel (II) cobalt (III) cyanide, formed by apple-green $Ni(H_2O)_6^{2+}$ and colourless $Co(CN)_6^{3-}$ (sometimes also containing impurities of yellow $Co(CN)_5(H_2O)^{2-}$). Though this might correspond to an anomalously increased Δ of the Ni(II) hexa-aquo ion, it is more an example of a *co-operative complex* as discussed on p. 197 for $Ni(CN)_2,NH_3,C_6H_6$, probably having nitrogen ends of cyanide groups as ligands for the nickel ion. Such co-operative complexes are even known in aqueous solution. Thus, Werner prepared addition compounds of Ag^+ with complexes such as $Coen_2Cl_2^+$ and $Coen_2(NCS)_2^+$, not giving immediate reactions for Ag^+ (precipitate with Cl^-, etc.). Schäffer (1959) demonstrated that Δ is much larger in the yellow $Cren_2(NCS)_2Hg^{3+}$ than in the red $Cren_2(NCS)_2^+$, and that the cherry-red $Cr(NCS)_6^{3-}$ gives a yellow precipitate with an excess of mercuric perchlorate. Waggener, Mattern and Cartledge (1959) report a similar increase ~ 5 per cent of Δ of $Co(NH_3)_5$ NCS^{2+} and *cis*-$Coen_2(NCS)_2^+$ with Hg^{2+}. The dark violet crystals of $Hg(SCN)_4Co$ with HgS_4 and CoN_4 tetrahedra were recently studied by Cotton, Goodgame, Goodgame and Sacco.

In aqueous solution not only are partly colloidal solutions of mixed oxidation states known, as the dark blue molybdenum (V, VI) and wolfram (V, VI), but also dark coloured chloro complexes of mixed oxidation numbers are observed in concentrated hydrochloric acid. Thus, Davidson *et al.* (1949–1951) studied copper (I, II), tin (II, IV), antimony (III, V), and the present writer (1957) titanium (III, IV). These complexes are not known as the predominating species in any given solution, but are inferred from absorption in the visible, which does not obey Beer's law, but is proportional to the product of the concentrations of the metal with the low and the

---

* cf. also the recent study of Co(III), Fe(II), Fe(III) cyanides by Haim and Wilmarth (1961).

high oxidation number. The present writer (1959) has also studied the spectra of green $N\{Ir(SO_4)_2H_2O\}_3^{4-}$ and the red-brown $N\{Ir(SO_4)_2OH\}_3^{7-}$, prepared by Delépine in 1909. These ions are now recognized to contain one Ir(III) and two Ir(IV) bound to the central nitrogen atom. They are diamagnetic and were previously supposed to be Ir(III) complexes, but they can be quantitatively reduced to a pale yellow complex with three Ir(III) by $V(H_2O)_6^{2+}$. The absorption spectra of these complexes of Ir(III, IV, IV) have bands at wave numbers as low as 15 kK, lower than expected for a mononuclear complex—even of a rather reducing ligand such as N^{3-}. Probably, the transitions are connected with the formation of M.O. from the 2p(z)-orbital of the nitrogen with some of the d-electrons from the iridium atom. Co-operative effects of colour in solution are also found for metal ions in the same oxidation state, such as the green, highly polymeric form of Pu(IV), as compared to the greyish yellow monomeric aquo ion, while it is not certain that the dark brown U(IV) polymers do not involve a slight oxidation (at least, pale green $U(OH)_4$ oxidizes to a brown colour before the formation of yellow uranates).

Yamada and Tsuchida (1956) have studied the absorption spectra of many crystalline salts, showing co-operative effects. When the *dichroism in polarized light* is investigated, it is often found by these authors that the light is preferably absorbed with the electric vector oriented in a given direction, relative to the crystal axes and thus relative to the axes of the microsymmetry of a given metal atom. Some of the more spectacular effects are caused by *bonding between adjacent metal atoms* in the crystal, often stabilizing the solid material to a great extent. Thus, the red precipitate of nickel (II) dimethylglyoximate has a special absorption above 17 kK in one of the directions of the polarized light, which does not occur in the yellow solutions (in $CHCl_3$, etc.), which have the first, rather weak shoulder at 20 kK and high bands at 27 kK and higher wave numbers. It is also known that solid platinum (II) tetracyanides often are strongly coloured ($BaPt(CN)_4$, $4H_2O$ is yellow, $MgPt(CN)_4$, $7H_2O$ orange-red, etc.) and fluorescent, while their aqueous solutions are invariably colourless, only showing bands in the ultra-violet (Table 22). Actually, the solids show strong absorption features above 22 kK.

Yamada and Tsuchida (1956) also studied another surprising deviation from the principle of addition of ionic colours, the green modification of *Magnus' salt* $[Pt(NH_3)_4][PtCl_4]$, precipitating from the mixed solutions of colourless $Pt(NH_3)_4{}^{2+}$ and strawberry-red $PtCl_4{}^{2-}$. There exists also a green $[Pt(NH_3)_4][PtBr_4]$. In all these cases, and in nickel (II) dimethylglyoximate, the X-ray crystallographic study has shown that the planar entities are piled in infinite "sandwich" structures with bonding between adjacent platinum or nickel atoms.

In other cases, the interaction between the metal atoms may go over *anion bridges*. Thus, Yamada, Shimura and Tsuchida (1953) studied the brown Co(III) peroxo complex $[(NH_3)_5CoOOCo(NH_3)_5]^{4+}$ and the green Co(III, IV) peroxo (or Co(III, III) superoxo) complex $[(NH_3)_5CoOOCo(NH_3)_5]^{5+}$ in crystals and observed a special band at 15·3 and 15·0 kK, respectively. The compound $Pt(NH_3)_2Br_3$ was shown by Brosset (1948) by X-ray analysis to be $[Pt(NH_3)_2Br_2]$ $[Pt(NH_3)_2Br_4]$, consisting of infinite chains, alternatively of the planar $Pt(NH_3)_2Br_2$ and the octahedral $Pt(NH_3)_2Br_4$. Similar results are valid for $[Pd(NH_3)_2Cl_2][Pd(NH_3)_2Cl_4]$, where the isolated Pd(IV) complex is very unstable. These substances are all strongly coloured; thus $[PtenCl_2][PtenCl_4]$ has a strong absorption edge above 19 kK. Similar compounds with infinite chains, consisting of $ClM_1ClM_2ClM_1ClM_2Cl$ . . . (where M_1 and M_2 are two different metals), are known such as $Cs_2Ag(I)Au(III)Cl_6$, the chloride bridges from $AgCl_2{}^-$ ions being perpendicular to the square-planar $AuCl_4{}^-$.

The square-planar d^8-systems seem specifically apt to form these co-operative complexes, with strong absorption bands not found in the monomeric entities. If only two d^8-metal ions interacted, we might explain the phenomenon by formation of M.O. by linear combinations of the empty p(z) and the filled $d(z^2 - \frac{1}{2}x^2 - \frac{1}{2}y^2)$ orbitals, directed along the z-axis. Since the infinite chain probably has a centre of inversion in each central ion, we cannot accept this explanation (except by action of configuration intermixing, due to the large K-integral between the two orbitals), but must find refuge in the empty d(t3)- or s-orbitals for M.O. construction.

We shall not here discuss in detail the *chemical bonding in metals*, which is a much too complicated subject for this book. The common properties of high electric conductivity, metallic reflection of light, etc., are undoubtedly caused by the presence of mobile electron clouds, surrounding the metal ions. The existence of amalgams of $NH_4{}^+$ and $N(CH_3)_4{}^+$ and the paramagnetic behaviour of rare-earth metals, suggest strongly that the ions with positive charge are actually present, and that it is reasonable to enquire about the number of partly filled shell and core electrons left on the ions. Thus, many properties of the alkali metals suggest that they have lost one electron entirely to the common fluid of electrons, leaving Na^+, etc., as embedded entities. This can be described by the extensive delocalization of, for example, the 3s-electrons of Na, forming chemical bonds between rather distant atoms. In the alkaline earth metals, such as Mg, the original 3s- and 3p-orbitals must be distributed in such a way that the energy ranges after the intermixing strongly overlap, decomposing the closed shell $3s^2$-structures of the atoms. Metallic bonding may exist in compounds, such as the borides and carbides of many metallic elements, and in the interstitial hydrides, where hydrogen atoms are placed in the lattice (e.g. in the non-Daltonian compound with the approximate composition Pd_2H). The chemical bonding in the complexes mentioned above involves to some extent a one-dimensional metallic bonding along the lines where the distances between the metal atoms are small. The simple behaviour of complexes not showing co-operative effects, as compared to that of the metals, is caused by the large distances generally found between the metal atoms in the former compounds.

When we consider in general the effects of bringing more metal atoms together, we discover discouraging numerical difficulties for the quantum-mechanical treatment of more atoms with a partly filled shell. For instance, if we consider the 120 states of three equivalent d-electrons, as found in Cr(III), two chromium ions produce 14,400 states and q chromium ions 120^q states. We meet again the problem of finding a reasonable approximation of microsymmetry to this function of an immense number of parameters for a macroscopic piece of matter. An interesting case is the interaction of the *total spin* S_1 and S_2 *of two component atoms*, producing a value S of the diatomic entity. Thus, two copper (II) ions have $S_1 = S_2 = \frac{1}{2}$ as monomeric d^9-systems. Most Cu(II) complexes exhibit the paramagnetism (proportional to $S(S + 1)$, see p. 182) expected for $S = \frac{1}{2}$. However, the acetate $Cu(CH_3COO)_2$, H_2O is known to have a relatively small distance between two copper atoms (2·64 Å, as compared to 3·21 Å between Pt-atoms in Magnus' green salt, and 3·25 Å between subsequent Ni-atoms in the dimethylglyoximate). Each Cu(II) is surrounded by four oxygen atoms in a plane (from carboxyl groups), one oxygen atom from a water molecule, and another copper atom on the perpendicular axis. Actually, Cu(II) acetate is diamagnetic at low temperature. Accurate measurements of the variation of the paramagnetism with temperature have shown that the ground state of the diatomic entity has $S = 0$, while the excited state with $S = 1$ has 0·31 kK higher energy. We see that for high temperatures, the paramagnetism will be the same as for the monomeric entities, since the dimeric entity has three states with $S = 1$ and one state with $S = 0$ (p. 67) and thus the probabilities $\frac{3}{4}$ and $\frac{1}{4}$ for having the two values of S, if their energy difference can be neglected:

$$\tfrac{1}{2} \times \tfrac{3}{2} + \tfrac{1}{2} \times \tfrac{3}{2} = \tfrac{3}{4}(1 \times 2) + \tfrac{1}{4}(0 \times 1) \tag{168}$$

We might argue that the *lower energy of the singlet state* is a breakdown of Hund's rule. However, we do not have here a single configuration. Actually, the interaction between the two Cu(II) atoms has created two new M.O., as linear combinations of the two (half-filled) orbitals $(x^2 - y^2)$, one of each copper atom. If we denote these new orbitals by

$$\psi_b = \frac{1}{\sqrt{2}}\{\psi_1 + \psi_2\} \text{ and } \psi_a = \frac{1}{\sqrt{2}}\{\psi_1 - \psi_2\} \tag{169}$$

with energies $E_b < E_a$, the energy levels will be distributed on three M.O. configurations (ψ_b^2 with a singlet state, $\psi_a\psi_b$ with a triplet and a singlet level and ψ_a^2 with a singlet state only) containing six states. It might be argued that the original configuration $\psi_1\psi_2$ contains only four states. However, the two others are supplied by the "charge-transfer" configurations (containing Cu(I), Cu(III) in the example chosen) ψ_1^2 and ψ_2^2. The energy levels are:

$$\left.\begin{array}{l} \text{triplet of } \psi_a\psi_b\text{:} \quad E_a + E_b - K(ab) + J(ab) \\ \text{singlet of } \psi_a\psi_b\text{:} \quad E_a + E_b + K(ab) + J(ab) \end{array}\right\} \tag{170}$$

while the singlets of ψ_a^2 and ψ_b^2 are the eigen values of

$$\left| \begin{array}{cc} 2E_a + J(ab) - E & K(ab) \\ K(ab) & 2E_b + J(ab) - E \end{array} \right| = 0 \tag{171}$$

The energy levels observed of Cu(II) acetate is the lower eigen value of eqn. (171) and the triplet of eqn. (170). The affiliation of the energy levels to the case of two isolated copper (II) atoms can be seen by setting $J(ab) = K(ab)$ (eqn. (169) if ψ_b and ψ_a has no overlapping region in space). In that case, E_a is approximately equal to E_b (completely equivalent in the Helmholz–Wolfsberg model), and the triplet of eqn. (170) and the lowest eigen value of eqn. (171) correspond to the four ordinary states from $\psi_1\psi_2$, while the singlet of eqn. (170) and the higher eigen value of eqn. (171) has the highly excited energy $2K(ab)$ above $\psi_1\psi_2$. Thus, we see from eqn. (171) that for small interactions between ψ_1 and ψ_2, separating E_a and E_b to a small extent only, the configurations of the singlet ground state are highly intermixed, being slightly above 50 per cent of $\psi_b{}^2$ and slightly below 50 per cent of $\psi_a{}^2$. Hence, we cannot expect small deviations from spherical microsymmetry by the perturbation (of linear symmetry) from another distant atom to be sufficient to produce pure M.O. configurations of the linear symmetry. In particular, the observed splitting between the lowest singlet and the triplet is much smaller than the M.O. energy difference $E_a - E_b$ and to a first approximation, it is only

$$(E_a - E_b)^2/2K(ab) \tag{172}$$

Since $K(ab)$ is about half the J-integral of ψ_1 with itself, $(E_a - E_b)$ may very well be 5 kK in Cu(II) acetate.

The phenomenon that the energy levels of half-filled orbital sets of the given microsymmetry are not split to first order by a distortion to a lower symmetry, where the orbitals are no longer equivalent, have been discussed in the case of octahedral complexes by Tanabe and Kamimura (1958). Thus, the energy levels such as $^4\Gamma_2$ and $^2\Gamma_3$ of $\gamma_5{}^3$, or $^3\Gamma_2$ and $^1\Gamma_3$ of $\gamma_5{}^6\gamma_3{}^2$ (use p. 134 for estimate of the nephelauxetic effects) all correspond to pure M.O. configurations containing the same amount of the higher and lower energy orbitals, formed from γ_5 and γ_3, e.g. in tetragonal symmetry. And levels like $^2\Gamma_4$ and $^2\Gamma_5$ of $\gamma_5{}^3$, containing two different M.O. configurations $\gamma_{t4}{}^2\gamma_{t5}$ and γ_{t5} in their two components $^2\Gamma_{t5}$, are forced by the baricentre rule to be nearly 50 per cent intermixed for small energy differences between γ_{t4} and γ_{t5}, since their complementary components, $^2\Gamma_{t2}$ and $^2\Gamma_{t4}$, respectively, both have the pure configuration $\gamma_{t5}{}^2\gamma_{t4}$. Consequently any splitting will be proportional to $(E_4 - E_5)^2$, as in the example eqn. (172).

By analogy with eqn. (168) we always have q times as much paramagnetism (as measured by $S(S + 1)$) with q ions as with one ion, if the energy differences between the different values of S in the composite systems can be neglected. Thus, two chromium (III) ions with $S = \frac{3}{2}$ have sixteen states, producing an average value of $S(S + 1) = 15/2$, twice the monomeric value:

$$
\left.
\begin{array}{cccc}
S = & e \text{ states} & S(S + 1) & eS(S + 1)/16 \\
3 & \text{seven} & 12 & 21/4 \\
2 & \text{five} & 6 & 15/8 \\
1 & \text{three} & 2 & 3/8 \\
0 & \text{one} & 0 & 0
\end{array}
\right\} \qquad (173)
$$

If the different states of total S of the system differ in energy, we have deviations from the usual paramagnetism. Notably, the term *antiferromagnetism*, used to describe the state when the energy level with lowest S has the lowest energy, and *ferromagnetism*, when the level with highest S has the lowest energy. It was seen above that most weak interactions between metal atoms in compounds have the tendency to induce antiferromagnetic behaviour. The formation of M.O., delocalized over more metal atoms, favour the low-energy M.O. configurations with low S. Moreover, diamagnetism can generally be produced by an influence separating one energy level (of an even number of atoms or electrons) sufficiently apart from other energy levels. The opposite phenomenon, ferromagnetism, occurs in a few metals (iron, cobalt, nickel, gadolinium and several alloys) below a characteristic Curie temperature. The reasons for this co-operative effect are too complicated to be reviewed here.

As discussed on p. 182, the effects that are important in the magnetic properties of compounds may be without a great influence on the absorption spectra, since the corresponding energy differences are small, compared to band widths above 1 kK, though large compared to the Boltzmann energy kT at low temperature. Hence, the magnetic anomalies of polynuclear complexes are not usually accompanied by spectacular effects in the colour. However, Schäffer (1958) has found that the blue "basic rhodo" ion $(NH_3)_5CrOCr(NH_3)_5^{4+}$ has an unusual spectrum with four narrow bands in the near ultraviolet, while the red "acidic rhodo" ion $(NH_3)_5Cr(OH)Cr(NH_3)_5^{5+}$ has a quite ordinary appearance. The former ion is almost diamagnetic at low temperature.

CHAPTER 12

CHEMICAL BONDING

THE previous chapters have summarized the known types of absorption bands of complexes and used the molecular orbital (M.O.) theory for the description of the energy levels involved. We shall in the following discussion attempt to show that the M.O. theory is an adequate instrument to use, while the hybridization theory and the valence-bonding description, as conceived by most chemists, are unsatisfactory and cannot be brought into agreement with the absorption spectra.

In the same way as the one-electron system H and the two-electron system He are the only two in spherical symmetry, where the wave function has been calculated accurately from Schrödinger's equation to give the observed energy value of the ground state, the systems H_2^+ and H_2 are the only ones treated exactly in linear symmetry. In M.O. language, the bonding orbital ψ_b and the antibonding orbital ψ_a can be constructed to a first approximation as the normalized linear combinations with positive and negative sign of the two 1s-orbitals, one on each hydrogen atom. We may refine this crude L.C.A.O. model by variation of the radial function of 1s from the value in the gaseous ion. The lowest configuration ψ_b^2 contains in that case only one energy level, $^1\Sigma_g$, consisting of one state (cf. p. 207). We may use all the mathematical apparatus of eqns. (169) to (171) (except for the neglect of the overlap integrals) for describing the configurations, identifying ψ_b with even σ and ψ_a with odd σ of the linear symmetry. Notably, we have a configuration intermixing according to eqn. (171), leading to a strong mixing of σ_g^2 and σ_u^2 (g = even, u = odd) for large internuclear distances between the two hydrogen nuclei. It is sometimes stated in textbooks that M.O. theory is unsatisfactory, because it predicts the hydrogen molecule to have 50 per cent probability of dissociating into H^- and H^+ with so much higher energy, 104 kK, than the other possible dissociation products H and H. Actually, this sentence should be restated as: the assumption of *pure* M.O. configurations is unsatisfactory at large internuclear distances, because the ground state goes smoothly over to two neutral hydrogen atoms according to eqn. (171). On the other hand, the situation near to the equilibrium value of the interprotonic distance, 0·742 Å, corresponds to a wave function, which is nearly as pure (∼98 per cent) a configuration σ_g^2 as $1s^2$ of the ground state of He. The rest of the wave function consists of many small contributions from excited σ_g^2, from σ_u^2, π_g^2, π_u^2, etc. The correlation effects discussed (p. 44) have been successfully described in the helium atom by use of a *correlated wave function*

$$\Psi = \psi_1(1s)\psi_2(1s)\{1 + \alpha r_{12}\} \tag{174}$$

explicitly containing a constant α multiplied by the distance r_{12} between the two electrons. In the same way, the ground state of the hydrogen molecule can also be described to a high degree of accuracy by equ. (174), choosing a suitable M.O. as $\psi_1 = \psi_2$. However, in systems containing more than two electrons, it is practically impossible to introduce correlation effects similar to equ. (174), and we must rely on the simpler concept of pure M.O. configurations and their possible intermixing.

When James and Coolidge (1933) completed their calculations on Ψ of the ground state of H_2, the chemists extrapolated these calculations and thought that probably any covalent bond could be explained in the same way as an electron pair common to two atoms. This is a somewhat optimistic extrapolation, and actually, the study of absorption spectra teaches us that the freedom inherent in M.O. theory of delocalization over more atoms in a rather arbitrary way is necessary for a useful classification, while the substitution of mere electron pairs for the classical valence lines does not describe the energy levels in an adequate way.

We must emphasize the fundamental differences in M.O. theory between what we *can* do in a significant way, and what we *cannot* do. A list of some opposed cases of this type can be given:

It is *possible*	It is *not possible*
(1) To use M.O. configurations as a classification of the energy levels.	(1) To believe that the actual wave functions are antisymmetrical Slater-determinants.
(2) To use the high microsymmetry as an approxima-	(2) To use group theory in the strict sense, since the actual

tion, involving the group-theoretical quantum numbers γ_n for orbitals and Γ_n for the energy levels.

(3) In the cases of most molecules outside the transition groups to talk about a closed-shell structure of the M.O. configuration of the ground state, since it is usually a considerable distance away from the first excited electronic level, as characteristic for the inert gases in atomic spectroscopy.

(4) To consider the M.O. as delocalized over two or more atoms. In the L.C.A.O. description, the coefficients of linear combination may assume any values between 0 and 100 per cent. However, this is probably a rather rough approximation, if the M.O. in each atom is not allowed to deviate somewhat from the shape of any individual orbital in atomic spectroscopy. We must especially arrange for the radial function to adjust to a fractional "effective ionic charge".

(5) To classify transition group complexes according to a definite number of electrons in the partly filled shell. Hence, we may use the names d^q and f^q from atomic spectroscopy, though the M.O.'s, containing the q

macrosymmetry is always too low.

(3) To indicate the number of electrons in each atom as an integer (which is already difficult because no sharp frontiers between the atoms exist in molecules, though the heavier atoms have a strongly concentrated core) and to think about a molecule as composed of atoms or ions with a definite charge.

(4) To consider the bond order of each bond and the bond number of a given atom as necessarily being an integer. It is customary to think about the continuous progression from electrovalent to covalent character in a given bond as the variation of one of the coefficients in L.C.A.O. from 0 to 50 per cent. However, due to the effects of overlap integrals, this is probably a rather crude picture.

(5) To use the electrostatic model, assuming that the partly filled shell is exactly that of the gaseous ion, having the same ionic charge as the oxidation number of the complex. It is probably not even possible to use the

electrons, deviate from $l = 2$ or 3. Only in polynuclear complexes of mixed oxidation numbers, q does not seem well defined in each central atom. It is possible to define oxidation numbers from q as $Z - q - Z_{core}$, where Z is the atomic number and Z_{core} the number of electrons in the central ion closed shells.

(6) To use orbital energy differences, introduced by the deviation of the microsymmetry from the spherical one, for description of the transitions in the partly filled shell. Notably, Δ in octahedral complexes varies in the spectrochemical series in a very regular way as function of the central ion and the ligands, and it is often possible with high accuracy to interpolate Δ of a new complex from other known values.

(7) To obtain a very fine agreement of the spectra with the Tanabe–Sugano diagram by assuming the expanded radial function model for octahedral d^n-complexes. The variation of β, the ratio between the interelectronic repulsion parameters in the complex and in the gaseous ion, is very regular in the nephelauxetic series as func-

expanded radial function model, assuming separability of angular functions for $l = 2$, for describing all properties, e.g. the temperature-independent paramagnetism, of the complex.

(6) To calculate Δ in a numerically significant way, as an electrostatic first-order perturbation energy, or as a combined first-order and second-order perturbation, e.g. in the Wolfsberg–Helmholz model. It is even not possible as yet quantitatively to explain the variation of Δ with ligands and with the central ion.

(7) To calculate the interelectronic repulsion parameters in a given complex, since even in gaseous ions, the variation of radial function in different energy levels of the same configuration and the intermixing of configurations produce term distances some 20 per cent smaller than calculated from the Slater model. It is not either pos-

tion of central ion and the ligands and can often be interpolated with great success. It is qualitatively correct for comparison in a series of isoelectronic systems to consider the interelectronic repulsion parameters as a measure of the average value of the reciprocal distance of the partly filled shell electrons from the central nucleus.

(8) To estimate M.O. energy differences from the electron transfer spectra. In particular, it is possible in octahedral complexes to use the linear microsymmetry, acting on the lone-pair electrons of the ligands, as a preliminary classification. The energy differences between the π-orbitals thus formed and the partly filled shell (γ_5 and γ_3) vary in a very regular way with the ligands, with the oxidation number and with the electron configuration of the central ion. It is even possible to make a linear interpolation formula in these quantities (eqn. 145). The intermediate coupling effects in heavy halide ligands can reasonably well be accounted for.

(9) To classify some M.O. as having a well-defined l to some approximation. In that

sible at the moment to predict the variation in complexes, though the nephelauxetic effect is qualitatively connected with the reducing character of the ligand and the oxidizing character of the central ion. It is expected that the difficulties for calculation of interelectronic repulsion parameters are nevertheless smaller than for calculation of M.O. energy differences.

(8) To calculate the M.O. energies *a priori*. Further, most of the excited configurations of the electron transfer transitions should have a very complicated structure of many energy levels, characterized as $^{2S+1}\Gamma_n$, or in intermediate coupling, as Γ_J. Actually, this structure predicted from group theory, is not resolved in the observed absorption spectra.

(9) To postulate that all M.O.'s to a good approximation can be considered as linear com-

way, not only the internal d^n- and f^n-transitions can be classified, but also $f^n \rightarrow f^{n-1}d$ and $s^2 \rightarrow sp$, and perhaps also $d^n \rightarrow d^{n-1}s$ and $1s \rightarrow nl$ (X-ray absorption) transitions. However, the electron transfer spectra such as filled M.O. $\rightarrow$ d or d $\rightarrow$ empty M.O., or internal ligand transitions, cannot generally be described in this way.

binations of a few n,l-shells of the same or different atoms.

(10) To make the observed energy differences plausible in terms of M.O. energies and interelectronic repulsion parameters. These quantities are not deduced from purely theoretical arguments, but derived from the measured absorption spectra.

(10) To use the variation principle on systems containing more than a few electrons. It is a condition for its validity that the approximate energy is below that of the first excited state of the same S, Γ_n, and this is not the case for any trial wave function known for transition group complexes.

(11) To use the linear microsymmetry of the bond between two atoms for classification of the bonding electrons into σ, π, δ, etc. These electrons may participate in the formation of M.O. extended over the whole molecule, and it cannot be argued that they are evenly shared between the two bonded atoms.

(11) To use the hybridization theory for a description of actual complexes.

All of these features have been discussed in the preceeding chapters. Some difficulties, as mentioned in points (1) and (7), are found universally in chemistry and in atomic spectroscopy; others, such as points (5) and (6), are specific for transition group complexes and have been largely discussed in the previous chapters. A rather fundamental problem for all chemical compounds, not only the transition group complexes, is point (3), that we cannot generally assume the molecules to be composed of atoms or ions with a definite charge though the oxidation number may be a well-defined integer, either from the conventions (p. 3), or for transition groups, point (5) above. The opinion that H_2O is stabilized by an intermixing between the structures $O^{2-}(H^+)_2$, $O^-(H)(H^+)$, and OHH, or that the wave function of CH_4 is a mixture of the numerous states C^{4+} $(H^-)_4$, $C^{3+}(H)(H^-)_3$, $C^{2+}(H)_2(H^+)_2$, $C^{2+}(H^-)(H^+)_3$, . . ., $C^{4-}(H^+)_4$ is expressed as the "resonance between ionic and covalent structures" in the *valency-bond description*.* However, we shall now by analysis of this rather confusing concept show that the M.O. configuration closed-shell picture of ordinary molecules such as H_2O or CH_4 is much more appropriate, but that it is compatible with nearly any overall charge distribution on the atoms in the molecule (apart from considerations of pure symmetry, e.g. that the charge of each hydrogen atom in CH_4 is expected to be the same). In molecules with internuclear distances near to the equilibrium values, the purity of the M.O. configurations has the same order of magnitude, 90–99 per cent, as in atomic spectroscopy, while the contributions from the ionic charge–wave functions discussed above would be in the range 10–30 per cent.

---

* Since every atom or positive ion has an infinite number of states, we always have the formal problem when atoms are assembled together in a chemical compound that a permutation of several infinities of wave functions participate in the resulting wave function. However, from chemical intuition, we have a rather clear idea that a finite number of the lowest states are of main importance in the formation of the complex. Thus, we may regard SO_4^{2-} as consisting of five cores (S^{6+} and four O^{6+}), each having the room for at most eight further, well-bound electrons. In this way, thirty-two electrons are to be distributed in forty available places. However, this simple operation produces $(40!)/(32!)(8!) = 76{,}904{,}685$ states. Since the sulphate ion is colourless and has no excited electronic states within the first 50 kK from the ground state, it is somewhat artificial to think about its ground state being a mixture of all these states, allocating each an integral number of electrons to each of the five atoms.

If we consider the possibilities O, O^- and O^{2-} only for the charge of the oxygen atom, but retain the variation between S^{6+} and S^{2-}, we obtain 12,870 states (of eight electrons in sixteen places). Of these, 1764 are singlet states with $S = 0$. If intermediate coupling is neglected, the ground state of the sulphate ion would be a mixture of the levels $^1\Gamma_1$ found among this number. The values of $M_s = \pm 4$, ± 3, ± 2, ± 1 and 0, are represented by 1, 8^2, 28^2, 56^2 and 70^2 states, respectively.

The first case we examine is the *halogen molecules* F_2, Cl_2, Br_2 and I_2. We shall simplify the discussion to consider only the most loosely bound p-shell of the two halogen atoms. The number of states of such two atoms is 66, viz.

$$
\begin{array}{llll}
(F)(F) & (p^5)(p^5) & 6\times6 = 36 & \\
(F^+)(F^-) & (p^4)(p^6) & 15\times1 = 15 & \quad(175) \\
(F^-)(F^+) & (p^6)(p^4) & 1\times15 = 15 &
\end{array}
$$

including the ionic structures (cf. p. 207 and eqn. 11). In linear symmetry, the p-orbitals combine to form M.O. with well defined $\lambda = 0(\sigma)$ or $\lambda = 1(\pi)$ and parity (even, g) or (odd, u). The spectra of diatomic molecules have demonstrated that the order of orbital energies are*:

$$
\left.
\begin{array}{lcl}
\text{unnormalized } \psi: & e = & \\
z_1 + z_2 & \sigma_u & 2 \\
(x_1 - x_2) \text{ and } (y_1 - y_2) & \pi_g & 4 \\
(x_1 + x_2) \text{ and } (y_1 + y_2) & \pi_u & 4 \\
z_1 - z_2 \quad .. & \sigma_g & 2
\end{array}
\right\} \quad (176)
$$

The p-orbitals x, y, z of each halogen atom 1 and 2 are combined in eqn. (176) taking the linear axis to be the z-axis. The chemical bonding separates the σ_g and σ_u much more than the π_g and π_u, because the orbitals z_1 and z_2 overlap much more than the x- and y-orbitals of different atoms. The bonding orbital σ_g has a density maximum between the two halogens, produced by the positive sign of z_1 and negative sign of z_2 in the internuclear domain. The bonding orbitals π_u have a ring of maximum density, where x_1 overlaps to the greatest extent with x_2. The parity of the bonding σ_g and bonding π_u is different, because different conditions are dictated by the creation of density maxima; the cross-products in ψ^2, viz. $2x_1x_2$ and $-2z_1z_2$, are both positive in the region between the two nuclei.

We have ten electrons to put in the M.O. of eqn. (176), and accordingly two holes, since the sum of the degeneracy numbers e is 12.

---

* Actually, owing to interaction with the 2s-shell, σ_g of eqn. (176) is above π_u, as can be seen from the fact that the groundstate of C_2 is $^1\Sigma_g^+$ of π^4, followed by $^3\Pi_u$ of $\pi^3\sigma_g$ at 0·61 kK higher energy (Ballik and Ramsay, 1959).

We may denote the M.O. configurations by the M.O. of the holes, producing again (of course) sixty-six states:

$$
\begin{array}{lrcl}
\sigma_u{}^2 & 1 & = & 1 \\
\sigma_u\pi_g & 2 \times 4 & = & 8 \\
\sigma_u\pi_u & 2 \times 4 & = & 8 \\
\sigma_u\sigma_g & 2 \times 2 & = & 4 \\
\pi_g{}^2 & 4 \times \tfrac{3}{2} & = & 6 \\
\pi_g\pi_u & 4 \times 4 & = & 16 \\
\pi_g\sigma_g & 4 \times 2 & = & 8 \\
\pi_u{}^2 & 4 \times \tfrac{3}{2} & = & 6 \\
\pi_u\sigma_g & 4 \times 2 & = & 8 \\
\sigma_g{}^2 & 1 & = & 1
\end{array}
\right\} \quad (177)
$$

Of these sixty-six states, the absorption spectrum in the visible and the near ultra-violet is determined by the energy levels $^3\Pi_0$, $^3\Pi_1$, $^3\Pi_2$, and $^1\Pi_1$ formed by the eight states of the first excited configuration, the holes $\sigma_u\pi_g$ of eqn. (177), i.e. $\sigma_g{}^2\pi_u{}^4\pi_g{}^3\sigma_u$. By intermediate coupling effects, the two levels $^3\Pi_1$ and $^1\Pi_1$ with the same $\Omega = 1$ may intermix, but the main part of the intensity is concentrated in the transition to $^1\Pi_u$ from the ground state $^1\Sigma_g$. Though these transitions are Laporte-allowed, the intensities are rather low in the gaseous halogens:

$$
\begin{array}{lcc}
 & \sigma_0 \text{ (kK)} & \epsilon_0 \\
F_2 & 35 \cdot 1 & 6 \\
Cl_2 & 30 \cdot 9 & 60 \\
Br_2 & (20 \cdot 3),\ 24 \cdot 1 & 150 \\
I_2 & (13 \cdot 7),\ (17 \cdot 9),\ 19 \cdot 8 & 800
\end{array}
\right\} \quad (178)
$$

The low intensity in F_2 exemplifies a very interesting case, well-known from the dinuclear transition group complexes: it is not a sufficient condition for a band to have high intensity to be Laporte-allowed, it is also necessary to supply a considerable electric dipole moment of the transition (eqn. 114). It is seen from eqn. (176) that if the orbitals x_1 and x_2 overlap only to a very small extent, then the one-electron transition $\pi_g \to \sigma_u$ does not have any dipole moment, but resembles an internal $p \to p$ (e.g. $x_1 \to z_1$) transition in each halogen atom. Consequently, the increased intensity of the absorption bands of Br_2 and I_2 in eqn. (178) must be connected with either larger overlap between the π-components of each atom or with larger extent of configuration intermixing. Mulliken (1940) discussed how much stronger absorption is expected of the $\sigma_u \to \sigma_g$ transitions in the far ultra-violet.

It is also interesting to note that the light quanta of the absorption maxima, given in eqn. (178), contain sufficient energy for dissociating the molecule completely to two halogen atoms at a large distance. The dissociation energies are $\sim$13 kK for F_2, 20·0 kK for Cl_2, 15·9 for Br_2 and 12·4 kK for I_2. However, the Franck–Condon principle does not allow such a direct optical transition to a state with much larger internuclear distance. In the same way, the dissociation energy of H_2 is 36 kK, and the broad emission band in the ultra-violet, utilized for the production of continuous spectra in hydrogen lamps, is caused by the decay of an excited state without a minimum on its potential curve.

It was one of the early victories of M.O. theory that the triplet ground state (and hence, the paramagnetism) of the *oxygen molecule* was explained. If the eight electrons available from the two 2p-shells are put in the lowest M.O.'s of eqn. (176), the configuration is $\sigma_g^2\pi_u^4\pi_g^2$. This is the simplest chemically well-known molecule having a *partly filled M.O. sub-shell in its ground state*, and consequently states of different total spin S of the ground configuration, viz.:

$$^3\Sigma_g^-, \quad ^1\Delta_g \text{ at } +K(\pi_g, \pi_g'), \text{ and } ^1\Sigma_g^+ \text{ at } +2K(\pi_g, \pi_g') \quad (179)$$

with the excitation energy in the pure M.O. configuration indicated. Hence, O_2 is the simplest example of the phenomenon that has been much elaborated in detail for transition group complexes that the sub-shell configuration may correspond to several energy levels. Actually, the spin-forbidden (and Laporte-forbidden) transitions from $^3\Sigma_g^-$ to $^1\Delta_g$ and $^1\Sigma_g^+$ (the line at 14·6 kK contains a vibrational contribution 1·43 kK) are analogous to the narrow "ruby lines". Since ζ_{2p} is so small in oxygen (0·15 kK), the spin-forbiddenness is very effective, and the later transition was originally discovered as one of the Fraunhofer lines of the solar spectrum, shown to be of terrestrial origin, since its intensity varied at sunset with the thickness of the layer of the earth's atmosphere traversed by the light. The internuclear distance r_L is very much the same for the first three levels (eqn. 179) of O_2, while it is larger for the next two, belonging to the excited configuration $\sigma_g^2\pi_u^3\pi_g^3$:

	r_L (Å)	energy (kK)	energy (kK)	
$^3\Sigma_g^-$	1·2074	0	0	
$^1\Delta_g$	1·2155	7·92	7·9	
$^1\Sigma_g^+$	1·2268	13·20	13·2	(180)
$^3\Sigma_u^+$	1·6	28·7	42	
$^3\Sigma_u^-$	1·604	33·96	50	

where the excitation energy is given for the minima of the potential curves, and also for the internuclear distance equal to r_L of the ground state (as would be observed in the absorption spectrum of a transition group complex).

Moffitt (1951) studied the particular case of O_2 for elucidating his idea of *atoms in molecules* that it is possible to use M.O. configurations as classification and that the chemical bonding can be considered as a small perturbation on the atoms. Hence, it is permissible to borrow information about energy level differences, n,l-orbital energies and ionization potentials from atomic spectroscopy, while the approximate wave functions estimated are only used for calculation of the interatomic perturbation. In that way, Moffitt removed some very large discrepancies with the experimental energies in eqn. (180) present in all the previous quantum-mechanical treatments of O_2. The concept of atoms in molecules has been criticized, e.g. by Pauncz (1958) showing that it is not applicable to H_2. However, this is not quite informative, since the chemical bonding in H_2 is not a small perturbation, while it only stabilizes O_2 0·13 per cent compared to the total binding energies of two oxygen atoms.

In the same way as the sixty-six states of two halogen atoms, two O have 495 states:

$$\left.\begin{array}{llll}
(O)(O) & (p^4)(p^4) & 15 \times 15 = 225 \\
(O^+)(O^-) & (p^3)(p^5) & 20 \times 6 = 120 \\
(O^-)(O^+) & (p^5)(p^3) & 6 \times 20 = 120 \\
(O^{2+})(O^{2-}) & (p^2)(p^6) & 15 \times 1 = 15 \\
(O^{2-})(O^{2+}) & (p^6)(p^2) & 1 \times 15 = 15
\end{array}\right\} (181)$$

The dissociation products of an oxygen molecule have highly different energy. For the free atoms, the ionization energy of O to O^+ is 109 kK and from O^+ to O^{2+} 282 kK; further, the multiplet terms ^1S and ^1D of neutral O have the energy 34 and 16 kK higher than the ground state ^3P. It is obvious that when an oxygen molecule increases its r_L to a large value, the classification in terms of M.O. configurations breaks down, and the classification according to multiplet terms and ionic charges of the two oxygens becomes appropriate. This was called by Moffitt transition from *valency-coupling* to L,S-coupling. For most states, the breakdown of the valency coupling happens for such large values of r_L (an increase by some $\frac{1}{2}$ Å) that the bonding energy is much less than for the bottom of the potential curve. For a group theorist, it is nonsense to say that we go from linear to spherical microsymmetry; for him, two oxygen atoms at a mutual distance of 1 km constitute a system having linear symmetry. We shall use the concepts of microsymmetry in another way, that we consider relatively insignificant such perturbations of lower symmetry, producing smaller orbital energy differences than the spreading of energy levels by interelectronic repulsion parameters of a given configuration. The tetragonal and rhombic perturbations of Jahn–Teller-stable, octahedral complexes with different ligands, discussed on p. 125, exemplify such insignificant perturbations, we still have octahedral microsymmetry to a large extent. Of course there will be intermediate cases, and the whole discussion of weak-field or strong-field determinants (eqn. 85) is actually a question as to whether the diagonal elements of energy are most conveniently given in spherical or octahedral microsymmetry with subsequent second-order perturbations from either the octahedral ligand field or the interelectronic repulsion, respectively.

Two nitrogen atoms, considered in the manner of eqn. (181), have 924 states. Actually, the ground state $^1\Sigma_g{}^+$ of the molecule N_2 has a nearly pure M.O. configuration $\sigma_g{}^2\pi_u{}^4$, and the first excited state, of $\sigma_g{}^2\pi_u{}^3\pi_g$, appears at very high energy, about 60 kK. It would not at all be convenient to describe the ground state of N_2 as a linear combination of many of the 924 states of two nitrogen atoms, while it is expedient to consider it as a closed-shell M.O. configuration. For forming this closed-shell, some promotion energy (e.g. of interelectronic repulsion, giving ^2P and ^2D of $2p^3$ higher energy than ^4S) must be subtracted from the bonding energy delivered by the M.O. energy differences. Mulliken proposed the name *valence-state energy* for these contributions from a mixture of multiplet terms, not corresponding to a stationary state of the gaseous atom. The diagonal elements of interelectronic repulsion in strong-field determinants, are actually valence-state energies of pure sub-shell configurations $\gamma_5{}^a\gamma_3{}^b$. The occurrence of high-spin rather than low-spin behaviour among many octahedral d^4, d^5, d^6, d^7-complexes indicates the importance of interelectronic repulsion parameters for disturbing the order of energy levels, predicted by simple M.O. theory.

While it is undoubtedly difficult or impossible to indicate the boundaries of each atom in a molecule, it is on the other hand demonstrated by X-ray crystallography that each reasonably heavy nucleus (at least with Z above 10) is surrounded by an individual core, a dense electron cloud. As a first approximation to the chemical bonding, both in complexes and in every other type of molecule, it is reasonable to maintain a picture resembling Fajans' "quanticule" model: we neglect the possibility of delocalization in M.O. of a certain number of inner-shell electrons near to each nucleus and we consider the rest of the electrons as a more mobile pool, available for formation of M.O. to the whole community of atoms in the molecule. As the hard cores of spherical microsymmetry, it would be prudent to reserve no electrons on hydrogen nuclei, $1s^2$ on Li, Be, B, C, N, O, F; $1s^2 2s^2 2p^6$ on Na, Mg, . . .; [A] in K, Ca, the first transition group, etc. We then have the full freedom of disposing the remaining eight electrons in CH_4, NH_3, H_2O, and HF in four more or less polarized, covalent bonds, leaving the charge of the C, N, O, F atoms more or less negative. So far we are only interested in the electron density distribution in a given state, e.g. the ground state, this is quite a satisfactory picture. It is only when we excite one electron from such a pool, consisting of bonding electrons, to another orbital that this simple model fails seriously, because it can tell us neither the shape of the hole left in the bonding orbitals nor the shape, i.e. delocalization, of the excited orbital. This failure in a simple molecule as CH_4 is of no interest to us, because we are not generally concerned with the absorption bands in the far ultraviolet region of methane spectra. If these were as well known as the absorption bands of transition group complexes, we would have recognized that the pool of the loosest electrons consists of the orbital γ_1 and the three equivalent γ_5 (eqn. 132). If these were only to be considered as linear combinations of σ-bonding electron pairs, they would have the same energy; but the properties of the carbon atom (or any other nucleus with Z some 5 to 9) in the centre of the molecule influence γ_1 and γ_5 to a different extent, since among the orbitals in spherical symmetry, 2s can only mix with γ_1 and 2p with γ_5 of tetrahedral symmetry. We should therefore observe transitions from γ_1 and γ_5 to a given acceptor orbital at different wave numbers.

If we do not consider absorption spectra, but only electron densities in one definite state, the simple *cores and pool* picture above makes interesting conclusions possible. Gillespie and Nyholm (1957) elaborate Sidgwick's idea that most stereochemical facts can be explained by the assumption that lone pairs and bonding pairs, connecting the central atom with a ligand, demand about the same angle in space, and that all such electron pairs repel each other. It is obvious that saturated molecules such as CH_4 and NH_4^+ assume regular tetrahedral symmetry with almost any type of repulsion between the hydrogen atoms and bonding between the carbon and the hydrogen atoms. It is much more remarkable that NH_3 and H_2O have about the angles, corresponding to the regular tetrahedron. In a sense, the one and two lone pairs need the same space as the hydrogen atom.* This concept was dramatically verified by the resolution into optically active antipodes of sulphur-containing cations $R_1R_2R_3S^+$ (Pope, Peachey and Smiles, 1900) and of sulphoxides R_1R_2SO (Kenyon and Phillips, 1925). In these cases, the lone pair occupies a long-lived position as a fourth ligand of the tetrahedrally co-ordinated sulphur atom. Gillespie and Nyholm remark that N(III), S(IV), Cl(V), Se(IV), Br(V), Sn(II), Sb(III), Te(IV), I(V), Pb(II) compounds such as NO_2^-, SO_3^{2-}, ClO_3^-, $TeCl_4$, IO_3^-, PbO, etc., generally behave as if a lone pair prevented a ligand from occupying a position in a regular co-ordination sphere. Even the strange shapes of some molecules, such as ClF_3, may be interpreted as a five-co-ordinated (bipyramidal) chlorine atom, having three ligands and two lone pairs. (cf. Table 34 in chapter 16).

From a group-theoretical point of view, this concept of *equivalent orbitals* (in the ground state of a molecule, as discussed by Lennard-Jones (1949) by a transformation of the M.O. with well-

---

* However, a difficulty common to both M.O. theory and the hybridization theory is to explain why the lone pairs assume hemihedric positions relative to the central atom rather than being symmetrically distributed on both sides of the nucleus. If the p-shell was very isolated in energy, relative to the other shells (and especially the corresponding s-shell), M.O. theory would explain the angles 90° in ML_3 and ML_2 in essentially the same way as the hybridization theory by participation of three or two orthogonal p-orbitals as the most effective binding. However, this argument is only valid if a given p-orbital cannot bind two ligands twice as well as one ligand. The asymmetry introduced in the electron density, being closer to the proton in the examples H_2O and NH_3, may be explained partly with the intermixing with even orbitals and partly by the overlap conditions, making the overlap with two ligands less effective than twice the overlap with one. Actually, the potential maximum for transforming NH_3 to a planar molecule is very low, some 2 kK, while the order of magnitude (according to information, kindly given by Professor Børge Bak) of energy necessary to stretch H_2O to a linear molecule can be estimated to 10 kK. Consequently, the directive valence forces are rather weak compared to the component of spherical symmetry (since the planation of NH_3 affords only 2 per cent and the linearization of H_2O some 12 per cent of the total bonding energy from the atoms N, O and H). However, the electrostatic forces of the protonic repulsion act obviously in the opposite way of the directive valence forces. cf. Walsh (1953).

defined γ_n to linear combinations) is inconvenient, when the properties of the central ion interact with the bonding orbitals of the pool. Thus, a single σ-bonding, localized orbital, say σ_5, can be constructed from the σ-orbitals of an octahedral complex in Table 4 by undoing the delocalization:

$$\sigma_5 = \{(\text{even } \gamma_1) + \sqrt{(2)} (\text{even } \gamma_3 a) + \sqrt{(3)} (\text{odd } \gamma_4 a)\}/\sqrt{6} \quad (182)$$

which might seem a highly artificial way of writing the orbital σ_5. However, it is remembered that σ_5 has not a group-theoretical quantum number γ_n, and hence, it is not invariant under the symmetry-operations of O_h. Actually, by rotating the co-ordinate system with a physically equivalent result, it is possible to transform σ_5 into any of the other $\sigma_1, \sigma_2, \ldots, \sigma_6$. It is only the combined density of all six σ-orbitals, which have the total symmetry even Γ_1 and which is invariant for such a transformation. Hence, we cannot expect to observe absorption bands, where one electron is excited to another orbital, leaving σ_5 of eqn. (182) singly occupied. The hole created in the σ-pool will have one of the symmetries even γ_1, even γ_3, or odd γ_4, and its energy depends on the interaction of the s-, d- and p-orbitals of the central ion with these sets of orbitals.

It is true that a definite one of the three odd γ_4 orbitals, e.g. $\sigma_1 - \sigma_2$, is also directed in a certain way in space and can be transformed to $\sigma_3 - \sigma_4$ or $\sigma_5 - \sigma_6$ by a change of names of the co-ordinate axes. However, these changes occur inside the set of three equivalent orbitals. This is also true in the less obvious case of the even γ_3 orbitals in Table 4, which are also transformed into a linear combination of two equivalent orbitals. In both cases, we have irreducible representations in the group-theoretical language, while the set of all six σ orbitals corresponds to a reducible representation, since it breaks down to three independent, irreducible representations, by application of all the possible symmetry operations.

Pauling's hybridization theory is a special case of L.C.A.O. description assuming equivalent M.O.'s consisting of a mixture of n,l-orbitals of the central atom, having the appropriate symmetries γ_n (and parity), with the σ- or π-bonding orbitals of the ligands. It is assumed that the hybridized n,l-orbitals of the central atom have approximately the same energy and the same radial function, because the linear combinations are written as an addition of the angular functions, multiplied by a common radial function. This is a rather dangerous assumption already in atoms such as C, N, O, where 2s- and 2p-orbitals have quite different energy in the neutral atoms according to atomic spectroscopy. Consequently, it is necessary to calculate valence state energies for the hybridization states, symbolizing a promotion energy to the somewhat unusual situation of the central ion wave function. The demand of equal coefficients of L.C.A.O., i.e. $1/\sqrt{2}$ if overlap integrals are neglected, in the hybridized M.O. from central atom and ligand orbitals is generally weakened by the admission of intermixed "ionic structures", where the M.O.'s are concentrated on one atom alone.

With the two amendments, of valence state energy and admixed ionic structures, it is generally possible to use this hybridization language for molecules of the short-period elements, though a superiority in intelligibility over M.O. theory has never been proved except by the psychological effect of habit on the mind of many organic chemists. There can never occur a superiority in content, since the hybridization theory is by definition a (very) special case of M.O. theory.

However, in the transition group complexes, the assumptions are impossible to maintain in the case of d^2sp^3-hybridization in octahedral complexes. Primarily, we can imagine an effective ionic charge between 0 and $+1$, where (in the first transition group) 3d- and 4s-electrons have the same orbital energy, but we cannot get this hybrid to coincide with the highly excited 4p-electrons. Secondly, we can compare the predictions of d^2sp^3-hybridization with the absorption spectra. It is equivalent to identical energies of the lower set of σ-bonding orbitals even γ_1, even γ_3, and odd γ_4 of eqn. (54), which should be filled. Hybridization-minded chemists do not usually speculate on the corresponding, antibonding orbitals, but these six orbitals should also have the same energy, lying above the sub-shell even γ_5. Hence, d^3-complexes such as Cr(III), and generally all covalent d^q-complexes with q < 6, should exhibit strong, Laporte-allowed transitions in two wave number ranges (odd $\gamma_4 \rightarrow$ even γ_5 and even $\gamma_5 \rightarrow$ antibonding odd γ_4) at nearly the same position as the Laporte-forbidden transitions usually described by ligand field theory. Moreover, d^6-complexes such as Co(III) should have one range of Laporte-allowed transitions (even $\gamma_5 \rightarrow$ antibonding odd γ_4) at nearly the same wave number as the Laporte-forbidden transitions. The observation of two weak, Laporte-forbidden bands in many d^3- and low-spin d^6-complexes at much lower wave numbers than the Laporte-allowed bands disproves the presence of pure d^2sp^3-hybridization. It might now be argued that the actual complexes exhibit an intermediate behaviour of, say, x per cent of the electrostatic model and $(100 - x)$ per cent of the hybridization model. This cannot be directly disproved, but it seems a rather artificial restriction on the M.O. theory to assume that the three parameters, describing the L.C.A.O. coefficients for each of the three sets of orbitals should necessarily be identical. It would not be surprising, if we were near to the 50 per cent mixing of even γ_1 from ligands and 4s, and of odd γ_4 and 4p, at the same time as we have only 10 per cent mixing of even γ_3 from the ligands and 3d. This would be described as a sp^3-hybridization with superposed electrostatic bonding. It is remarked that these parameters of L.C.A.O. do not necessarily change abruptly by going from high-spin to low-spin behaviour, this is only a question of the relative size of the energy difference Δ between the two equivalent, antibonding even γ_3-orbitals, and the three equivalent even γ_5, as compared to the interelectronic repulsion parameters (eqn. 76). It may be questioned why text-book writers so much prefer d^2sp^3-hybridization in $Cr(H_2O)_6^{3+}$ and SF_6 for sp^3-hybridization, though they do not cite experimental evidence. Probably it is felt that q orbitals are necessary for the efficient binding of q ligands. This is a fallacious argument in view of M.O. theory, since the bonding capacity of each orbital symmetry type γ_n must be judged independently.

For securing the hybridization theory in transition group complexes, it has been customary to assume *promotion of excess electrons to high* (5s and 4p) *orbitals*. This is directly against experience and is caused by the neglect of the two σ-antibonding even γ_3-orbitals as an available accommodation for the excess electrons. Many authors have tried to distinguish between "outer-shell" hybridization $4s4p^34d^2$ and "inner-shell" hybridization $3d^24s4p^3$. However, the M.O. description implies in a certain sense that we have always one of the two types of hybridization. It is a somewhat ambiguous question, whether the bonding even γ_3 is nearest to be 4d and the antibonding even γ_3 nearest to be 3d, or the reverse. The spatial extension suggests that the bonding orbital should be named 4d and the antibonding 3d, while the number of nodes suggest the opposite order, 3d and 4d, and the energy scale perhaps might induce the imaginative notation 2d and 3d. Actually, in M.O. theory, it is clear that the electrons are filled in the M.O.'s according to increasing energy, and that in any complex, the bonding even

γ_3 is filled in the ground state, and the next even γ_3 takes care of the "excess" electrons, if present. The difference between complexes, where the 3d-electrons do not participate in the covalent bonding and complexes, where the 3d-electrons are involved in such bonding, is a qualitive difference between the behaviour of both the bonding and antibonding even γ_3-orbitals, smoothly evolving delocalization effects.

Magnusson justly remarked in his review paper (1957) that ligand field effects, connected with the presence of a partly filled shell, are always small corrections, superimposed on the much greater effects of chemical bonding in general in the complex. To a certain extent, we may divide the "pool" of non-core electrons in a transition group complex into a comparatively more localized and stable set of bonding electrons and a quite loosely bound set of partly filled shell electrons, occupying the orbitals formed by distortion and expansion of the shells (3d, 4d, 5d, 4f, 5f) from spherical symmetry. This is analogous to the division of the pool of electrons in aromatic compounds into a skeleton of fixed σ-electrons and a mobile fluid of π-electrons. In both cases, the distinction is only approximate, and much less strictly valid than, for example, the distinction from the true core electrons. On the other hand, most transition group complexes can be fairly well represented as having a number of electron pairs, equal to the co-ordination number, being the lone pairs of the ligands, directed against the central ion and being deformed by polarization and partly invading the core domain of the central ion. In this respect, the transition group complex has the same structure as molecules outside the transition groups, viz. $Zn(NH_3)_4^{2+}$ or $PbCl_6^{2-}$. Superimposed on this skeleton, delivering the effective core field of the complex, are the electrons of the partly filled shell. The spectrochemical series is connected with the energy difference between γ_5-electrons, interacting to a lesser extent with this core field, and γ_3, being directly antibonding to the σ-bonding components of the skeleton. The nephelauxetic effect is caused by the change of this core field, as compared to the central field of the corresponding gaseous ion. It is emphasized that the concept of dative electron pair bonds from the ligands and the concept of "the corresponding gaseous ion" are not too well founded in the general M.O. theory. We are not accustomed to think about NH_3^+ or H_2O^+ as possible ligands, while Cl and Br might seem possible, as compared to the usual formulation NH_3, H_2O, Cl^-, and Br^-.

However, we could not *a priori* know whether $Co(NH_3)_6^{3+}$ was best formulated as $Co^{3+}(NH_3)_6$ or $Co^{3-}(NH_3^+)_6$, and the qualitatively very attractive idea of electroneutrality in the complexes expressed the judgment that the truth might be somewhere between, $Co^0(NH_3^{0.5+})_6$ or $Co^{1.2+}(NH_3^{0.3+})_6$. The freedom of this variation, maintaining the nearly pure M.O. configuration of the ground state, has the same reason as in H_2O or CH_4, that the electron-pair bonds may have asymmetric delocalization on different atoms. The property of $Co(NH_3)_6^{3+}$ of having the unique qualification of the oxidation number $+3$ of cobalt (when we discuss the absorption spectrum) is that six of the mobile electrons occur in the complex, as in $3d^6$ of gaseous Co^{3+}. In the excited states of electron transfer bands of, for example, $Co(NH_3)_5Cl^{2+}$, there occur seven mobile electrons, and a hole is created in the more firmly bound orbitals. On the other hand, the topic of ligand field theory is the possible energy levels of the six mobile electrons remaining in a fairly constant core field of the other electrons and the nuclei.

If we use the "Aufbau" principle of atomic spectroscopy of filling electrons into orbitals according to increasing energy, we may also understand the behaviour of a molecule or a crystal, containing many nuclei and, especially, why the firmly bound core electrons are so localized and why the most loosely bound electrons tend to delocalize. If we fix the nuclei at their equilibrium positions in the finished system (it will need violent forces to begin with before all electrons are placed), each nucleus is surrounded by an electric potential hole, like a deeply carved whirl which occurs in a bath tub during emptying, being the function $-Ze^2/r$. When we allow the first electrons to enter the system, they are captured by the nuclei with highest Z, forming nearly hydrogen-like, doubly occupied 1s-orbitals with the energy $-Z^2$ ry. In the finished particle of matter, this is possible for nearly all nuclei, except for the shallowest potential hole of hydrogen with $Z = 1$. The following electrons admitted to the system are captured in nearly hydrogen-like 2s- and 2p-orbitals. However, their energy is not as low as expected, $-Z^2/4$ ry, since the 1s-electrons have a screening effect. As is known from X-ray spectra, this is nearly equivalent to a value of Z some 2 to 3 units lower than the atomic number. For the following electrons, we must not only reckon with the *internal screening*, the positive contribution $+e^2/r$ to the central field from an electron completely at smaller values of r than that considered, but also the

external screening, since the potential is $+e^2/r_0$ for points with
$r < r_0$ inside a sphere, completely surrounded in spherical symmetry
by one electron, and having the radius r_0. Hence, when the Z_{core}
electrons have been added to a given nucleus, the electric potential
outside the electron cloud will be given by $-(Z - Z_{core})e^2/r$ (neglecting external screening from other atoms, nearly cancelling due to
the simultaneous presence of nuclei and electrons), and we talk about
the ionic charge $Z - Z_{core}$. During the whole of this process, the orbitals of the strongly bound electrons have rearranged to some extent,
corresponding to the self-consistent behaviour in the central field.

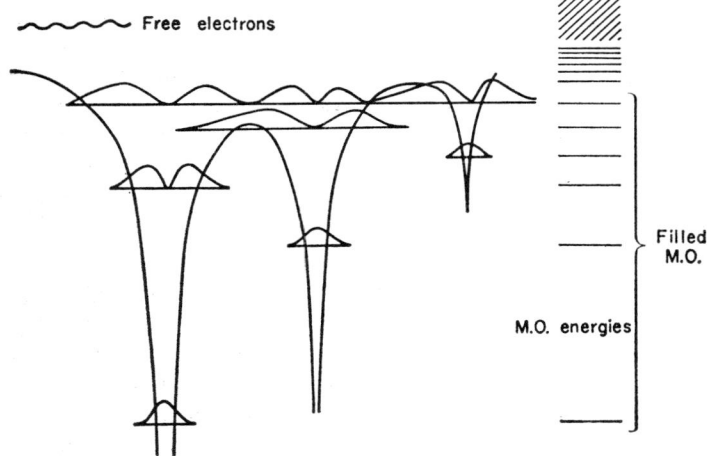

Fig. 28. Qualitative representation of the spatial extent of M.O. as
function of the potential around three atoms with different nuclear
charge. It is seen how the M.O.'s become more delocalized at the top of
the figure, just below the continuous wave functions of free electrons.

When we have filled the core electrons into the system, we are left
with the boundary regions between the nuclear potential holes (now
even more sharply descending in the middle near each nucleus,
where alone the value $-Ze^2/r$ is approached). These boundary
regions do not exhibit the spherical microsymmetry of the atoms, but
have queer forms determined by the superposition of potential holes
from more atoms. Here is the place for the delocalized orbitals. If
these M.O.'s end up by being confined in entities, consisting of some
few atoms, we have the usual situation of a molecule or a monomeric
complex. On the other hand, if the most loosely bound orbitals

Q

are delocalized in long, linear channels, we have the situation described in the section of co-operative effects in strongly coloured,
low-spin d^8-complexes (p. 204). If the loosest M.O. are delocalized
over the whole three-dimensional lattice of a solid, we may have
different situations: if the ground state of the system has a fairly
pure closed-shell M.O. configuration, the result is generally a strongly
coloured compound without well-defined oxidation number of the
transition group atoms possibly present; and the excited states of
the visible absorption correspond to a rearrangement of the pool
of loosely bound orbitals. If the excited states are very low, and
distributed almost continuously from 0 to 100 kK, and if consequently
the M.O. configuration of the ground state is not defined, we have
metallic bonding. Many compounds such as PbS or InSb are semi-
conducting which does not demand a very large number of available conduction electrons, as found in the genuine metals such
as Na and Ag. It is customary for solid-state physicists to talk about
these broad ranges of M.O. energies as *energy bands* of the crystal,
assuming the normal behaviour of monomeric entities, when the
highest filled energy band is completely filled and there is a considerable distance to the following, empty energy band. The intermediate cases of semi-metallic behaviour occurs when two energy
bands, one filled and one empty, are slightly overlapping, or, in the
case of many semiconductors, some few holes exist in the highest,
filled energy band or a few electrons occur in the lowest otherwise
empty band. The properties of the solids may in this case depend
very strongly on the presence of very small traces of other elements.
Finally, the genuine metal is said to have a highest energy band, which
is approximately half-filled. This language of solid-state theory is
often useful, though it may be misleading. For instance, it is thought
by many physicists that the half widths of absorption bands, e.g.
of Tl(I) in KCl, are simply measuring the width of the excited
energy band. This is a somewhat inadequate description, since the
vibrational amplitude of the ground state is of obvious importance
to the action of the Franck–Condon principle on the excited
state. When it is stated that the absorption bands of KBr at
53 kK is caused by the excitation of an electron from the highest
filled energy band to the lowest empty one, it is not quite as
informative as the analogous statement in M.O. theory (cf. pp.
167 and 241).

For many chemists, there is a fundamental difference between the discussion of electronic levels, as applied to visible and ultra-violet spectra, and of the vibrational levels, as applied to infra-red spectroscopy. It is felt that M.O. theory is appropriate for discussion of the large energy differences, while it is not related to the concepts of force constants, anharmonicity, etc. However, it must be realized that the potential curves of molecules are simply the variation of electronic energy levels as function of the internuclear distances. When we discussed the construction of a particle of matter above, we placed the nuclei at their equilibria positions. If one nucleus is slightly removed from such a position, its repulsion from the other nuclei and the energy of all the electron clouds undergo such changes as to increase the energy slightly. If we expand the energy increase in a Taylor's series, we will not obtain any term of first degree in the nuclear displacement, since the necessary condition for the original position to be a point of equilibrium is that the differential quotient of the energy with respect to the nuclear displacement is zero. Hence, the first, non-vanishing term is of second degree. There is no mysterious tendency towards Hooke's law forces and a special applicability of the harmonic oscillator as a model for nuclear motion in a molecule. However, the demand of orthogonality between the different M.O.'s and the proportionality of kinetic energy of an electron cloud in quantum mechanics to its average radius in minus second power for a given shape, produce other types of resistance towards compression than are usually met with in an electrostatic, classical model of matter. This low compressibility, due to core-repulsion, is not a specific property of molecules; also isolated atoms should collapse in classical electrodynamics.

The macrosymmetry of *crystal lattices* is studied in crystallography. There is no immediate relation between the type of chemical bonding in a compound and its crystal lattice. Notably, the continuous development between electrovalent and covalent bonding does not necessarily correspond to changes of the crystal structure. Thus, it cannot be argued that the chemical bonding is strictly analogous in NaCl, KBr, AgF, AgCl, AgBr, MgO, CaO, BaO, MgS, CaS, BaS, CdO, NiO, MnO, CrN, PbS, PbSe, VC, ZrN and ZrC, though they all crystallize in the NaCl lattice.

Among the elements, most metals crystallize in dense packings, where the co-ordination number of each atom* is 12. This is the

---

* Often with cubic symmetry, the neighbours at the middle of each of the twelve edges of a cube.

situation expected when the bonding between equivalent atoms is not specifically directed. The number 12 is somewhat related to the fact that the surface of a sphere is $4\pi r^2$, i.e. $12 \cdot 566 r^2$. However, some metals have rather low macrosymmetries, e.g. Mn. The non-metallic elements prefer in general two other types of crystal structures: the *molecular lattice*, where the molecules such as H_2, N_2, Cl_2, Br_2, P_4, S_8 are conserved as distinct entities in the lattice, and the *covalent atomic lattice* (as distinguished from solid inert gases, which form atomic lattices somewhat analogous to the molecular lattices), e.g. of C in diamond, where each carbon atom is bound by four electron pair bonds to four other carbon atoms. This example shows the bulkiness of structure induced, when the chemical bonding is strongly directive.

A similar bulky structure can be caused by *hydrogen bonding*. This type of chemical bonding has not yet been discussed in this book, because it is without great importance to the visible absorption spectra, though it has spectacular effects on the vibrational levels studied in the infra-red. While the usual covalent bonds evolve an energy between 15 and 40 kK (see eqn. 184) per electron pair in-volved, the presence of a hydrogen nucleus between two lone pairs, each from different atoms (symmetrically between the lone pairs or mainly embedded in one of them) corresponds generally to a stabi-lization around 1 kK only. However, the hydrogen bonding effects are sufficiently strong to be of importance in solution, and also in the biochemical processes in organisms. The structure of ice is known to be a special molecular lattice with each oxygen atom sur-rounded by four protons, two belonging to the water molecule itself and the other two, to other water molecules, in which the protons are attracted by the two lone pairs. In consequence, the specific density is very low, only about half that of the neutral atoms. Hydrogen bonding is also the cause of the high boiling point of water and the large heat of vaporization, $3 \cdot 6$ kK per molecule, corresponding to the breakdown of two hydrogen bonds.

Some examples of lattices of binary compounds, AB, are:

CsCl-lattice ($r_A/r_B > 0 \cdot 73$). Each A surrounded by eight B in a cube (and vice versa), e.g. CsCl, CsBr, CsI, TlCl, TlBr, TlI.

NaCl-lattice ($0 \cdot 73 > r_A/r_B > 0 \cdot 41$). Each A surrounded by six B in a regular octahedron (and vice versa), e.g. see above.

Wurtzite-lattice ($0 \cdot 41 > r_A/r_B$). Each A surrounded by four B in a regular tetrahedron (and vice versa), e.g. CuCl, CuBr, CuI, AgI,

BeO, ZnO, ZnS, CdS, GaP, GaAs, GaSb, AlN, Si, diamond. Sometimes, the macroscopic symmetry is only hexagonal (and not cubic, as in zincblende).

Some examples of binary compounds, AB_2, are:

CaF_2-lattice ($r_A/r_B > 0.73$). Each A surrounded by eight B in a cube, e.g. CaF_2, BaF_2, CdF_2, HgF_2, ZrO_2, CeO_2, ThO_2 and "antifluorites" Li_2O, Na_2O, K_2O. Rutile-lattice ($0.73 > r_A/r_B > 0.41$). Each A surrounded by six B in a regular octahedron, e.g. MgF_2, ZnF_2, TiO_2, MnO_2, SnO_2, PbO_2.

Cristobalite-lattice ($0.41 > r_A/r_B$). Each A surrounded by four B in a regular tetrahedron, e.g. BeF_2, SiO_2, Cu_2O, Ag_2O.

There is one common case of ternary compounds, ABC_3:

Perovskite-lattice: Each B surrounded by six C in a regular octahedron, e.g. $BaTiO_3$, $KMgF_3$, $KNiF_3$, $CsNiCl_3$, $LaAlO_3$, $RbIO_3$. Small distortions often occur (Orgel's "ferroelectric rattling", 1958).

It has been evident for many years that *isomorphous* crystals (i.e. having the same lattice symmetry) may be formed by compounds of different oxidation numbers of the components, e.g. of $CaCO_3$ and $NaNO_3$. It is seen from the list given above that it is more the relative number of the atoms of which the molecule is composed, as in the perovskites ABC_3, than the oxidation number which determines the lattice. Actually, some metallic alloys with Daltonian composition such as AgMg, AgZn, CuBe, CuPd, NiAl and TlBi crystallize in CsCl lattice and SnAs and SnTe in NaCl lattice.

Undoubtedly, the pronounced difference in electrical charge among the constituents of ionic lattices usually prevents more than eight anions from being bound to a cation (though anhydrous UCl_3 is nine-co-ordinated with U(III)). However, the most important factor, governing the occurrence of different lattices, seems simply to be the relative size of the cations and the anions. The limiting values of the ratio r_A/r_B between the ionic radii A and B given above are calculated from the model of *close packing of hard spheres*. They agree well with the ionic radii usually assumed. It is well known that within some limits, a constant increase $+x$ could be assumed of all cationic radii, if all anionic radii were assumed to be $-x$ smaller, since the quantity observed by X-ray crystallographers is actually the size of *unit cells*, and for atoms inside the unit cells, the distance between the centres of the electron clouds of atoms practically coincides with the nuclei. The unit cell is the entity, which reproduces the crystal by translation an integral number of times along each of the three axes of the (oblique or orthogonal) co-ordinate systems. *Covalent radii* are often mentioned in the literature; they are derived from similar principles as the ionic radii, but utilize the existence of homonuclear diatomic molecules, and the equilibrium internuclear distance is equal to twice the covalent radius:

$$\begin{array}{lllll} \text{covalent radius (Å)} & \text{F } 0\cdot64 & \text{Cl } 0\cdot99 & \text{Br } 1\cdot14 & \text{I } 1\cdot33 \\ \text{ionic radius (Å)} & 1\cdot36 & 1\cdot81 & 1\cdot95 & 2\cdot16 \end{array} \right\} \quad (183)$$

Though the ionic radii are $0\cdot8$ Å larger than the covalent radii, it must not be concluded that the electrovalent bonds are on average $0\cdot8$ Å longer than the covalent bonds. Actually, while the corresponding ionic radii, e.g. of Cr(III) or Fe(III), are around $0\cdot6$ Å, the covalent radii of these metals are given about $1\cdot4$ Å in the literature. Consequently, the internuclear distances calculated are nearly the same; it is not possible in general to estimate the extent of covalent or electrovalent bonding in a complex from the measured internuclear distances.

In most cases, it may be assumed that strong covalent bonds cannot be much longer than usual, while electrovalent bonds are weakened less by stretching; however, we cannot draw the opposite conclusion from a moderate value of the bond length. Genuine salts stabilize their electrovalent bonds by compression, with the result that no strong covalent bonding is ever observed for r_L ranging from a value (at high pressure) well below the equilibrium value at zero pressure up to high values.

The *bond-lengths* (or more specifically the radii of eqn. (183)) and the *bond-energies* are often mentioned as *additive properties of bonds*. It is surprising to workers who study quantum mechanics that these additive properties so often persist, though no direct theoretical explanation can be given. Actually, these bond energies, attributed to single bonds, are (in kK):

$$\begin{array}{llll} \text{H—H } 36 & \text{C—C } 29 & \text{C—O } 29 & \text{F—F } 13 \\ \text{H—F } 47 & \text{C—H } 35 & \text{C—F } 38 & \text{F—Cl } 21 \\ \text{H—Cl } 36 & \text{O—H } 39 & \text{C—Cl } 27 & \text{Cl—Cl } 20 \end{array} \right\} (184)$$

Most organic compounds have heats of formation (from the gaseous atoms) which agree with eqn. (184) within 1 per cent. In the case of multiple bonds, attempts have often been made to define *bond-order from additive properties*. Thus, the bond lengths are especially useful for estimating the bond order (the calibration scale is 1·33 in graphite, 1·50 in benzene, 2·00 in ethylene, etc., for carbon–carbon bonds) and the bond energies for the detection of resonance stabilization by delocalization of the more mobile M.O. (thus, benzene is stabilized 14 kK, compared to the total value 444 kK calculated from eqn. (184) for C_6H_6, if this molecule is assumed to contain three ordinary double bonds with the $C=C$ bonding energy 49 kK).

The bond energies in salts do not form a regular set of additive properties, as expressed in eqn. (184) for the covalent bonds. The heats of formation of crystalline salts such as NaCl or KBr are comparable to the heats of formation of gaseous molecules of organic compounds. The situation is quite different when we consider the heats of formation of the dissociated ions, such as Na^+ and Cl^-, in aqueous solution. It will be argued below that the hydration of the gaseous ion Na^+ produces some 30 kK, and of highly charged ions much more, some 360 kK of Fe^{3+}. However, only a part of the energy needed for producing these ions in the gaseous state is received back in this process.

In the Haber–Born cycle description of ionic salts, it is assumed that the electrostatic *Madelung energy*

$$az^2e^2/(r_A + r_B) = az^2(115 \text{ kK})/(r_A + r_B \text{ Å}) \qquad (185)$$

is the most important part of the bonding energy of the crystal. z is the smallest ionic charge of those occurring in the lattice, r_A and r_B the ionic radii of AB, and a is the Madelung constant, being 1·748 for the NaCl lattice, 1·762 for CsCl, 5·039 for CaF_2 and 4·82 for the rutile lattice. Actually, eqn. (185) gives values which are some 10 per cent larger than the observed heats of formation of crystalline alkali halides (and many other salts) from gaseous ions. This discrepancy is easily explained by the potential of the repulsive forces between the closed-shell ions and by other, secondary effects.

Since the Madelung energy (being 72 kK for NaCl and 385 kK for MgO) is proportional to z^2 and increases with small ionic radii, large ionic charges can have quite large amounts of Madelung energy available for forming the highly charged ions in gaseous phase. Thus, it was argued that TiO_2, MnO_2 and PbO_2 all contain the ions Ti^{4+}, Mn^{4+} and Pb^{4+}, since eqn. (185) applies to these compounds, crystallizing in rutile lattice. On the other hand, Rabinowitch and Thilo (1930) proved that CCl_4 may hardly contain C^{4+}

and four Cl⁻, as once suggested by Kossel. However, as discussed, p. 212, there is no serious reason for believing that these compounds are either entirely electrovalent or covalent, and the approximate validity of eqn. (185) only shows that quite high ionic charges may defy the electroneutrality principle, if the structure is sufficiently

TABLE 24. HYDRATION ENTHALPIES OF ELEMENTS IN kK

	$-\Delta H$ atoms $H_h = 0$	$-\Delta H$ atoms $H_h = 40$	$-\Delta H$ ions $H_h = 40$	r_{ion} A	$-\Delta H_{ion} \cdot r_{ion}/z^2$
Li(I)	39·8	−0·2	43·2	0·67	29
F(−I)	34·2	+74·2	45	1·36	61
Na(I)	29·2	−10·8	30·6	0·98	30
Mg(II)	51·4	−28·6	153·4	0·78	30
Al(III)	70·2	−49·8	380	0·57	19·5
S(−II)	15·2	+95·2	—	1·74	—
Cl(−I)	24·1	+64·1	34·1	1·81	62
K(I)	28·6	−11·4	23·5	1·33	31
Ca(II)	62·0	−18)0	126·8	1·06	33·5
Sc(III)	85·0	−35·0	321	0·83	29·5
Mn(II)	42·4	−37·6	148·6	0·91	34
Fe(II)	41·3	−38·7	156	0·83	32·5
Fe(III)	38·0	−82·0	359	0·67	21·5
Co(II)	42·5	−37·5	165	0·82	34
Ni(II)	41·0	−39·0	169	0·78	33
Cu(I)	24·1	−15·9	46·1	0·96	44·5
Cu(II)	23·2	−56·8	169·4	0·80	34
Zn(II)	23·8	−56·2	163·8	0·83	34
Ga(III)	40·9	−79·1	383	0·62	21·5
Rb(I)	27·8	−12·2	21·4	1·48	31·5
Sr(II)	59·5	−20·5	113·8	1·27	36
Y(III)	95·2	−24·8	291	1·06	34
Ag(I)	33·2	−6·8	54·0	1·13	61
Cd(II)	16	−64	145	1·03	37
In(III)	31	−88	337	0·92	28
I(−I)	16·2	+56·2	29·7	2·20	65
Ba(II)	59·9	−20·1	102·0	1·43	36·5
La(III)	92·7	−27·3	275	1·22	37
Pr(III)	91·4	−28·6	—	—	—
Nd(III)	90·6	−29·4	—	—	—
Sm(III)	90·2	−29·8	—	—	—
Yb(III)	87·1	−32·9	—	—	—
Hg(II)	−10	−90	145	1·12	41
Tl(I)	15·1	−24·9	24·1	1·49	36
Tl(III)	6·1	−115·9	338	1·05	32
Pb(II)	16·1	−63·9	116	1·32	38
U(III)	87·1	−32·9	—	—	—
U(IV)	95·9	−64·1	—	—	—

compact with small ionic radii. The latter condition limits the possibilities of electrovalent bonding, because the anions expand with increasing negative charge and overlap considerably with the small cations.

It cannot be argued that the hydration energies of the individual ions in aqueous solution should rather be calculated for the gaseous ions than for the gaseous atoms. The reason is that the bond energies of eqn. (184) can arbitrarily be made much larger by consideration of gaseous ions. Thus, the heat of formation of gaseous HCl from H^+ and Cl^- in the gaseous state would be 116 kK rather than 36 kK from neutral H and Cl atoms.

Table 24 gives the *hydration energies* (or more exactly, the enthalpies, cf. eqn. (193). The value $-\Delta H$ is given, since conventionally, a positive ΔH of a reaction corresponds to an absorption of heat) calculated from the very accurate thermodynamic data available. However, these must be corrected for ΔH, here called H_h, of the reaction

$$\tfrac{1}{2}H_2 + H_2O_{aq} = H_3O^+{}_{aq} + e^-{}_{vacuo} \tag{186}$$

Thermodynamicists generally assume, for convenience, that $H_h = 0$, as is ΔG of the hydrogen electrode.

Unfortunately, a certain experimental device has introduced the term ionization potential for the *ionization energies* of gaseous atoms. In aqueous solution, we may still talk about ionization energies of the chemical entities. Probably, the absolute value of ΔG of the reaction eqn. (186) cannot be defined as a consistent quantity, since it involves the transfer of an electric charge over the boundary between two phases, the aqueous solution and the empty space at large distance. But the energy and heat differences of eqn. (186) are valid quantities, though they are not accessible for a very accurate measurement.

Arguments* can be given for H_h being between 35 and 40 kK. Latimer (1955) remarked that the hydration heats of anions, calculated for the gaseous ions ($-\Delta H_{ions}$ in Table 24) are very close to the electrostatic energy of a charged sphere $z^2e^2/2r = z^2 \times 58$ kK/Å, if the choice $H_h = 40$ kK is made. This corresponds to the dielectric effects of water being sufficiently large for removing all the electrical energy of the anionic charge. This is not the case with the cations; according to Latimer, r should be increased some 0·85 Å above the

* See the review by Randles (1960).

usually adopted ionic radii. This could hardly be interpreted as a physical model of the aquo ions. Table 24 shows the values of the proportionality constant of $-\Delta H_{ions} \times r_{ion}/z^2$ which is seen to vary between 19·5 and 44·5 kK for the positive ions.

The reason why $-\Delta H_{atoms}$ (with $H_h = 40$ kK) can be strongly negative for the cations in aqueous solution, and thus the bond

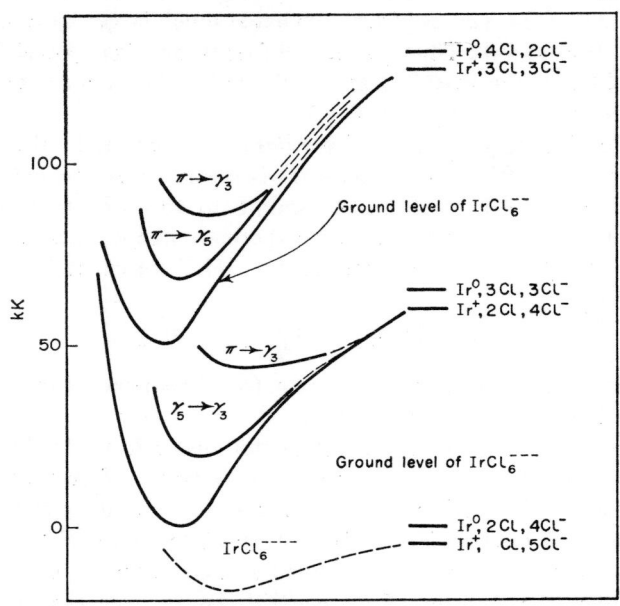

FIG. 29. The potential curves of Ir(IV), Ir(III) and a hypothetical Ir(II) hexachloro complex as function of the internuclear distance r_L (assuming regular, octahedral symmetry except in the limiting cases of infinite separation). It is seen that the wave numbers of both the ligand field bands $\gamma_5 \to \gamma_3$ and the electron transfer bands $\pi \to \gamma_5$ and $\pi \to \gamma_3$ decrease strongly with increasing r_L. The relative position of the set of Ir(IV) curves, where the system contains one electron less than in the set of Ir(III) curves, does not immediately tell us anything about the relative stability of $IrCl_6^{2-}$ and $IrCl_6^{3-}$, since electrons are not available with zero energy in ordinary, chemical systems (cf. p. 239). It is interesting to compare with Fig. 20, p. 165, the potential curves for a NaCl crystal.

energies also negative, though there is no tendency to separation of the free metal atoms, is of course that electrons are not available with zero energy, but only with much lower energies, some -40 to -70

kK, depending on the electrochemical potential U of the solution (eqn. 197), corrected for H_h. However, we cannot directly compare these values with orbital energies obtained from optical transitions, since the equilibrium reactions involved in the establishment of electrochemical potentials allow the rearrangement of internuclear distances in the oxidized and the reduced form of a given complex, contrary to the Franck–Condon principle for an optical transition.

Fig. 29 gives a qualitative picture of the potential curves of the systems $IrCl_6^{3-}$ and $IrCl_6^{2-}$, which are known to have the standard oxidation potential $U_0 = +0.962$ V $= 7.8$ kK, giving an energy difference between the minima of the two potential curves (occurring at different r_L) approximately 47.8 kK with our assumption. It is seen that the ground state A of $IrCl_6^{3-}$ does not necessarily have the lowest energy obtainable, if a free variation in the number of electrons is allowed. Thus, the ground state of a hypothetical iridium (II) complex $IrCl_6^{4-}$ might well occur some 20 kK below the minimum of A. However, the usual high energy (some 50 kK) required to provide an electron, prevents the latter complex from being observed; hydrogen evolution would occur with water, since the electron affinity of the latter substance, forming OH and H_2, is some 45 kK per electron.

We should not expect the potential curve of $IrCl_6^{2-}$ to increase for large r_L to the very high energy, some 800 kK above the minimum, of Ir^{4+} (nearly in the same state as in gaseous systems) and 6 Cl^-_{aq}. As shown in Fig. 29, the complex dissociates rather to a neutral Ir atom, four neutral Cl, and two Cl^-_{aq}, demanding only some 100 kK energy. The similar phenomenon is well known for gaseous alkali halide molecules, where the ground state dissociates, for example, to neutral Na and Cl atoms, though the ground state for values of r_L near to the equilibrium value rather corresponds to an ion-pair Na^+Cl^-.

Dunn (1959) supported the principle of electroneutrality by pointing out that a trivalent aquo ion as $Fe(H_2O)_6^{3+}$ is expected to dissociate at high r_L to Fe^+, $2H_2O^+$, and $4H_2O$, because the ionization energy of H_2O, forming H_2O^+, is only 101 kK *in vacuo*, smaller than the ionization energies of gaseous Fe^+ (130 kK) and Fe^{2+} (248 kK). This argument, though qualitatively convincing, is not a certain indication of the properties of the ground state (cf. gaseous NaCl). Of course, even the dissociation to H_2O^+ does not represent the lowest energy path possible, since the heat of $4H_2O^+$, reacting with $2H_2O$ to form $4H_3O^+$ and one O_2 molecule can be estimated to 200 kK. Consequently, the bond energy of each metal–water co-ordination can only very roughly be estimated as some 10 to 15 kK for the "cheapest" dissociation products. The energy required to remove one water molecule from hexa-aquo ion is much smaller than, say, a sixth of the total dissociation energy, since it does not need to carry positive charge away during this unsymmetric distortion of the complex. We are actually here asking for the activation energy (see eqn. 223) of the special reaction of forming a penta-aquo ion without rearrangement of the five water ligands and with a completely empty, sixth co-ordination place. The latter energy can be very low and is probably only high in cases such as $Cr(H_2O)_6^{3+}$, where it has been measured by Taube using tracer methods (the oxygen isotope 18) to be some 8 kK. However, this activation energy is sufficient for a half-life of the water-exchange, amounting to several days at room temperature.

As we did not obtain very definite bond energies for complexes in solution, the most invariant property seems to be the (somewhat fictitious) orbital energies, derived from the absorption spectra. If we combine eqn. (145) and other information given above on the electron transfer spectra with the ligand field theory, we get a rather uniform energy scale in kK of the most loosely bound electrons in different complexes:

F	-100	$Re(IV)\gamma_5$	-36	$Re(IV)\gamma_3Cl_6$	-8
H_2O	-87	$Os(IV)\gamma_5$	-43	$Os(IV)\gamma_3Cl_6$	-15
Cl	-72	$Ir(IV)\gamma_5$	-50	$Ir(IV)\gamma_3Cl_6$	-22
Br	-66	$Pt(IV)\gamma_5$	-57	$Pt(IV)\gamma_3Cl_6$	-29
I	-56	$Pb(IV)\gamma_5$	-85	$Pb(IV)\gamma_3Cl_6$	-57
$Mo(III)\gamma_5$	-28	$Mo(III)\gamma_3Cl_6$	-8	$Pb(IV)\gamma_1Cl_6$	-40
$Ru(III)\gamma_5$	-42	$Ru(III)\gamma_3Cl_6$	-22	$Sn(IV)\gamma_1Cl_6$	-27
$Rh(III)\gamma_5$	-49	$Rh(III)\gamma_3Cl_6$	-29	$Rh(III)\gamma_3(NH_3)_6$	-15
$Ir(III)\gamma_5$	-42	$Ir(III)\gamma_3Cl_6$	-18	$Ir(III)\gamma_3(NH_3)_6$	-1
$Ru(IV)\gamma_5$	-50				

$$(187)$$

It can successfully be assumed that the most loosely bound electrons of the ligands have the same energy to this approximation (cf. the discussion pp. 166 and 154). Further, it is assumed in eqn. (187) that the energy of the γ_5-orbitals is independent of the ligands for a given central ion, while the variation of Δ is expressed by the energy of γ_3, being a function of both central ions and ligands. The zero point of the energy scale in eqn. (187) is estimated from different sources of evidence, especially H_h discussed above.

However, two serious objections arise against the objective validity of eqn. (187) as a list of M.O. energies. The high absorption bands of crystalline alkali metal halides imply in all cases an energy of the acceptor orbital -10 kK, assuming eqn. (187). This might seem quite reasonable, if it was not for the electrons of the partly filled shell of trivalent ions (cf. the γ_5-electrons of Mo(III) and Ir(III)) which should be excited to this acceptor orbital at even lower wave numbers than the ligand field bands, caused by $\gamma_5 \rightarrow \gamma_3$ transitions. This disagreement with experimental results is probably caused by the neglect of interelectronic repulsion energy. The electron transfer and ligand field transitions, forming the basis for eqn. (187), correspond to a certain standard amount of interelectronic repulsion energy, which is smaller than that corresponding to the highly varied electron density in a process where an electron is removed to a distant acceptor orbital. The actual expression for the energy difference between the excited configuration $Ca^{n-1}b^{m+1}$ above the ground configuration $Ca^n b^m$ (where C are the filled orbitals common to both states) contains the contribution

$$-(n - 1)J(aa) + mJ(bb) + (n - m - 1)J(ab) \qquad (188)$$

neglecting the K-integrals and assuming all J-integrals of one of each of the three types to be identical. A physical explanation of the rather complicated eqn. (188) is that it is the difference between the ionization energy $a^n b^m \rightarrow a^{n-1}b^m$ (giving the contribution $-(n - 1)J(aa) - mJ(ab)$) and the electron affinity $a^n b^m \rightarrow a^n b^{m+1}$ (contributing $mJ(bb) + nJ(ab)$) corrected by the quantity $-J(ab)$. The latter quantity* represents the attraction between the hole a and the new electron b in the excited state. It probably amounts to some -30 kK in ordinary electron transfer levels (such as $\pi \rightarrow d$) and is much nearer 0 kK in the Ir–py bands discussed below. In a case like NiO (crystallizing in NaCl lattice), it would be expected on account of primitive ideas of M.O. energies that since the γ_3 electrons on two different nickel atoms have the same energy, it would need much less energy

---

* If both a and b occur in the ground configuration, the excitation energy of eqn. (188) equals the difference between the ionization energy of a and of b, corrected by the quantity $J(bb) - J(ab)$. Actually, $J(bb)$ is the difference between the ionization energy ($b^m \rightarrow b^{m-1}$) and electron affinity ($b^m \rightarrow b^{m+1}$) of b.

to transfer an electron from one Ni to another, than the usual electron transfer from the filled M.O. mainly localized on the oxygens to the half-filled γ_3. However, this argument is completely false, since the creation of Ni^+ and Ni^{3+} in NiO would need much energy. Eqn. (188) informs us that the energy difference between the ionization energy and the electron affinity of two different nickel ions would be expected to be J(3d, 3d), a quantity $\sim$80 kK .This demonstrates a complete breakdown of the energy band description of NiO, since this substance would be said to contain a half-filled energy band. Actually, it shows no metallic conductance, when pure (as show the mixed oxidation numbers in black $NiO_{1.01}$ or in Fe_3O_4).

We may doubt that the ionization energies of halogen electrons and central atom γ_5 and γ_3 electrons, if they were observed, would exhibit the differences displayed in eqn. (187), as derived from electron transfer spectra. We would expect the quantity J(3d, 3d) $-$ J(3d, π) from the footnote on p. 241 to be some 30 kK, while the similar J(5d, 5d) $-$ J(5d, π) would be much smaller, due to the comparable spatial extension of the 5d- and the π-orbitals. Hence, in the first transition group, the 3d-electrons are much nearer to have the same energy as the ligand electrons, e.g. in Fe(III) and Cu(II) chloride complexes, than estimated directly ($\sim$30 kK) from the electron transfer spectra.

Another surprising phenomenon is the high wave number of "inverted" electron transfer bands. The energy of the most loosely bound electrons of pyridine must be about -80 kK in eqn. (187). Since the pyridine $\pi \to \pi^*$ bands occur at 38 kK (p. 191), we might expect the empty π^*-orbital to have the energy -42 kK. Hence, the Ir–py bands (p. 193) were expected to occur at 0 kK, since the γ_5-electrons of Ir(III) were also estimated to have the energy -42 kK. Actually, the Ir–py bands have wave numbers between 30 and 35 kK (Table 20). The explanation must be that the excited state of py in $\pi \to \pi^*$ has a much lower energy, when the number of electrons in the molecule is not changed, than the excited state of $\gamma_5 \to \pi^*$, where the π^* electron is moving in the core field of a neutral pyridine molecule. We may express this fact either by saying that the π^*-energy is highly different in the two cases (as is the d-shell energy in gaseous ions with different charge, for example) or by assuming a large attraction $\sim$20 kK between the π^* electron and the π-hole with the arguments given above. In this way, we may generally treat the problem that transitions from the partly filled shell of the central ion to empty orbitals of the ligands were expected from eqn. (187) to occur at much smaller wave numbers than the observed values.

This is a general restriction on the absolute validity of such concepts as M.O. energies. A strict analysis shows that the interelectronic repulsion is a very large quantity, screening away nearly all the attraction by the nuclei for the most loosely bound electrons. In most cases, by far the largest part of it can be abstracted away in an effective one-electron operator quantity, the contribution of the "core field" to the M.O. energy. However, when describing an excitation in the visible or ultra-violet absorption spectrum as a transition of an electron from one M.O. to another, we cannot expect the variation of the core field to be small, compared to the other contributions to the relatively small energy difference. Hence, there are not only numerical difficulties for considering M.O. energies as anything else than a convenient classification of energy levels, but there are also severe theoretical difficulties. A special aspect of this difficulty is our ambiguous use of the term "M.O. energy", either for denoting the ionization energy (varying strongly with the ionic charge) or taking the energy of n electrons in a shell as the sum of n times the M.O. energy plus $n(n - 1)/2$ times a representative expression for the interelectronic repulsion (cf. eqn. 149). The latter point of view, which is obviously a closer approximation to the truth, leaves us with many undetermined parameters for the description of a given ionization process, since the M.O. energy and the interelectronic repulsion integrals cannot, unfortunately, be expected not to vary. According to Koopman's theorem, however, the ionization energy is given to a certain approximation as minus the M.O. energy in a self-consistent field of the original system, independent of the fact that the M.O. energies may be strongly changed in the ionized system. It is remarkable that it is the contribution of the interelectronic repulsion to the ionization energy $a^n \rightarrow a^{n-1}$, viz. $-(n - 1)J(aa)$, and not the average value per electron in the configuration a^n, viz. $-(n - 1)J(aa)/2$, which is of importance for the actual M.O. energy. In general, the sum of all one-electron ionization energies ($Ca^n \rightarrow Ca^{n-1}$) of a system (with opposite sign) would be larger, $-E(f_1) + 2E(g_{12})$, than the actual energy $-E(f_1) + E(g_{12})$, because the two-electron operator quantities g_{12} would be taken twice. The one-electron operator quantities f_1 are the nuclear attraction and the kinetic energy of the electrons.

CHAPTER 13

DETERMINATION OF COMPLEX SPECIES IN SOLUTION AND THEIR FORMATION CONSTANTS

SEVERAL chemical problems are involved, if we want information on absorption spectra of complexes in solution. The principal question is the composition of the coloured species, and the relative concentration of different species, if a mixture occurs. We may either have solutions, establishing their equilibria very quickly at room temperature, or *robust complexes*, reacting slowly. In the former case, many methods are available for determining the composition and the equilibrium mass-action *formation constants* of the complexes, as discussed below. In the latter case, the reaction rates of different paths may be so dissimilar that our solutions have compositions very far from the thermodynamic equilibrium. In the quickly reacting, *labile* systems, we can only prepare the thermodynamically stable complexes, and there is usually not much point in isolating the pure solids in order to study their solution chemistry, because they react with the solvent to form the same mixture as their original solution. Most complexes outside the transition groups, e.g. mercury (II) and lead (II), and some inside, e.g. the high-spin manganese (II) and cobalt (II), are of this type. Some low-spin complexes are also quite labile, e.g. most complexes of gold (III) and palladium (II), which is very convenient for the study of their solution chemistry. On the other hand, most reactions of rhodium (III) are very slow, and the preparations often difficult. Thus, most amines (such as ethylenediamine) in aqueous solution are used by this metal as a source of OH^- rather than as a ligand, and the yellow precipitate of $Rh(OH)_3$ redissolves only very slowly, by boiling, to the colourless, thermodynamically very stable amine complexes. Only in the cases where a good catalyst is found, such as J. Bjerrum's active charcoal for cobalt (III) complexes or Delépine's case of small amounts of alcohol catalysing the reaction between $RhCl_6^{3-}$ and pyridine, may the speed of the preparation be increased much. Iridium (III) forms the most robust complexes known, and its chemistry is rather "organic", demanding high temperatures for most reactions. This has the advantage that it is possible to isolate geometrical isomers and measure the spectra of complexes which could not exist at all in equilibrium conditions.

If the central ion M forms complexes with the ligand L, and if we neglect to begin with the number of solvent molecules in the first co-ordination sphere, we may sometimes have the simple case of *consecutive formation* of a series of *mononuclear* (with only one M) complexes ML, ML_2, ML_3, . . ., ML_N (we are not indicating a possible ionic charge). Then, the *maximum co-ordination number* is N, i.e. in the case of monodentate ligands 6 for octahedral, 4 for tetrahedral and square-planar, and 2 for linear complexes. J. Bjerrum (1941) pointed out that the first complexes of such a series sometimes have much higher formation constants than the last members. He then defined the *characteristic co-ordination number* N_c as the number of such strongly bound, monodentate ligands. Thus, mercury (II) has $N_c = 2$, but may raise N to 4 or possibly 6, since $Hg(NH_3)_2^{2+}$ may take up further ammonia with much smaller constants to form $Hg(NH_3)_4^{2+}$, and $HgCl_2$ forms $HgCl_3^-$ and $HgCl_4^{2-}$ with an excess of chloride. This phenomenon probably corresponds to a change from linear to tetrahedral co-ordination.

The formation constants are determined in a solvent, where the mass-action law is valid in concentration units. If we write the *molar concentrations* [M], etc., still leaving the possibility of co-ordinated solvent molecules open, the nth consecutive *step formation constant* K_n is

$$K_n = [ML_n]/[ML_{n-1}][L] \qquad (189)$$

for the step reaction

$$ML_{n-1} + L = ML_n \qquad (190)$$

while the nth *total formation constant* β_n is

$$\beta_n = [ML_n]/[M][L]^n \qquad (191)$$

related to the total reaction

$$M + nL = ML_n \qquad (192)$$

A warning may be given that many authors use a different nomenclature; thus, K_n is often called k_n, and our β_n called K_{1-n} or K_n.

R

The assumption of the validity of the mass-action law on concentration units is often a rather dubious one. It is related to thermodynamics by the following procedure: The heat content or *enthalpy* H is the sum of the energy E and the mechanical term PV (pressure times volume). The *change of free energy* ΔG (often called ΔF in America) by a reaction is defined as

$$\Delta G = \Delta H - T\Delta S \qquad (193)$$

where T is the absolute temperature and ΔS is the *entropy change* by the reaction. Since the entropy S is a measure of the logarithm of the probability of the state (its sum of degeneracy numbers, its "state sum", in statistical thermodynamics), eqn. (193) expresses the fact that the chemical reactions do not simply lead to an equilibrium, where only the substances with the lowest energy are present, but that for $T > 0$, the tendency towards disorder demands smaller or larger concentrations of substances of higher enthalpy in equilibrium. Consequently, some chemical reactions proceed spontaneously, due to a large ΔS though they absorb heat, e.g. the dissolution of many salts in water. Now, the simplifying assumption, expressed in the *mass-action law in concentration units*, is that the free energy of the substance C is related to the concentration [C] by

$$G = G_0 + RT \ln [C] \qquad (194)$$

R being the gas constant, 1·986 cal/mole. If eqn. (194) is valid for the reacting substances

$$a_1A_1 + a_2A_2 + \ldots + a_nA_n = b_1B_1 + b_2B_2 + \ldots + b_nB_n \qquad (195)$$

the condition of equilibrium, $\Delta G = 0$, is obtained for

$$RT(b_1 \ln [B_1] + b_2 \ln [B_2] + \ldots + b_n \ln [B_n] - a_1 \ln [A_1] -$$
$$- a_2 \ln [A_2] - \ldots - a_n \ln [A_n]) = a_1 G_0(A_1) + a_2 G_0(A_2) +$$
$$+ \ldots + a_n G_0(A_n) - b_1 G_0(B_1) - b_2 G_0(B_2) - \ldots - b_n G_0(B_n)$$
$$= - \Delta G_0 = RT \ln K \text{ of reaction eqn. (195)} \qquad (196)$$

It is well known from Nernst's law for *electrochemical potentials*

$$U = U_0 + \frac{RT}{nF} (\ln [B] - \ln [A]) \qquad (197)$$

for the reaction

$$A = B + n \text{ electrons} \qquad (198)$$

(using the "European" sign of the standard oxidation potentials U_0 with $Na/Na^+ - 2 \cdot 71$ V rather than the Lewis sign $+2 \cdot 71$ V, and setting $[A] = 1$ for a pure metal) that for redox systems in equilibrium, the oxidizing character of the solution can be characterized by a single quantity, U of eqn. (197), and that all reactions of the type (198) then have their relative concentrations $[B]$ of the higher and $[A]$ of the lower oxidation number established. In the same way, it is possible to define a chemical potential for hydrogen ions in a given solution. The pH-scale is a linear function of this chemical potential, shifting one power of ten for each $0 \cdot 059$ V of the latter at $T = 300°K$. The electrochemical measurements of a hydrogen–platinum or glass-electrode potential gives pH plus a constant to be derived from a measurement of a solution with a known hydrogen ion concentration. Thus, for the equilibrium between the acid HB and the base B, it is valid

$$-\log_{10}[H^+] = pH = pK - \log_{10}[HB] + \log_{10}[B] \qquad (199)$$

with $pK = -\log_{10}K$, where K is the acid dissociation constant $[H^+][B]/[HB]$, still without consideration of the solvation of the proton. Since all acid–base reactions, involving mononuclear complexes (as contrasted to many redox reactions in eqn. (198)) are very quick, the acidity of a given solution is determined by a unique value of pH, establishing all equilibria according to eqn. (199). On the other hand, the acid–base reactions, decomposing or forming oxygen bridges, as in the bases $P_2O_7^{4-}$, $S_2O_8^{2-}$, $ox_2Cr(OH)_2Crox_2^{4-}$, $Cl_5RuORuCl_5^{4-}$, and many polymeric hydroxyl complexes, may be very slow.

It is easily shown that the mass-action law on concentration units is not generally valid. Thus, the *solubility product* $[Na^+][Cl^-]$ (which is a special case of the mass-action law eqn. (196), having constant free energy of the solid phase) is $30 \cdot 2$, if derived from a saturated solution of NaCl at room temperature, having the molar concentration $5 \cdot 5$. However, if an equal volume of $12 \cdot 5$ M hydrochloric acid is added to such a saturated salt solution, nearly all the NaCl crystallizes out, though the product of $[Na^+] = 2 \cdot 25$ moles/l. and $[Cl^-] = 9$ moles/l. does not supersede the solubility product.

It is customary, in many text-books, to remove this difficulty by assuming the mass-action law valid for *activities* a_n, i.e. the molar concentrations multiplied by the *activity coefficients* f_n. It must be emphasized that if we define the activities by substituting $G = G_0 + RT \ln a$ for eqn. (194), we are reducing the mass-action law to a mere tautology, which is only valid, because we have defined it to be so. This is an attitude somewhat too superficial to apply to the mass-action law, since the electrochemical measurements (eqn. 197) and the pH-scale (eqn. 199) actually indicate that the mass-action law is a useful approximation in aqueous solution, at least in a qualitative way, for discussing powers of ten of the concentrations of oxidized/reduced and acid/basic forms of pairs of reaction products. However, it is only in some cases that it is possible to perform plausible measurements of mass-action constants to a higher accuracy than a power of ten. Such special cases are *constant ionic media in aqueous solution*. In water, most of the deviations from the mass-action law (in concentration units; we omit this predicate in the following) are caused by the strong electrostatic interactions between ions. If we maintain a large and constant concentration of a salt, not participating to a large extent in the complex equilibria to be studied, the mass-action law is valid for the variation of the concentration of the ionic complexes, when these concentrations are only small, compared to the "neutral salt" concentration. If the complex concentration is not quite negligible, it is sometimes argued that the concentration of the main salt component should be decreased, leaving the *ionic strength* μ, the sum of $\frac{1}{2}z^2[C]$ with $z =$ the ionic charge, constant of the solution. However, while the ionic strength fairly well describes the behaviour of dilute salt solutions as function of concentration, it is not completely certain that it is relevant to the variation of the properties of strong solutions.

It is highly interesting empirically to note that the formation constants for complexes having the same ionic charge, i.e. of neutral ligand molecules, are not strongly dependent on the total salt concentration. In the same way, the activities of neutral molecules (ammonia, urea, glucose, etc.) are not generally very dependent on salt concentration, and especially, there is no dramatic change of the slope of the activity vs. salt concentration near the limit, where the latter vanishes. On the other hand, the reactions decreasing the number of free ionic charges ($M^{n+} + L^- = ML^{n-1}$) are generally much enhanced in dilute solution, while a strong salt solution is much more indifferent to the presence of slightly more ionic charge. Here it is found that measurements of activities do not always correspond to spectrophotometric measurements of changes of absorption spectra. Thus, the equilibrium

$$Co(NH_3)_6{}^{3+} + SO_4{}^{2-} = Co(NH_3)_6, SO_4{}^+ \qquad (200)$$

is strongly shifted to the right-hand side at low salt concentrations. However, the invariant visible spectrum clearly indicates that we do not have sulphate ions entering the first co-ordination sphere of Co(III), occupied by the six NH_3, but that *ion-pairs* are formed. (Actually, the ultra-violet light absorption of the ion pair is increased.) In the same way, the physicochemical measurements of activities (e.g. in ion-exchange resins) of cerium (III) chloride solutions indicate the formation of $CeCl^{2+}$ in appreciable concentration even below 1 M HCl. On the other hand, the spectral changes of $4f \rightarrow 5d$ bands of Ce(III) and internal $4f^n$-transitions in other trivalent lanthanides do only occur above 5 M HCl, the spectrum being constant in the range 0–5 M. In such cases, it is believed that only ion pairs are formed at low chloride concentrations, while actual chloro complexes are formed at high chloride molarity.

Assuming full ionic dissociation of the salts, Kohlrausch's law indicates the conductivity of a solution as the sum of the ionic conductivities, multiplied by the ionic concentrations. The deviations from Kohlrausch's law were originally ascribed to formation of ion pairs. *Debye* and *Hückel* pointed out that more continuous salt effects might be responsible, and also predicted the variation of activity coefficients as function of the ionic strength μ

$$\log_{10}f_n = -kz^2\sqrt{\mu} \qquad (201)$$

where the constant k (at room temperature and assuming the dielectric constant of water $= 80$) accidentally is near to 0·5. For higher concentration than some 0·05 M, specific salt effects begin to introduce differences between individual salts, and the Debye–Hückel theory is no longer valid. For monovalent ions A^+B^-, the activity coefficients f_n decrease smoothly from 0·75 for 0·1 M to some 0·55 for 1 M solutions, and increase for higher concentrations. For salts such as $A_2{}^+B^{2-}$ and $A^{2+}B^{2-}$, the activity coefficients are much less regular, and decrease in the latter case to some 0·04 in

1 M solutions. For concentrated acids and salt solutions, the specific salt effects are very important, and it is no longer possible to substitute for, e.g. 10 M HCl, either 10 M LiCl or other solvents as a constant salt medium. We may say that the properties of the solvent are now too far removed from the properties of water to make a simple extrapolation possible.

Actually, our description of activity effects in salt solutions suffers from two large difficulties. As mentioned (p. 237), it is not possible to define the free energy or the activity of a charged species alone; it is necessary to consider at the same time the ions of opposite charge, making up a neutral entity. Hence, only *activity products* $f(A^+)f(B^-)$ of ions, and not the activity of eqn. (201), are measurable quantities. Consequently, the pH-scale cannot be strictly defined without reference to the anions present in the solution, and pH-measurements can only be compared in a constant salt medium. This is an obvious inconvenience to the practical use of the pH concept and it is actually possible to use it in a looser way, i.e. not relying on more accuracy than, say, half a unit in comparing different solutions. For comparing strongly acidic solvents, e.g. almost anhydrous or SO_3-containing H_2SO_4, the Hammett-function has been introduced, comparing, for example, the colour shifts of pH-indicators. Thus, a quantity somewhat analogous to pH is -16 in pure H_2SO_4. In aqueous solution pK values cannot be measured below 0, because such a strong acid reacts with the base water, forming H_3O^+, and not either above 14, because such a strong base reacts with H_2O, now functioning as acid, and forming OH^-. We consider that the pK-values of H_2SO_4 (forming HSO_4^-), $HClO_4$, HCl and HBr are situated in the range -20 to 0, while HNO_3 and CF_3COOH, for example, are weaker acids with pK near 0. At the other end of the scale, pK of $(C_6H_5)_3CH$, forming $(C_6H_5)_3C^-$, is assumed to be some $+33$, according to measurements in diethyl ether, while the order of magnitude of pK for $NH_3 \rightarrow NH_2^-$ (assuming pK of $NH_4^+ = 9$) is $+40$ in liquid ammonia. The dissociation of acids is much weaker in alcohols than in water, permitting the observation of the much smaller acidity of HNO_3 than of HCl. However, many individual variations may occur. Thus, HNO_3 and picric acid have the same pK in C_2H_5OH, while the difference is at least 1·5 units in water. In general, the *solvents of low dielectric constant* decrease acidities of the type $HA \rightarrow H^+ + A^-$, forming more ionic charge (tending to attract each other to form ion pairs), while the acidities of the type $HA^+ \rightarrow H^+ + A$, e.g. amines and nitrogen groups of amino-acids, are much less affected. In many alcohols, NH_4^+ tends to be as strong an acid as CH_3COOH, preventing NH_3 and CH_3COOH from reacting quantitatively to ammonium acetate, analogous to NH_3 and HCN in aqueous solution.

These solvent effects constitute the other great difficulty for the activity description. It is not realistic to think about metal ions as unsolvated entities. Actually, the free energy of hydration of tri-valent metal ions (having the same order of magnitude as the change of enthalpy) is some 1000 kcal/mole, corresponding to a total forma-tion constant (equ. 191) β_6 about 10^{700} from unsolvated M^{3+} and water. (However, cf. p. 239). This is much larger than the formation constants relative to aquo ions (e.g. J. Bjerrum determined β_3 of $Coen_3^{3+}$ from $Co(H_2O)_6^{3+}$ and en to be 10^{48}). The solvation effects are of no importance to the discussion of ΔG, as long as the solvent is not varied. However, by comparing alcohol–water mixtures, rather curious effects occur.

The *Katzin effect* is that strong spectral changes can be ascribed to anion complexes formed by dissolving transition group salts in anhydrous alcohols. These complexes disappear by the addition of some 2 to 10 vol. per cent water, and cannot be recovered by addi-tion of a large amount of, for example, ammonium nitrate or chlor-ide. This might be understood if the free energy of bonding the ligands decreased strongly in the series water $\gg$ anions $\gg$ alcohols. However, it is known from systems (perchlorate) where the anion has no effect that water is only bound some ten times more strongly than alcohol molecules to the metal ions in alcohols. The Katzin effect is caused rather by the fact that the pure alcohol solvates are very unstable against uptake of anions, while the aquo (and amine) complexes are much less interested in exchang-ing one ligand with an anion. The electrical conductivities of transi-tion group complexes as of other salts decrease strongly in many solvents. However, this seems to be caused more by ion-pair forma-tion than by exchange in the first co-ordination sphere. Thus, the aquo ion $Co(H_2O)_6^{2+}$ has nearly the same spectrum in a mixture of 50 (vols.) per cent CCl_4, 43 per cent C_2H_5OH, 7 per cent H_2O as in water, while the dielectric constant of the former solvent must be very low. Rather than this macroscopic quantity, we might talk about a *microscopic polarity* of the solvent, increasing the activities of ions in the direction: strong salt solutions in H_2O < weak salt solutions in $H_2O < C_2H_5OH < CH_3COCH_3 < C_6H_5NO_2 < C_6H_6 < CHCl_3 < CCl_4$. It is remembered that the activity of water in such organic solvents is also rather surprising, since the saturated solutions of H_2O in CCl_4 (0·01 vol. per cent) or C_6H_6 (0·1 vol. per cent) have the same activity as pure liquid water; and hence, the activity coefficient of H_2O in these solvents is 10^4 and 10^3.

In the opposite direction of the organic solvents of low polarity, we may study the complex formation in molten salts. The *electrochemical series* of metals, arranged according to their standard oxidation potentials with respect to the most common oxidation number for the aquo ions, is somewhat different, if studied for metals in salt melts. This is comprehensible, since many ions in aqueous solution, such as Cs^+ and Ba^{2+} are probably stabilized by macroscopic dielectric effects only, not co-ordinating a definite number of water molecules to the central ion, while other ions, such as $Cr(H_2O)_6^{3+}$ and $Ni(H_2O)_6^{2+}$ are definite complexes, having an internal bonding stabilization. In salts, the co-ordination of, for example, Cl^- is expected to vary in a slightly different way with the central ion than with H_2O in aqueous solution.

We have now recognized that accurate measurements of formation constants are only possible in some special cases, such as constant salt media in aqueous solution. For determining K_n, we should use the variation of the concentrations of complexes $[ML]$, $[ML_2]$, . . . as function of $[M]$ or $[L]$. Since we know also, from preparation or analysis of the solutions, the *total concentrations* of central ion M and ligands L

$$C_M = [M] + [ML] + [ML_2] + \ldots + [ML_n]$$
$$C_L = [L] + [ML] + 2[ML_2] + \ldots + n[ML_n] + [HL]$$
$$+ [H_2L] + \ldots \quad (202)$$

(where the possibility of acid–base reactions of L is indicated), it may sometimes be sufficient to know only $[M]$ or $[L]$, or, in cases discussed below, only C_M and C_L together with some properties of the solution. The concentration of metal aquo ions $[M]$ may be found from electrochemical measurements according to eqn. (197) (actually, most of the very small solubility products known, e.g. 10^{-50} of HgS, are not derived from direct measurements of concentrations, amounting to some atoms per litre, but are free energy measurements) and $[L]$ by vapour pressure measurements, e.g. of ammonia over aqueous NH_3 with and without metallic salts present. However, the most common method applied for determination of formation constants has been the measurement of *displacements of hydrogen ions from ligands by central ions*

$$M^{+n} + LH_k^{m+k} = ML^{n+m} + kH^+ \qquad (203)$$

which is for instance the principle of *complexometric titrations*, where the reaction

$$M^{+n} + H_2enta^{2-} = Menta^{+n-4} + 2H^+ \qquad (204)$$

is quantitative within a controlled range of pH with the result that a titration of the metal content is transformed into an acidometric titration. For study of complex formation, eqn. (202) may be used in two different ways. J. Bjerrum used a large, constant concentration of the hydrogen ion addition product of the ligand, e.g. NH_4^+. It is seen from eqn. (199) that a measurement of pH indicates the ratio between $[NH_3]$ and the known $[NH_4^+]$ in this case. We may also titrate a mixture of, say, $0 \cdot 1$ M LH_k and $0 \cdot 1$ M metal with OH^-, noting the *titration curve*, pH as function of the base or acid added. If compared with the titration curve of pure LH_k, the formation constants can be calculated. G. Schwarzenbach has discussed the complications arising, when the complex ML itself can also take up protons.

J. Bjerrum pointed out that for mononuclear complexes, the extent of formation is only dependent on [L], and not on the total C_M for a given [L]. This can be seen from the mass-action law, and the average number of ligands co-ordinated per central ion, the *formation number* ñ can be defined:

$$ñ = ([ML] + 2[ML_2] + \ldots n[ML_n])/C_M \qquad (205)$$

J. Bjerrum defines the *formation curve* as ñ as function of $log_{10}[L]$. This is a monotonically increasing function, levelling off for ñ → N for large [L], and sometimes also levelling off, showing relatively low slope for some intermediate values of ñ near the characteristic co-ordination numbers N_c. If N is one, we have the expression

$$ñ = \frac{K_1[L]}{1 + K_1[L]} \qquad (206)$$

equivalent to the Langmuir-isotherm, if we defined the formation curve as ñ as function of [L]. $\bar{n} = \frac{1}{2}$ for $[L] = 1/K_1$. In general, for most formation curves found in practice, it is a rather good approximation to the nth step formation constant

$$K_n \cong 1/[L] \text{ for } \bar{n} = n - \tfrac{1}{2} \qquad (207)$$

In the special case of a *statistical formation curve*,

$$\bar{n} = \frac{NK_{av}[L]}{1 + K_{av}[L]} \qquad (208)$$

the individual step constants are related to the average value K_{av} by

$$K_n = (N - n + 1)K_{av}/n \quad \text{and} \quad \beta_N = K_{av}{}^N \qquad (209)$$

and the formation curve is isomorphous to eqn. (206) with a scaling of ñ. The theory of formation curves does not only apply to the complexes, but also to titration curves of acids, since they are also characterized by a set of eqn. (206) with ñ representing the average number of protons bound per base, and with $[L] = [H^+]$.

J. Bjerrum determined the ammonia complex formation constants by adding, if possible, sufficiently little NH_3 to a solution, containing much NH_4NO_3 (serving at one time as salt medium and as pH-buffer according to eqn. 199) and a small concentration of the heavy metal ion to be studied. In this case

$$C_L = \bar{n}C_M + [L] \qquad (210)$$

(not assuming $[NH_4{}^+]$ a part of C_{NH_3}, as done in eqn. (202)). ñ can be determined from C_M, C_L and [L], if the two latter quantities are not nearly identical. When using the hydrogen ion displacement reactions of the type eqn. (203), it is necessary to control the possible formation of amido or hydroxyl complexes, because reactions of the type

$$Hg^{2+} + 3NH_3 = Hg(NH_2)NH_3{}^+ + NH_4{}^+$$
$$Al^{3+} + 4H_2O + 4NH_3 = Al(OH)_4{}^- + 4NH_4{}^+ \qquad (211)$$

contribute to the decrease of $[NH_3]$ to the same extent as actual formation of ammonia complexes. A distinction can be made by variation of the concentration of NH_4^+, since the consequent change of pH for a given $[NH_3]$ influences the extent of the reactions eqn. (211), but not the uptake of NH_3.

The formation constant of a given complex does not give all the pertinent information on the thermodynamical stability we might want. Thus, we may sometimes be more interested, when comparing the different efficiency of complexing agents, to know the hydrogen ion displacement constants for eqn. (203) rather than for the formation eqn. (190) and (192). In the former case we consider, as a source of the ligand, that species LH_k which is dominant at the pH we are interested in. Thus, many amine complexes have larger values of K_n than amino acid or other anion complexes of the same central ion, but they decompose by acidification at a higher pH than the latter complexes. This is connected with the fact that the amines have higher pK values of their proton additives than the usual acids, of which the anions form complexes. In some cases we cannot determine pK of LH_k in aqueous solution, because it is well above 13 (e.g. many phenolates, amido groups, the oxide ion, etc.) and in these cases, only the hydrogen ion displacement constants can be determined.

Another important fact is that K_n of eqn. (189) has the dimension of a reciprocal concentration. Consequently, a change of the units for concentration influence all K_n by the same factor. This is related to the *chelate effect*, much discussed in the literature. Polydentate ligands seem to have higher values of their formation constants than monodentate ligands. However, this is partly an effect of our choice of concentration scale. It is not quite so interesting to discuss the choice between the usual molarity scale, where the concentration of the solvent is equal to 1 and the other concentrations are expressed in moles/litre, or molality scale, whether the other concentrations are expressed in moles/kg solvent, or molar fraction scale, since we could not in any of the three cases correct for the intrinsic change of solvent, when one of the solute concentrations is largely increased. However, it is obvious that a concentration scale, setting all solutes and the solvent on equal footing, would influence the numerical value of the chelate effect strongly. Thus, if the water concentration on the molarity scale is set equal to 55 M, the value of β_n for replacement of q water molecules by a polydentate ligand will be increased by the factor 55^q. If we consider generally the chelate effect as the ratio between, say, K_1 of the polydentate ligand P and β_q of q similar, monodentate ligands [L], it is $[L]^q/[P]$ for the situation $[MP] = [ML_q]$. Hence, the chelate effect is dependent on the $(q - 1)$th power of the concentration units. If we define the activities of the pure ligands (amines, water, etc.) to be one, this contribution nearly vanishes. On the other hand, by dilution, the polydentate ligands form complexes much more resistant towards dissociation than the monodentate ligands.

The determination of formation constants from titration curves can be made independently of whether the metal complex is coloured or not. On the other hand, this method is not always possible in practice, if $\bar{n}C_M$ is small compared to [L] in eqn. (210). Thus, it is from absorption spectra that it is possible to obtain evidence that $Ni(NH_3)_6{}^{2+}$, for example, has the same composition even in concentrated aqueous ammonia or in liquid ammonia. As discussed on p. 90 the study of absorption spectra is one of the most versatile methods of identifying the coloured species, since it may be used for determination of many components, when the light absorption of a solution is known at many wavelengths. J. Bjerrum mainly applied the formation curves, determined by the pH-measurements and other physicochemical methods, for calculating the *distribution function*, the values of $[M]/C_M$, $[ML]/C_M$, $[ML_2]/C_M$, . . . for a given value of $\bar{n}$. Generally, the complex ML_n is by no means the only one occurring in solution, even when $\bar{n}$ is an integer. If the constants do not deviate strongly from the statistical case eqn. (209), or if the deviations go in the direction of even larger differences between consecutive K_n, it is a fair approximation to state that for $\bar{n} = n + \frac{1}{2}$, the main components of the solution is ML_n and ML_{n+1}, while for $\bar{n} = n$, the main components are equal amounts of ML_{n-1} and ML_{n+1} and a larger proportion of ML_n. The corresponding *disproportionation constant* δ_n is

$$\delta_n = [ML_{n-1}][ML_{n+1}]/[ML_n]^2 = K_{n+1}/K_n \qquad (212)$$

In the case of statistical formation (eqn. 209), the disproportionation is rather extensive for $\bar{n} = n$, according to eqn. (212) for $N = 6$

$$[ML_{n-1}]/[ML_n] = [ML_{n+1}]/[ML_n] = 0\cdot646 \text{ for } n = 1 \text{ and } 5$$
$$= 0\cdot731 \text{ for } n = 2 \text{ and } 4 \qquad (213)$$
$$= 0\cdot750 \text{ for } n = 3$$

Hence, for $\bar{n} = 3$ in the statistical case with $N = 6$, the distribution on the single complexes $[ML_n]$ is

$$\left. \begin{array}{l} n = 0 \quad 1 \quad 2 \quad 3 \quad 4 \quad 5 \quad 6 \\ 1 \, : \, 6 \, : \, 15 \, : \, 20 \, : \, 15 \, : \, 6 \, : \, 1 \end{array} \right\} \qquad (214)$$

showing that only 5/16 of C_M occurs as ML_3. In such cases, J. Bjerrum (1931, 1941) applied the distribution function to the absorption spectra of the solutions, having different $\bar{n}$, and separated the *individual spectra* (except *cis–trans* isomers for $n = 2, 3, 4$, which cannot be distinguished from the formation curve) by solving linear equations of the type eqn. (99). This of course involves a large amount of calculation, and introduces large experimental uncertainties for the intermediate complexes. For a study of nickel (II) complexes with a mixed set of ligand atoms, the present writer (1956, 1957) has preferred to use polydentate ligands, where it can often be argued that one definite complex strongly predominates in the coloured solution.

J. Bjerrum introduced the *principle of corresponding solutions* that two solutions, involving only monomeric complexes, and having the same absorption spectrum must have the same $\bar{n}$ and the same $[L]$. If we mark the concentrations of the second solution by a dash, eqn. (210) implies

$$\bar{n} = (C_L' - C_L)/(C_M' - C_M) \qquad (215)$$

making it possible to determine the formation curve, if the difference $C_L' - C_L$ can be determined sufficiently accurately for corresponding solutions (e.g. by interpolation in the variation of the spectra). This principle can be generalized to some extent. If we start to study an unknown system, it will be reasonable to measure the spectra of a series of solutions with $C_L/C_M = 0, 0.5, 1, 1.5, 2, \ldots$ up to a number somewhat higher than the expected maximum co-ordination number N, and at two metal concentrations C_M differing by a factor at least 2, and preferably 5 (it is often useful to obtain the same optical density by using 1 cm and 5 cm cells). We may have two limiting cases: if the spectrum for a given ratio C_L/C_M is independent of C_M, then this ratio is simply $\bar{n}$, and we cannot obtain $[L]$, because the formation constants are too large (larger than some five times the highest value of $1/C_M$ applied). The other limiting case has a spectral evolution, which is a function of C_L only, and then, C_L is very near to [L]. In this case, $\bar{n}$ cannot easily be estimated. In the intermediate cases, eqn. (215) can be fruitfully applied. The presence of polynuclear complexes will then be discovered by a variation of $\bar{n}$ with C_M for the same [L]. This is very common in the hydrolysis reactions, studied by L. G. Sillén, such as the reaction

$$2Fe^{3+} + 2OH^- = Fe(OH)_2Fe^{4+} \text{ (or } FeOFe^{4+}) \tag{216}$$

where the apparent pK for the acid/base equilibrium decreases with increasing metal concentration. Rather than apply eqn. (215) to each set of experimental points, we may also plot C_L/C_M as function of the reciprocal metal concentration $1/C_M$:

$$C_L/C_M = \bar{n} + [L]/C_M \tag{217}$$

For $C_M^{-1} \to 0$, the ratio $C_L/C_M \to \bar{n}$. The slope of the line in this graph of corresponding solutions gives $[L]/C_M$.

Even though the distribution of the single species ML_n can be calculated in principle from the formation curve, it will often be more exact to solve the equations of the effective molar extinction co-efficients (eqn. 99), obtaining the concentration of the individual species, since some of the pure spectra (of M and of ML_N) can often be directly obtained or easily extrapolated from solutions containing one prevailing component. Thus, for usual values of K_n/K_{av}, the spectrum of ML can be obtained by extrapolation in a series with $\bar{n} = 0.25, 0.5, 0.75$ and 1. In an actual study, it is always useful to recognize the possible influence of pH changes in solutions of different C_M, having the same C_L/C_M. The property that most ligands have enabling them to act as bases, is nearly always accompanied by an ability to form metal complexes. It cannot be argued that the affinity to hydrogen ions and to metal ions is exactly proportional; pyridine and heterocyclic diimines (dip and phen), and the halide ions (except F^-) are attracted more to metal ions than to protons, while on the other hand tertiary amines R_3N are generally very strong bases, but do not form stable complexes. For practical purposes of com-plexing a metal M at a given pH, it is interesting to seek a ligand for which the *complexing efficiency*

$$-\log_{10}([M]/C_M) \cong \log_{10}\beta_N - N \log_{10}[L] \tag{218}$$

is as large as possible. To a good approximation (the error in maximum being 0·3 for one acid dissociation) for $C_L > NC_M$

$$\log_{10}[L] = \log_{10}(C_L - NC_M) - \Sigma_{pos.}(pK_n - pH) \quad (219)$$

where the summation is only to be taken for positive contributions from such pK_n (acidity constants of the ligands) larger than pH. However, eqn. (218) and eqn. (219) are only good criteria for the complexing of M, if another metal is not present in a concentration comparable to C_L. In this case, the difference of $\log_{10}\beta_N$ for the two metals will be of importance.

As an actual case of a study of complex formation, supported by spectrophotometrical evidence, we may discuss the chloro complexes of palladium (II). The salt K_2PdCl_4 is known from X-ray analysis to contain the square-planar group $PdCl_4{}^{2-}$. Samuel and Despande (1933) found that the absorption spectrum of an aqueous solution of K_2PdCl_4 changes by addition of KCl. Several authors (Sundaram and Sandell, 1955; Droll, Block and Fernelius, 1957) interpret this phenomenon as the formation of $PdCl_5{}^{3-}$ and $PdCl_6{}^{4-}$ with step formation constants K_5 and K_6 slightly above 1, while the first four K_n are very large, at least 200. Actually, by applying eqn. (217) to a set of absorption spectra of 0·01 M and 0·003 M Pd(II) in 1 M $HClO_4$ (serving as constant ionic medium and suppressing hydrolysis reactions), the present writer found that $PdCl_4{}^{2-}$ actually dissociates at moderately low chloride concentrations,

$$K_4 = [PdCl_4{}^{2-}]/[PdCl_3(H_2O)^-][Cl^-] = 6 \quad (220)$$

while K_2 is larger than the reciprocal Pd-concentrations applied. On the other hand, $PdCl_4{}^{2-}$ has the same absorption spectrum in 12 M HCl as the predominating species in solutions above 2 M HCl (eqn. 220), and consequently, the formation of higher chloro complexes is not probable.

The aquo ion $Pd(H_2O)_4{}^{2+}$ cannot easily be prepared pure; the usual way of precipitating black "Pd(OH)₂" from $PdCl_4{}^{2-}$ plus a carefully controlled amount of $2OH^-$, and (partly) redissolving the filtered precipitate in boiling $HClO_4$ is unsatisfactory; the solutions have too high ϵ and precipitate Pd or $Pd(OH)_4$ after some days. However, it can quantitatively be prepared by the addition of an excess of mercury (II) perchlorate in $HClO_4$:

$$PdCl_4{}^{2-} + 4Hg^{2+} + 4H_2O \rightarrow Pd(H_2O)_4{}^{2+} + 4HgCl^+ \quad (221)$$

Since K_1 and K_2 of Hg(II) chloro complexes are $10^{6\cdot74}$ and $10^{6\cdot48}$, respectively, this reaction proves that K_1 and K_2 of Pd(II) chloro complexes must be below 10^6. They might probably be estimated by competition with Bi(III) on the free chloride ions.

In the case of robust complexes, the spectra may be compared with those of freshly prepared (or extrapolated back in time) solutions of solid compounds of known composition. However, the monomeric acid/base reactions of robust complexes are very quick, necessitating the control of pH of the solutions, e.g. of $Co(NH_3)_5H_2O^{3+}/Co(NH_3)_5OH^{2+}$ or $Pt(NH_3)_6^{4+}/Pt(NH_3)_5NH_2^{3+}$. Generally, the mixed amine–halide complexes exchange their halide more quickly with solvents or other ligand molecules (especially with OH^-) than they lose amine molecules. *Reaction kinetics* studies the evolution in time (at a constant temperature) of such reactions, and to the first approximation indicates the reaction as *Zeroth, first* or *second order* in the reactants, according to the reaction rate being invariant, or proportional to $[C]$ or $[C]^2$ of a given reactant. The interpretation of these kinetic data in terms of a *reaction mechanism* is not easy, and hardly ever unique, because many different assumptions on the intermediate reaction products (often present in immeasurably small quantities) may lead to the same evolution in time. A special experimental difficulty is the possible influence of small, almost indetectable impurities, as the acid decomposition of $Fe(CN)_6^{4-}$ to to Fe^{2+} and HCN, studied by G. Emschwiller, where the presence of very small traces of Hg^{2+} is sufficient for catalysing the reaction to a great extent. It is often attempted to measure *activation energies* A from the Arrhenius equation, assuming the reaction rates proportional to

$$\exp(-A/kT) \qquad (222)$$

However, heterogeneous catalysis from the walls of the container for the solution, etc., may disturb such a determination to a great extent. Most chromium (III) complexes seem to react with homogeneous reactions, having A about 25 kcal/mole ($= 9$ kK).

At the present time, ligand field theory is not able to indicate quantitatively specific reasons for high or low reaction rates, i.e. lability or robustness of a given complex. However, some connexion seems to exist with the ligand field stabilization $\rho\Delta/5$ and the degeneracy number e of the ground state in octahedral symmetry. Thus, systems with $e = 2$ or 3 in the first transition group do not usually form robust complexes, while the low-spin d^4- and d^5-systems with $e = 3$ in the next two transition groups are much less robust than the corresponding d^6-systems with $e = 1$. On the other hand, the *systems with $e = 1$ and large values of $\rho\Delta$* form robust complexes. The filled shell and high-spin d^5-systems have $e = 1$, but $\rho = 0$ and do not form robust complexes according to this criterion. If the robustness is directly a monotonic function of $\rho\Delta$, we expect it to increase in the series

$$Ni(II) < V(II) < Cr(III) < Mo(III) < Re(IV) < Co(III)$$
$$< Rh(III) < Ir(III) < Pt(IV) \quad (223)$$

This is in qualitative agreement with experience. The apparent exceptions Mo(III) and Pt(IV) may be connected with the easy change of oxidation numbers in intermediate complexes, e.g. to Mo(V) and Pt(II). The heterogeneous catalysts, such as active carbon and palladium sponge, may establish equilibria with the labile oxidation numbers such as Co(II).

The explanation of the criterion ($e = 1$ and large $\rho\Delta$) may be that a distortion of the complex as an intermediate step in the reaction is made easy by the degenerate ground states, able to show Jahn-Teller distortions; and that the removal of a single ligand has more serious consequences for the chemical bonding of the complex in the cases where $\rho\Delta$ is large.

S

CHAPTER 14

SURVEY OF THE CHEMISTRY OF THE
HEAVY, METALLIC ELEMENTS

BEFORE the next chapter, which gives a summary of the absorption spectra of complexes, it may be useful to make a few remarks on the chemistry of these complexes. Unfortunately, most chemical textbooks treat the metalloids at very great length, followed by the alkali and alkaline earth metals, while transition group metals, perhaps with exception of chromium, manganese, iron and copper, are rather neglected. In discussing the overall chemical properties, we may divide the metallic elements into two groups, named A and B by Ahrland and Chatt (1958). The A characters are to form complexes, mainly determined by electrovalent bonding, and with the formation constants varying in the following way:

$$\left.\begin{array}{l} F^- \gg Cl^- > Br^- > I^- \\ OH^- \gg NH_3 \sim H_2O \sim CN^- \\ N \sim O > S > P \end{array}\right\} \qquad (224)$$

while the B characters have the opposite order of ligands:

$$\left.\begin{array}{l} F^- \ll Cl^- < Br^- \ll I^- \\ CN^- > NH_3 > OH^- \gg H_2O \\ S \gg N > P \sim O \end{array}\right\} \qquad (225)$$

The inert-gas configuration ions, such as Ca^{2+}, Sr^{2+}, Ba^{2+}, Al^{3+}, Sc^{3+}, Y^{3+}, Zr^{4+}, Th^{4+}, and the trivalent lanthanides and trivalent and tetravalent actinides all belong strictly to the A group. The d^{10}-systems with low oxidation number, such as Cu^+, Ag^+, Au^+ and Hg^{2+}, are the most characteristic members of group B.

The other ions are to some extent intermediate cases* between A and B:

---

* Poë and Vaidya (1961) criticize the concept of A and B elements. However, the bond-energies ahd enthalpies in solution, discussed by these authors, are quite difficult to define consistently (p. 240) and no doubt remains that the relative tendency is more B-like of Hg than of Zn.

A			B	
Fe^{3+}, V^{3+}, Cr^{3+}		Co^{3+}		
Mn^{2+}, Fe^{2+}	Co^{2+}, Zn^{2+}	Ni^{2+}, Cu^{2+}		(226)
Mo^{3+}	$Ru^{4+}, In^{3+}, Sn^{4+}$	$Ru^{3+}, Rh^{3+}, Cd^{2+}$	Pd^{2+}	
	Ir^{3+}	Pt^{4+}	Pt^{2+}, Au^{3+}	

Thus, the (high-spin) d^5-systems are nearly pure A elements, as are also the first members of each transition group. The d^{10}-systems with high oxidation number, and some octahedrally co-ordinated transition group elements are intermediate cases, while the square-planar, low-spin d^8-systems belong clearly to group B. The s^2-family Sn^{2+}, Sb^{3+}, Tl^+, Pb^{2+}, Bi^{3+} takes a special position, having B-character with respect to the bonding of I^-, S^{2-}, and S-containing ligands, but distinctly A-character with respect to NH_3 and CN^-, due to the reluctance to form σ-bonds, making the 5s- and 6s-electrons anti-bonding.

The high oxidation numbers $(+6, +7, +8)$ of all three transition groups are mainly known as tetroxo complexes. A special phenomenon is the dioxo complexes of the actinides $(+5$ and $+6)$. The usual metalloids form mainly trioxo anions in the 2p-group (C, N), tetroxo anions in the 3p- and 4p-groups (Si, P, S, Cl, As, Se) but hexa-hydroxo anions in the 5p-group (Sn, Sb, Te, I). The latter change from tetrahedral to octahedral co-ordination is probably caused mainly by the variation of relative *atomic radii*, which have the tendency to vary for the loosest shell of a given oxidation state

$$\left.\begin{array}{l} 2p < 3p \sim 4p < 5p < 6p \\ 3d < 4d \sim 5d \end{array}\right\} \qquad (227)$$

the approximate equality signs being caused by the increasing atomic number during the filling of the 3d-shell (cancelling the expansion of the 4p-shell, compared to the 3p-shell) and during the filling of the 4f-shell, the *lanthanide contraction* (making the chemical properties of 5d-elements as close to those of 4d-elements as to make the separations Zr–Hf, Nb–Ta, Mo–W, and even Rh–Ir, rather difficult).

The *oxidation numbers* increase and obtain a larger freedom of variation in the 4d- and 5d-groups, compared to the 3d-group, and in the 5f-group, compared to the 4f-group. This agrees (though not perfectly well) with the observed ionization energies in atomic spectroscopy, as also the opposite development, the increased stability of the 5s- and especially the 6s-shell, leading to decreasing oxidation numbers in the series Si, Ge, Sn, Pb and P, As, Sb, Bi.

Within the partly filled d-shells, the ligand field stabilization of certain electron numbers has a strong influence on the oxidation numbers observed, particularly in the 4d- and 5d-groups. Thus, among the octahedral complexes, the low-spin d^6-systems have the largest ligand field stabilization, while the low-spin d^8-systems form the most stabilized square-planar complexes.

Actually, in the first transition group, the adherence to a rather constant oxidation number of most complexes (except the tetroxo with high and some cyanide and carbon monoxide complexes with low oxidation numbers) is so firm that the M.O. energy differences do not have a large effect. Thus, among the divalent $3d^q$-hexa-aquo ions, all values of q between 2 and 10 are known, while the trivalent $3d^q$-hexa-aquo ions are known for q between 0 and 6, inclusive. In the next two transition groups, the selectivity for q is much more pronounced: $4d^1$ and $5d^1$ are known in some particular cases of strongly distorted structures, such as $MoOCl_5^{2-}$, but have a large tendency towards dimerization and also delocalization of the partly filled M.O. with badly defined oxidation number; $4d^2$ and $5d^2$ are not known in many well-defined complexes, except the types $Mo(CN)_8^{4-}$ and $RuO_2Cl_4^{2-}$, $4d^3$ and $5d^3$ are known in a series of complexes of Mo(III), Tc(IV), Ru(V), W(III), Re(IV), Os(V) and Ir(VI). $4d^4$ and $5d^4$ are known in hexahalide complexes of Ru(IV) and Os(IV), and oxo-bridged dimeric complexes, $4d^5$ and $5d^5$ are more well-known of Ru(III), Os(III), and Ir(IV). However, $4d^6$ and $5d^6$ are by far the best known configurations, involving Rh(III), Pd(IV), Ir(III), and Pt(IV). On the other hand, monomeric $4d^7$ and $5d^7$ is not certainly known in any complex, $4d^8$ and $5d^8$ are again well-known in the square-planar complexes of Pd(II), Pt(II), and Au(III). Recently, organic compounds have been prepared of Rh(I). The last odd electron number, $4d^9$ and $5d^9$, are not too well represented by the Ag(II) complexes and $Ir(NH_3)_5$.

If considering a given column in the periodical system, the preference of certain electron configurations and the tendency towards increasing the oxidation number is clearly observed. In the series cobalt, rhodium, iridium and nickel, palladium, platinum, the two

3d-members form in most cases high-spin, octahedral complexes of the oxidation number +2. In the cases (such as CN^- and some amines) where the ligands are able to induce larger M.O. energy differences than found in the aquo ions, Co(II) is very readily oxidized to a diamagnetic, octahedral Co(III) complex, while Ni(II) can do better than conserve its oxidation number forming a diamagnetic, square-planar complex. In the 4d-group, the low-spin behaviour is the general rule (except for the paramagnetic PdF_2), and the octahedral Rh(III) and square-planar Pd(II) represent the M.O. configurations expected. In the 5d-group, the tendency towards high oxidation number permits the formation of both low-spin d^5 Ir(IV) and low-spin d^6 Ir(III) and Pt(IV) (as already observed in some compounds of Pd(IV)), while low-spin d^8 Pt(II) is no longer the most stable oxidation number.

Some ligands, such as oxide mentioned above, permit the central ion to show very high oxidation numbers. Similar effects are found in the fluoride complexes, which are also those having the greatest tendency towards high-spin behaviour, as explained from the very small nephelauxetic effect. Thus, CoF_6^{3-} and PdF_2 are high-spin compounds, while $Co(H_2O)_6^{3+}$ and $Pd(H_2O)_4^{2+}$ are low-spin compounds. On the other hand, it cannot be argued that all fluoride complexes are high-spin ones; the d^6-systems NiF_6^{2-}, PdF_6^{2-} and PtF_6^{2-} are all diamagnetic, and RhF_6^{2-} and IrF_6^- low-spin.

Cyanide ($Ni(CN)_4^{4-}$, the red Ni(I) cyanide, $Pd(CN)_4^{4-}$), isonitrile (Mn(I), Rh(I)), dipyridyl* (V(−I), Co (I), Rh(I) etc.), and carbon monoxide complexes are particularly liable to form complexes of low oxidation numbers. Special electron configurations, such as d^{10} ($Ni(CO)_4$) and d^6 (V dip_3^-, $Cr(CO)_6$, $Mn(CN)_6^{5-}$, $Mn(CNC_2H_5)_6^+$) seem to have bond structures appropriate for this bonding, probably involving formation of M.O. from d-electrons and empty π-orbitals of the ligands. A special case is the "Sandwich" complexes of the *cyclo*pentadienide ion $cp^- = C_5H_5^-$, where a large number of low oxidation numbers are now known.

Returning to the more usual chemistry, the important difference between the 3d-group on the one hand and the 4d- and 5d-group on the other hand, is that the aqua ions of +2 and +3 are the usual

---

* However, the existence of Li dip($S=\frac{1}{2}$) and Li_2dip (diamagnetic) shows that this ligand may occur in anionic forms with additional electrons and the central ion configuration may hence be indeterminate, diamagnetic Ti dip_3 not necessarily being a d^4-system.

starting materials (excepting Co(III)) for a complex chemical study of the first transition group, while $Ru(H_2O)_6^{3+}$, $Rh(H_2O)_6^{3+}$ and $Pd(H_2O)_4^{2+}$ are the only aquo ions known of the following group, and they are not easily prepared pure. The usual starting material for study of the platinum group chemistry is the hexachloro complexes, prepared by the action of chlorine (or, according to M. Delépine, vapours of CCl_4) at red heat on a mixture of the finely powdered metals and alkali chloride. It is often believed that the chloro complexes of the platinum group have exceedingly high formation constants, as compared to the first transition group. Actually, this is due to the confusion between robustness and high thermodynamic stability. Thus, at room temperature, $RhCl_6^{3-}$ continues to lose chloride in 1 M $HClO_4$ for several weeks, while an equilibrium is established in 1 M HCl after some days, leaving $\bar{n} \sim 4 \cdot 5$. The sixth step constant K_6 is below 0·5. It was seen in eqn. (220) that K_4 of $PdCl_4^{2-}$ is only 6, and K_4 of $PtCl_4^{2-}$ is known to be 55. $IrCl_6^{3-}$ also aquates slowly in dilute HCl or $HClO_4$. The spectra in the ultra-violet demonstrate that many chloro complexes of heavy central ions are only completely formed in 12 M HCl, but not in 4 M HCl. It is not possible to indicate the mass-action law formation constants in concentration units, since the mass-action law is known not to be valid in that way in strong hydrochloric acid (see p. 247). We shall rather consider the approximate range of concentration of HCl, where the complex with the maximum co-ordination number (in the case of Hg(II), the characteristic co-ordination number) is almost completely formed, as indicated from measurements in the literature of the strong chloro complexes, or from the constancy of the absorption band positions as function of increasing concentration of HCl. The somewhat qualitative series is:

$$
\left.
\begin{array}{ll}
< 0 \cdot 1 \text{ M HCl:} & \text{Hg(II), Tl(III), Os(IV)?, Ir(IV)?, Pt(II),} \\
& \text{Pt(IV)?, Au(III)} \\
\sim 1 \text{ M HCl:} & \text{Cu(I), Pd(II), Ag(I), Cd(II), In(III), Pb(IV),} \\
& \text{Bi(III), Ru(IV)?} \\
\sim 8 \text{ M HCl:} & \text{Zn(II), Mo(III), Ru(III), Rh(III), Sn(IV),} \\
& \text{Re(IV)?, Os(III)?, Ir(III)?, Tl(I), Pb(II)} \\
\sim 12 \text{ M HCl:} & \text{Fe(III), Co(II), Cu(II), Ga(III), Sb(V),} \\
& \text{U(IV), Pu(IV)} \\
\gg 12 \text{ M HCl:} & \text{Al(III), V(III), Cr(III), Mn(II), Ni(II),} \\
& \text{Ce(III), Nd(III), Th(IV), U(III)}
\end{array}
\right\} \quad (228)
$$

It does not show a clear-cut division according to A and B elements (cf. eqn. 226) but emphasizes the continuous development of most such divisions of chemical properties.

The chloro complexes $SnCl_6^{2-}$, $SbCl_6^{-}$, $RuCl_6^{2-}$, and $IrCl_6^{2-}$ decompose at low concentration of HCl, probably not forming mixed chloro–aquo complexes, but rather hydroxo complexes. Actually, NaOH has a rather curious effect on several Ir(IV) and Pt(IV) complexes, since the corresponding Ir(III) hexahalide and Pt(II) tetrahalide complexes are formed at first and then, during some days, the halides decompose slowly to other compounds. This reduction must involve an equilibrium

$$2IrCl_6^{2-} + 2Cl^- = 2IrCl_6^{3-} + Cl_2$$
$$PtI_6^{2-} \quad\quad = PtI_4^{2-} + I_2 \tag{229}$$

where Cl_2 and I_2 are consumed by OH^- to form ClO^-, etc. This is analogous to the equilibrium between the violet-blue Ir(IV) and yellow-green Ir(III) bromo complexes

$$2IrBr_6^{2-} + 2Br^- = 2IrBr_6^{3-} + Br_2 \tag{230}$$

where the left-hand side reactants are only stable at an extremely low concentration of bromide ions.

In most electrostatic models, halide anions were expected to form much stronger complexes than water and ammonia. Eqn. (228) demonstrates clearly that the latter, neutral molecules have an unusually large ability of binding metal ions, at least in aqueous solution (cf. the Katzin effect, p. 251).

There is no simple, quantitative explanation of this fact. We may of course make the high dielectric constant $D = 80$ of water partly responsible for the weak binding of anions. If this macroscopic concept, actually expressing the polarizability in a homogeneous electric field, could be applied, we see how the interaction between a positive and a negative unit charge at a distance of 2 Å leads to the potential energy of only 0·72 kK, rather than 58 kK for $D = 1$. Table 25 gives a series of selected values for complex formation constants, mainly given for 0·1–1 M salt solutions in water. It is seen that most constants K_1 for monodentate ligands are less than 10^5. If we take the molar concentration 55 of water into account (cf. p. 255), the corresponding value of $10^{6·74}$ is equivalent to $\Delta G = 3·2$ kK (eqn. 97).

TABLE 25. THE COMPLEXITY CONSTANTS OF A NUMBER OF METAL COMPLEXES

(log K_n as defined in eqn. (189) for conditions not very different from 0·1–2 M salt media at 20–25°C; the table is only conceived as a qualitative review, for exact values and for literature references, the reader is referred to ref. 55, and for rather interesting discussions to ref. 48 and 520. The values marked β_n indicate log β_n as defined in eqn. (191). The values marked solub. indicate − log the equilibrium concentration of free ligand [L] needed for maintaining the free metal ion concentration [M] = 10^{-3} moles/l.)

Metal	OH^- log K_1	OH^- solub.	F^- log K_1	Cl^- log K_1	Br^- log K_1	I^- log K_1	SCN^- log K_1	S^{2-} solub.	NH_3 log K_1	NH_3 β_n	en log K_1	en log K_2	en log K_3	den log K_1	den log K_2	gly$^-$ log K_1	gly$^-$ log K_2	gly$^-$ log K_3	ata^{3-} log K_1	enta^{4-} log K_1
Mg(II)	2·5	3·9	1·2						0·2		0·4					3·4			5·4	8·7
Al(III)	9	9·7	6·1																	16·1
Ca(II)	(1)	1·3						0	(−0·2)							1·4			6·4	10·7
Sc(III)	9	(9)	6·2																	23·1
V(III)	11			0			2·0													25·9
Cr(III)	10	9·0	4·4	0	0		1·9	11·8	(1)											25
Mn(II)	3	4·9		0·4			(0)	15·4			2·7	2·1	0·9	4·0	2·8	2·9			7·4	14·0
Fe(II)	4	5·9		0·8				22·5			4·3	3·2	2·0	6·2	4·1				8·8	14·3
Fe(III)	11·5	11	5·2				2·3												15·9	25·1
Co(II)	5	6·0			0		1·3	20·8	2·1	(β_4 5·1)	5·9	4·8	3·1	8·1	6·0	4·6	3·8		10·6	16·2
Co(III)	12									(β_6 35·2)	(19)	(16)	(13·5)							36
Ni(II)	4·5	7	0·7	−1·5 (β_2 5·5)	(β_2 5·1)	(β_2 8·2)	1·2	40·7	2·8	(β_6 8·7)	7·6	6·4	4·5	10·7	8·2	5·8	4·8	2·5	11·3	18·6
Cu(II)	7·5	8·4	0·7	0·3	0·3		1·8 (β_2 10)	41·1	4·1	(β_4 12·7)	10·7	9·3	(−1)	16·0	5·3	8·2	7·0	3·6	12·7	18·8
Zn(II)	5·5	6·8	0·7	−0·2	−0·6	(−1)	0·5	19·9	2·4	(β_4 9·5)	5·9	5·1	1·9	8·9	5·5	4·8	4·1	(1)	10·4	16·3
Ga(III)	11	11	4·4	−0·6																20·3
Sr(II)																			5·0	8·6
Y(III)	7	6·6	(0)																11·4	18·0
Ag(I)	(2)	4·8		3·7	β_3 9·0	β_3 13·7	β_2 9·8	42·8	3·2	(β_2 7·0)	4·7	3·0		6·1		3·5	3·4			7·3
Cd(II)	5	5·5	0·5	1·5	1·8	2·1	1·4	25·5	2·6	(β_6 5·1)	5·6	4·6	2·1	8·4	5·4	3·9	3·2	(2)	9·5	16·5
In(III)	10	10·1	3·8	2·2	1·9	1·2	2·6	22												25·0
Ba(II)																0·8			4·8	7·8
La(III)	5	5·2	2·7																10·4	15·4
Nd(III)	6	6·2		−0·7															11·1	16·5
Hg(II)	11·5	11·7	1·0	6·7	9·0	12·9	β_2 22	50	8·8	(β_2 17·5)	(β_2 23·4)			21·8		10·3	8·9			21·8
Tl(I)	(0·5)			0·5	0·8	0·7	0·2	25	(−1)											
Tl(III)	13	13·6		6·2	8·9															
Pb(II)	7·5			1·1	1·5	1·4	0·6	24·5								5·4	3·4		11·8	18·3
Bi(III)	12·5			2·4	2·3	3	1·1	32												
Th(IV)	10	10·2	7·7	0			1·5													23·2
U(IV)	12·5		>9	0·3	0·2															

It is evident from Table 25 that most K_n increase strongly with the ionic charge z for a given ligand. Exceptions are certain monovalent group B metals, such as Cu(I) and Ag(I) (and *a fortiori* Au(I), where the aquo ion is not known). If we compare different ligands, it is surprising to compare the enormous thermodynamic stability of hydroxo complexes with nearly all other anion complexes. This may partly be caused by the use of water as solvent, the hydrogen bonding properties, etc., of hydroxo complexes being more favourable than of chloro complexes, for example. A complication arises from the fact that most hydroxo complexes are polymeric, either in solution or in heterogeneous precipitates. In Table 25, not only log K_1 for MOH, but also $-\log[OH^-]$ for $[M] = 10^{-3}$ M in the solubility product are given. However, these two numbers are generally much less certain than most other values of K_n. A similar situation exists for F^-, where the solubility products of the difluorides are sufficiently low to have the quantity $-\log[F^-]$ for $[M] = 10^{-3}$ M equal to 2·6 for Mg, 3·7 for Ca, 2·5 for Ni, and 2·8 for Sr, though substantially no complex formation of MF^+ and MF_2 occurs in the supernatant solution. We may consider these fluoride precipitates as a special case of polymeric complexes, the bonding between several fluorides and several metal ions stabilizing a lattice, which has no counterpart as entities in the solution.

One of the most conspicuous differences between A and B elements is the criterion, as to whether K_1 is largest of the fluoro or the chloro complex. It is seen from Table 25 that Al(III), Cr(III), Fe(III), Ga(III), Th(IV) and U(IV) distinctly are A elements, while Cu(I), Ag(I), Hg(II) and probably Tl(III) represent the typical B elements. Another characteristic difference is the solubilities of sulphides, as well-known from classical, qualitative analysis. As seen from Table 25, the corresponding values of $[S^{2-}]$ for $[M] = 10^{-3}$ M may be exceedingly small for B elements.

Unfortunately, the complexity constants are not known for a very wide selection of ligands and central ions. There is a certain number of ligands (NH_3, many organic amines and amino acids) and metal ions (mainly with the oxidation number $+2$) which traditionally are much studied, and for which K_n is known with the accuracy obtainable in constant salt media. On the other hand, the robust central ions such as Cr(III), Ru(III), Rh(III), Ir(III), and Pt(IV) are at the present nearly unknown in the tables of complexity constants.

Among the ligands, interesting cases such as CN^-, some heterocyclic amines, NO_2^-, isonitriles RNC, nitriles RCN, porphyrines, etc., produce such large experimental difficulties that the constants have not been determined. When we say that most numbers in Table 25 are small, we must not forget that many of the undetermined constants are presumably very large. We do not consider either those complexes for which K_n are so small that they cannot be prepared in aqueous solution. Many interesting preparations (e.g. complexes of R_3P, R_2S, R_2SO, ROH, R_3PO, $(RO)_3P$, PCl_3, etc.) are performed in anhydrous solvents.

The entropy difference ΔS of formation of most anion complexes is positive, i.e. ΔG is more negative than ΔH. This corresponds to the higher order (i.e. less probability) of the many water molecules, polarized in the second sphere of co-ordination around highly charged ions. This order can be released to a more probable state by forming the less charged complex. It is well known that the formation of most chloro complexes is enhanced by increasing temperature, i.e. heat is absorbed. Similar effects are found by the neutralization of H_3O^+ with OH^-, having ΔG corresponding to $- 6 \cdot 7$ kK, but ΔH only $- 4 \cdot 7$ kK. We cannot at the present hope to obtain any detailed understanding of the details of free energy of complex formation, as expressed in Table 25, as long as we do not have a rather complete theory for all the interactions of the solvent with the complexes. These long-range interactions have quite small influences on the energy differences, as measured from the absorption spectrum; and consequently, we may apply M.O. theory to the rather ideal case of an isolated complex ion or molecule. But it is obvious that the interaction of this entity, if charged, with other molecules of a solvent cannot be described by simple, electrostatic considerations. Even more than the thermodynamics of complex equilibria, the mechanisms of reactions are practically inaccessible to calculations at the moment.

The following is a short resumé of the complex chemistry of most of the heavy metals to be discussed in the next chapter, ppt. is "precipitate", and "amphoteric" indicates the formation of hydroxo complexes, having negative charge*.

Titanium.—The composition of the species of Ti(IV) in acidic solution is not known. At pH 1 to 14, hydrated $Ti(OH)_4$ is precipi-

---

* cf. Table 34 in Chapter 16 as a survey of co-ordination numbers and symmetry. Abbreviations of ligands are given in Table 10, p. 108.

tated. TiO_3^{2-} is extremely alkaline, but many solid salts are known. The peroxo complexes are much more amphoteric, being red cations in acid and colourless anions in alkaline solution. TiF_6^{2-} is very stable. In acid solution, Zn may reduce Ti(IV) to purple $Ti(H_2O)_6^{3+}$, forming blue Ti(III) chloro complexes in strong HCl, and dark brown Ti(III, IV) chloro complexes by partial oxidation. The Ti(III) solutions and black $Ti(OH)_3$ evolve hydrogen with water.

Vanadium.—In acid solution, yellow VO_2^+ (or $V(OH)_4^+$) and dark red polyvanadates represent $+5$, while colourless $V_3O_9^{3-}$ and HVO_4^{2-} occur in alkaline solution. The blue vanadyl ion is probably VO^{2+}, forming a series of complexes with four more ligand atoms, e.g. VO aca_2 and the blue VO ox_2^{2-}. At higher pH, brown ppt. redissolving in OH^- to brown V(IV) solutions. In HCl and H_2SO_4, V(III) forms green chloro and sulphate complexes. The violet $V(H_2O)_6^{3+}$ can be prepared in $HClO_4$. At higher pH, dark yellow hydroxo complexes in solution, and finally olive green $V(OH)_3$ ppt. The purple $V(H_2O)_6^{2+}$ is very unstable towards hydrogen evolution; a solid mixture $(NH_4)_2(V, Zn)(H_2O)_6(SO_4)_2$ can be kept in air.

Chromium.—Cr(VI) is yellow CrO_4^{2-} for pH > 5 and orange $HCrO_4^-$ below 4. In strong H_2SO_4, red CrO_3 can be prepared, but the liquid CrO_2Cl_2 does not seem to form cations. Cr(III) as violet $Cr(H_2O)_6^{3+}$ is very robust. The green $Cr(H_2O)_5OH^{2+}$, formed at pH ~ 4, polymerizes to dark green solutions, which do not react immediately with H_3O^+. The greyish green $Cr(OH)_3$ is soluble in excess OH^-, forming green anions. The green Cr(III) sulphate and chloro complexes, violet Cr ox_3^{3-}, purple Cr aca_3, purple $Cr(NCS)_6^{3-}$, and purple Cr $enta(H_2O)^-$ are easily prepared, at least in solution, while the monomeric ammine complexes as the cherry-red $Cr(NH_3)_5$ Cl^{2+} and yellow $Cr(NH_3)_6^{3+}$ can only be prepared from anhydrous $CrCl_3$ and liquid ammonia, or by oxidizing Cr (II). The dark red solutions of Cr(III) in aqueous NH_3 and other amines contain polynuclear "Jowitschitch" complexes. The blue Cr(II) aquo ion is very reducing, it has $N_C = 4$, since a purple Cr en_2^{2+} is formed.

Manganese.—The strongly coloured, purple MnO_4^- is well known. In alkaline solution, the green MnO_4^{2-} and in very strong OH^- the blue MnO_4^{3-} are stable, while they disproportionate to the brownish black MnO_2 and MnO_4^- at lower pH. Pale yellow MnF_6^{2-}, red Mn(III) fluorides, red Mn ox_2^- and the purple solutions of Mn(III) in strong H_3PO_4 and H_2SO_4 are the best known examples of $+4$ and $+3$, having $N_C = 6$ and 4, respectively. The pale pink $Mn(H_2O)_6^{2+}$

form complexes as the tan-coloured Mn enta^{2-} and greyish green Mn en$_3^{2+}$. However, in alkaline solution, these complexes are often easily oxidized by air, as also Mn(OH)$_2$. A few tetrahedral Mn(II) complexes are known, as the strongly fluorescent salts of yellow MnBr$_4^{2-}$.

Iron.—The dark purple FeO$_4^{2-}$ evolves slowly O$_2$, even in strong OH$^-$. Pure Fe(H$_2$O)$_6^{3+}$ is pale purple, as can be observed in strong HClO$_4$, but the brown hydroxo complexes formed at higher pH, or the strongly yellow chloro complexes, are better known to most chemists. In strong HCl, tetrahedral FeCl$_4^-$ occurs. Octahedral complexes of many anions, green Fe ox$_3^{3-}$, blue-green Fe mal$_3^{3-}$, white FeF$_6^{3-}$; and the pale blue Fe urea$_6^{3+}$ are known. The pale green Fe(H$_2$O)$_6^{2+}$ oxidizes very slowly in air in acid solution, but its complexes such as Fe en$_3^{2+}$ and Fe(OH)$_2$ are very reducing at higher pH. The oxidation products, containing mixed $+2$ and $+3$, as also the mixed cyanide complexes, are strongly coloured.

Cobalt.—The blue Co(H$_2$O)$_6^{3+}$ can be made by electrolysis or from the green Co(CO$_3$)$_3^{3-}$ in saturated NaHCO$_3$, added quickly to an excess of HClO$_4$. However, it evolves O$_2$ with water. Most Co(III) ammine complexes are made by oxidizing a solution of the pink Co(H$_2$O)$_6^{2+}$ and an amine with H$_2$O$_2$ or air:

$$2\text{Co(H}_2\text{O)}_6^{2+} + 10\text{NH}_3 + \text{O}_2 \rightarrow (\text{NH}_3)_5\text{CoOOCo(NH}_3)_5^{4+} + \\ + 12\text{H}_2\text{O} \quad (231)$$

The dark brown peroxo complex formed can be decomposed by warming (to raspberry-red Co(NH$_3$)$_5$OH^{2+}) or by acids, forming strawberry-red Co(NH$_3$)$_5$H$_2$O^{3+}, purple Co(NH$_3$)$_5$Cl^{2+}, violet Co(NH$_3$)$_5$Br^{2+}, etc. If the yellow Co(NH$_3$)$_6^{3+}$ or Co en$_3^{3+}$ is required, active carbon catalyses the formation in NH$_4^+$ or en H$^+$ solutions:

$$4\text{Co(H}_2\text{O)}_6^{2+} + 20\text{NH}_3 + 4\text{NH}_4^+ + \text{O}_2 \rightarrow 4\text{Co(NH}_3)_6^{3+} + \\ + 26\text{H}_2\text{O} \quad (232)$$

Other Co(III) complexes, as the yellow Co(NO$_2$)$_6^{3-}$, green-blue Co ox$_3^{3-}$, purple Co enta$^-$ and Co enta(H$_2$O)$^-$, are also formed by oxidation of the corresponding Co(II) complexes. Tetrahedral Co(II) occurs in very strong solutions of some ligands: blue CoCl$_4^{2-}$, green-blue CoBr$_4^{2-}$, olive-green CoI$_4^{2-}$, blue Co(NCS)$_4^{2-}$, and violet Co(OH)$_4^{2-}$.

Nickel.—The green Ni(H$_2$O)$_6^{2+}$ form instantaneously and with rather large affinity complexes with an excess of ligands, such as blue Ni gly_3^-, blue Ni $enta^{2-}$, violet $Ni(NH_3)_6^{2+}$, purple Ni en_3^{2+}, and red Ni $phen_3^{2+}$. This is connected with the fact that the (non-amphoteric) $Ni(OH)_2$ only precipitates at rather high pH $\sim$ 8. Lower values of Δ than of the aquo ion are presented by the yellow solution in concentrated H_2SO_4 and the solid salts yellow-green $KNiF_3$, yellow $CsNiCl_3$, and orange $pyHNiBr_3$. Tetrahedral high-spin complexes are rare; blue $NiCl_4^{2-}$ is known in salt melts and salts of large tetra (alkyl) ammonium ions. Square-planar, low-spin complexes are known of many complicated, organic ligands. Yellow $Ni(CN)_4^{2-}$ and Ni $trien^{2+}$ are the simplest cases. The latter compounds are often somewhat robust*

Copper.—The red-brown Cu(III) periodate complexes (low-spin) and pale green CuF_6^{3-} (high-spin) are among the rather scarce examples of +3. The blue $Cu(H_2O)_6^{2+}$ has with most ligands $N_c = 4$, as demonstrated by violet $Cu(NH_3)_4^{2+}$ and purple Cu en_2^{2+}. Cu(II) forms complexes with amino-acids, as the blue Cu gly_2 and Cu $enta^{2-}$, and with anions, as blue Cu ox_2^{2-}. However, the strong tendency of forming hydroxo complexes and precipitating $Cu(OH)_2$ counteract the complex formation with many amines. In strong chloride and bromide solutions, the dark yellow $CuCl_4^{2-}$ and dark purple $CuBr_4^{2-}$ are formed. Cu(I) aquo ions are very unstable towards disproportionation to metallic copper and Cu(II). Only complexes are well known, as the precipitates (much resembling Ag(I)) of white CuCl, CuBr, CuI, and the colourless solutions of $Cu(NH_3)_2^+$, $Cu(CN)_2^-$ and $Cu(CN)_4^{3-}$. Cu(II) is even reduced to +1 by I^- or CN^-.

Zinc.—The colourless aquo ion is probably $Zn(H_2O)_6^{2+}$. It forms amine complexes as $Zn(NH_3)_4^{2+}$ and Zn en_3^{2+}, amino acid complexes as Zn $enta^{2-}$, is amphoteric, forming $Zn(OH)_4^{2-}$, and forms the tetrahedral $Zn(CN)_4^{2-}$ and $ZnCl_4^{2-}$. However, these complexes are weaker than most of the corresponding complexes of Ni(II) and Cu(II). They do not show absorption bands in the measured range.

Molybdenum.—In alkaline solution, +6 occurs as colourless MoO_4^{2-}. At low pH, polymolybdates and molybdic acid is ppt. It re-dissolves in strong HCl. The latter solution can be reduced by I^- (catalysed by phosphates, forming the yellow hetero-polyanion $P(Mo_3O_{10})_4^{3-}$) to the green $MoOCl_5^{2-}$ (hydrolysing below 6 M HCl,

---

* Monomeric Ni aca_2 is red and low-spin, while the green high-spin Ni_3aca_6 recently has been shown to be a trimer with octahedral microsymmetry of six oxygen atoms.

forming brown colours) and by Zn or $SnCl_2$ in 12 M HCl to red $MoCl_6^{3-}$ and $MoCl_5(H_2O)^{2-}$. In weaker acids, or in H_2SO_4, Mo(V) occurs as brown ppt., and magnificent blue colours of Mo(V, VI) aɔɹ often observed. SCN^- and Sn^{2+} (which reduces the red Fe(III)) forms a blood-red Mo(V) thiocyanate complex, extractable in organic solvents.

Technetium.—Colourless TcO_4^- is much less oxidizing than MnO_4^- $TcCl_6^{2-}$, $TcBr_6^{2-}$ and TcI_6^{2-} have been prepared.

Ruthenium.—Orange, volatile, RuO_4 is more easily reduced than OsO_4. The green RuF_6^- and RuF_5 (or perhaps rather Ru_2F_{10}) are not stable in aqueous solution. In strong OH^-, the orange RuO_4^{2-} can be formed by melting Ru with NaOH and an oxidizing agent and extracting the mass with water. At low $[OH^-]$, disproportion to a black ppt. of RuO_2, xH_2O and greyish green RuO_4^- occurs. With strong HCl, the dark brown, diamagnetic Ru(IV) complex $Cl_5RuORuCl_5^{4-}$ is formed. The dark orange $RuCl_6^{2-}$ can be formed by oxidation of $RuCl_6^{3-}$ with Cl_2. The latter, strawberry-red anion, can be formed by prolonged boiling of the higher oxidation numbers with formaldehyde in strong HCl. At 4 M HCl, it aquates to forms such as $RuCl_5(H_2O)^{2-}$. Many complexes, e.g. the green Ru ox_3^{3-}, yellow $Ru(NH_3)_5Cl^{2+}$ and white $Ru(NH_3)_6^{3+}$ have been prepared. In strong HCl, reduction to dark blue Ru(II) is possible. In aqueous NH_3-NH_4Cl, zinc dust may produce the orange-yellow $Ru(NH_3)_6^{2+}$. Ruthenium red is a trimetric Ru(III, III, IV) amine–dioxo complex. The purple $Ru(NO)Cl_5^{2-}$ is one of the cases where the oxidation number of the central ion is not well-defined, it may be $+2$, $+3$, $+4$, according to whether the ligand, having also a partly filled M.O., is assumed to be NO^+, NO or NO^-.

Rhodium.—Yellow $Rh(H_2O)_6^{3+}$ can be formed by prolonged boiling of $RhCl_6^{3-}$ with strong $HClO_4$. It forms amphoteric $Rh(OH)_3$ with bases. The raspberry-red $RhCl_6^{3-}$ aquates slowly, even in 1 M HCl, to $RhCl_5(H_2O)^{2-}$ and lower complexes. Similarly, the equilibrium

$$RhBr_6^{3-} + H_2O = RhBr_5(H_2O)^{2-} + Br^- \qquad (233)$$

is at the left-hand side in 5MNaBr, but at the right-hand side in 1 M NaBr, according to the absorption spectra. The pale yellow $Rh(NH_3)_5Cl^{2+}$ and white Rh en_3^{3+} can fairly easily be prepared, while the other complexes known as yellow Rh ox_3^{3-}, orange $Rh(NH_3)_5I^{2+}$, yellow Rh $enta^-$, etc., require more preparative attention. A series of pyridine complexes comprise orange *cis*- and red *trans*-$Rhpy_2Cl_4^-$ and pale yellow *trans*-$Rhpy_4Cl_2^+$.

Palladium.—The dark red K_2PdCl_6 only produces $PdCl_6^{2-}$ in solution if some free Cl_2 is present. The aquation of yellow-brown $PdCl_4^{2-}$ was discussed on p. 259, as also the yellow $Pd(H_2O)_4^{2+}$. With OH^-, the black ppt. redissolves to colourless hydroxo complexes. The reaction with other ligands is very quick in most instances; however, the pink analogue $[Pd(NH_3)_4][PdCl_4]$ of Magnus' salt may be formed before colourless $Pd(NH_3)_4^{2+}$. The dark orange $PdBr_4^{2-}$, brownish-black PdI_4^{2-}, pale yellow Pd ox_2^{2-}, and all the colourless Pd gly_2, Pd en_2^{2+}, Pd py_4^{2+}, and $Pd(CN)_4^{2-}$ are such examples. A spectrochemical series can almost be established with the shift of wave numbers of the mixed complexes of the tridentate amine den:

$$Pd \ den \ I^+ < Pd \ den \ Br^+ < Pd \ den \ Cl^+ < Pd \ den \ (OH)^+$$
$$< Pd \ den \ (NH_3)^{2+} \quad (234)$$

Lanthanides.—The oxidation number is nearly always $+3$ in aqueous solution, except occasionally $+4$ for cerium (forming anion complexes such as $Ce(NO_3)_6^{2-}$ and $Ce(SO_4)_3^{2-}$ and having a very complicated hydrolysis behaviour in acid solution) and $+2$ for europium (resembling Ba^{2+}, except for the strongly reducing character). The trivalent aquo ions probably have $N = 9$, as in the solid $Nd(H_2O)_9(BrO_3)_3$. They react at pH well above 6 to ppt. the non-amphoteric hydroxides, and form complexes such as R enta$^-$. While RF_3 is very insoluble, as is $R_2 ox_3$, the chloro complexes are only formed above 6 M HCl, as judged from the absorption spectra. The tendency towards complex formation, also with OH^-, increases steadily through the series of increasing Z, corresponding to decreasing ionic radius, except for a possible effect of missing ligand-field stabilization at the half-filled shell $4f^7$ Gd(III).

Wolfram.—In alkaline solution, the colourless WO_4^{2-} occurs. The behaviour with strong acids resembles that of Mo(VI), except that it is more difficult to reduce. The hetero-polyanions such as $P(W_3O_{10})_4^{3-}$ are colourless. Most complexes of $+3$ known are dimeric, as the olive-green $W_2Cl_9^{3-}$.

Rhenium.—The colourless ReO_4^- has a property common to both Mo and As, i.e. it precipitates more easily (as black Re_2S_7) in strong than in weak HCl. It can be reduced in acid solution to various brown Re(V) and Re(IV) complexes. The mononuclear entities, pale blue ReF_6^{2-}, pale green $ReCl_6^{2-}$, orange $ReBr_6^{2-}$, and dark purple ReI_6^{2-}. At increased pH, they hydrolyse and form black ppt. The blue-green Re(I) hexacyanide and gery Re(III) tetra-hydrido complexes represent low oxidation numbers. Re(II) and Re(III) are also known in HCl solutions.

Osmium.—The colourless, volatile OsO_4 is used for the isolation of this element from the rest of the platinum group. It reacts with strong HCl to form the yellow $OsCl_6^{2-}$ (of which the $N(C_2H_5)_4^+$ salt is yellow, as expected from the additivity of ionic colours, while K_2OsCl_6 is dark red, due to a co-operative effect). Analogously, boiling with strong HBr produces orange $OsBr_6^{2-}$, while HI (I_2 can be removed with ascorbic acid) and KI forms the greyish purple OsI_6^{2-}. These halide complexes can be reduced to Os(III) with Ag powder:

$$OsCl_6^{2-} + Ag + 2Cl^- = OsCl_6^{3-} + AgCl_2^- \qquad (235)$$

but the pale yellow Os(III) complexes (which have been identified from the close analogy of spectra with Ru(III)) re-oxidize quickly to Cs(IV). The highest fluoride is the volatile OsF_6. Yellow-brown $OsO_2Cl_4^{2-}$ and $OsO_2(OH)_4^{2-}$ are remarkably stable.

Iridium.—The volatile IrF_6 exhibits an interesting $5d^3$-spectrum. The constitution of the purple solution of Ir(IV), prepared by boiling its compounds, e.g. with $HClO_4$, is not known. With OH^-, blue ppt. of IrO_2, xH_2O is formed, which is slightly amphoteric. The dark orange $IrCl_6^{2-}$ and dark violet blue $IrBr_6^{2-}$ are strong oxidizing agents (eqns. 229 and 230). Not many other Ir(IV) complexes are known, apart from the dark purple $IrpyCl_5^-$, *cis-* and *trans-*$Irpy_2Cl_4$, and the dark green $N\{Ir(SO_4)_2(H_2O)\}_3^{4-}$. The olive-green $IrCl_6^{3-}$ and $IrBr_6^{3-}$ aquate slowly at low halide concentration; the colourless amine complexes such as $Ir(NH_3)_5Cl^{2+}$ and Ir en_3^{3+} are very difficult to prepare, the latter has to be heated to 180°C for several days in an autoclave. A series of oxalate and pyridine complexes are known, e.g. the yellow *cis-* and the pink *trans-*$Irpy_2Cl_4^-$.

Platinum.—The best known compounds of $+4$ contain colourless PtF_6^{2-}, yellow $PtCl_6^{2-}$, tomato-red $PtBr_6^{2-}$ and dark red PtI_6^{2-}. Several amine complexes, such as the colourless $Pt(NH_3)_6^{4+}$, $Pt(NH_3)_5Cl^{3+}$ and Pt $en_3^{:+}$, can be prepared. They are acids with pK between 6 and 9, forming the yellow amido complexes as $Pt(NH_3)_5NH_2^{3+}$. The square-planar strawberry-red $PtCl_4^{2-}$, cherry-red $PtBr_4^{2-}$, and dark olive-green PtI_4^{2-} represent $+2$. They often react very slowly with amines, forming Magnus-salt intermediates of the type $[Pten_2][PtCl_4]$ before redissolving to colourless Pt $en_2{}^{2+}$. The co-operative effects of more platinum atoms in solid salts of this category and of $Pt(CN)_4{}^{2-}$ were discussed (p. 204). Definite complexes with H_2O as ligand are very little known, except of cases as *cis*- and *trans*-$Pt(NH_3)_2(H_2O)_2{}^{2+}$.

Gold.—The yellow $AuCl_4{}^-$ reacts with OH^-, forming $AuCl_3(OH)^-$, . . ., $Au(OH)_4{}^-$. The red $AuBr_4{}^-$ and $Au(SCN)_4{}^-$ have both a strong tendency of reduction to $+1$ (as the colourless Au(I) complexes, cf. $Au(CN)_2{}^-$) and, even more, to the metal. The colourless Au $en_2{}^{3+}$ is known to be an acid, forming yellow $Au(en-H)en^{2+}$. The amide ppt., forming in aqueous NH_3, is strongly explosive.

Mercury.—$HgCl_2$, $HgBr_2$ and HgI_2 (colourless in solution, not as the red or yellow crystals) are very slightly dissociated in aqueous solution. With an excess of halide, they form the tetrahedral ions such as the pale yellow $HgI_4{}^{2-}$. $Hg(SCN)_4{}^{2-}$ and the thio-anions are very strong complexes, and the ppt. of HgS is an even stronger complex. Hg(I) occurs as the dimeric, colourless $Hg_2{}^{2+}$, having a strong tendency to disproportionation to Hg and Hg(II) in the presence of complexing ligands. Only the solid Hg(I) halide ppt. are well-defined complexes. The unusually high formation constants of several Hg(II) complexes can be seen from the almost quantitative reactions:

$$Hg(H_2O)_2{}^{2+} + 2NH_4{}^+ = Hg(NH_3)_2{}^{2+} + 2H_3O^+$$
$$HgO + 4I^- + H_2O = HgI_4{}^{2-} + 2OH^- \qquad (236)$$

Thallium.—The colourless Tl(III) aquo ion is a strong acid, readily forming brown, non-amphoteric hydroxide ppt. The chloro- and bromo-complexes are in equilibrium with small amounts of Cl_2 and Br_2. Tl(I) resembles in some respects Ag(I), e.g. in its crystalline halides, and in other respects the alkali metals (as expected from the ionic radius), e.g. by its reluctance to form amine, hydroxo, or cyanide complexes.

Lead.—While brown PbO_2 and red Pb_3O_4 are well known, the solution chemistry of $+4$ is very meagre. The dark yellow $PbCl_6{}^{2-}$ decomposes quickly in HCl, forming Cl_2. Pb(II) forms a colourless aquo ion, being rather acidic, and very amphoteric to form $Pb(OH)_3{}^-$. The halide complexes, such as $PbCl_4{}^{2-}$, have fairly high formation constants. Pb $enta^{2-}$ is known, though amines alone form the hydroxo complexes.

T

Bismuth.—Sodium bismuthate (V) is very oxidizing. Bi(III) forms a colourless, very acidic aquo ion. Among the hydrolysis products seems to be $Bi_6(OH)_{12}^{6+}$, forming many insoluble salts. They are not amphoteric. The highest chloro complex is $BiCl_5^{2-}$ or $BiCl_6^{3-}$ and is formed already in 2 M HCl. The pale yellow bromo and orange iodo complexes also have high formation constants. Bi enta$^-$, but no amine nor cyanide complexes, are known.

Polonium.—The dark yellow $PoCl_6^{2-}$ is more chemically stable than $TeCl_6^{2-}$. Po(II) in its red chloro complexes form probably square-planar entities of the same type as iodine (III) in ICl_4^-.

Thorium.—In aqueous solution, only +4 is known. N of the colourless aquo ion is not known, but it is certain that the ionic charge is +4, and it is, oddly enough, weakly acidic (Th(OH)$_4$ ppt. at pH $\sim$ 4). The hydroxide is not amphoteric. The complexes, such as Th ox$_4^{4-}$, Th enta and Th aca$_4$, have relatively high formation constants. While ThF$_4$ is very insoluble, no evidence has been presented for the formation of chloro complexes.

Protactinium.—The ppt. of Pa_2O_5, xH_2O is very slightly soluble in acids or bases, with the exception of fluoride solutions, and the consequent chemistry of hydroxo complexes is very complicated. The reduction to Pa(IV) is possible in strong acids.

Uranium—does not resemble Mo and W much (more pronounced analogies are present between Nb, Ta, Pa). The yellow, fluorescent UO_2^{2+} forms uranyl complexes with many anions, such as Cl^-, NO_3^-, F^-, ox^{2-}, and especially SO_4^{2-}. With higher pH, yellow ppt. of the hydroxide and polyuranates, containing foreign cations, are formed. The yellow solution of $UO_2(CO_3)_3^{4-}$ obtained with CO_3^{2-} is decomposed by boiling. In acid solution, many reducing agents (e.g. photochemical reactions with alcohols) may form apple-green U(IV). The chemistry of +4 resembles that of thorium; the aquo ion form $U(OH)(H_2O)_x^{3+}$ with pK $= 1.5$ and hydroxide ppt. at slightly higher pH; in 1 M HCl, a monochloro complex is formed, while UCl_6^{2-} exists in 12 M HCl. U ox$_4^{4-}$ and U aca$_4$ have N $= 8$. The grey U(III) aquo ion, as found in $HClO_4$ and < 6 M HCl, is very strongly reducing, evolving H_2 rather quickly. In strong HCl, a dark red chloro complex is formed.

Neptunium—forms the pink NpO_2^{2+}, resembling the uranyl ion, though more oxidizing, and less acidic. The green-blue NpO_2^+ does not seem to form strong complexes and is not acidic below pH $= 6$. The yellow-green $Np(H_2O)_x^{4+}$ is closely analogous to U(IV). The purple $Np(H_2O)_9^{3+}$ is easily oxidized, but does not evolve H_2 with water.

Plutonium—exhibits only small differences from the chemistry of Np. The orange PuO_2^{2+} is very oxidizing. The disproportionation of the (presumably) pink PuO_2^+ and Pu(IV) is a slow reaction, since (as is the case for all actinides) the equilibria MO_2^{2+} : MO_2^+ and M^{4+} : M^{3+} are quickly established, but not the reactions changing the oxo ligands. Further complications arise from the polymerization of the yellow-brown Pu(IV) aquo ion to bright green hydroxo complexes (having an average charge $+0.15$ per Pu, as also known from Ce(IV) chemistry) and the reduction from H_2O_2, produced by the α-radioactivity. $Pu(NO_3)_6^{2-}$ and $PuCl_6^{2-}$ are well-established complexes as also Pu(IV) sulphates. The lavender-blue $Pu(H_2O)_9^{3+}$ is closely analogous to the lanthanides.

Americium.—The prominent oxidation number is $+$ 3, as found in the pink aquo ion. Solid AmF_4 (soluble with pink colour in 15M NH_4F), purple AmO_2^+, and brown AmO_2^{2+} are exceptions.

Curium.—In aqueous solution, only the pale yellow Cm(III) aquo ion is known. Solid CmF_4 has been prepared.

Berkelium.—Both $+3$ and $+4$ form fairly stable aquo ions. Their absorption spectra would be highly interesting.

The occurrence of a definite oxidation number of a given element depends obviously on the electrochemical potential U of the solution and its content of ligands, able to form strong complexes. Table 26 summarizes the information for elements in 1 M acid solution, for $U = -0.5, 0, +0.5, +1$ and $+1.5$ V, compared to the hydrogen electrode. Values of U do not often occur in practice outside this range in aqueous solution, since the over-voltage, preventing the equilibrium establishment of hydrogen or oxygen evolution is rarely very large. The oxidation numbers in parenthesis in the columns refer to chloro complexes of such elements, where the aquo ions are not well known. The numbers in parenthesis outside the columns indicate other oxidation numbers, often encountered in complexes. The distribution of oxidation numbers at a given U is often strongly dependent on pH. The formation of hydroxide precipitates and amphoteric hydroxo complexes in solution has a general tendency to favour higher oxidation numbers at high pH. Conspicuous effects of this kind are observed in V, Cr, Mn, Co, Ru, Ce, Np and Pu. In some cases, the equilibria are not generally established. Thus, the platinum group metals do not readily form their chloro complexes in hydrochloric acid (or dissolve in any other solvent), and tungsten and rhenium metal exhibit similar difficulties. In these examples, the stability ranges of oxidation numbers in Table 26 are only qualitatively certain. For illustrating the behaviour in another environment than aqueous solutions, the oxidation number achieved with elementary fluorine, either alone or in combination with KF or CsF is also indicated. In most cases, this

TABLE 26. THE OXIDATION NUMBERS OBTAINED IN 1 M NON-COMPLEXING ACID SOLUTION IN WATER (CHLORO COMPLEXES IN HCl GIVEN IN PARENTHESES) FOR VARIOUS ELECTROCHEMICAL POTENTIALS U AS DEFINED IN EQN. (197)
(After the column U = 1·5 V are indicated in parentheses many other important oxidation numbers. Finally, the oxidation number obtained in anhydrous state with elementary fluorine and an alkali fluoride is indicated.)

Z	Element	−0·5V	0	+0·5	+1·0	+1·5	Anhydrous, F_2 + KF (or CsF)
1	H	0	0, 1	1	1	1	1
2	He	0	0	0	0	0	0
3	Li	1	1	1	1	1	1
4	Be	2	2	2	2	2	2
8	O	−2	−2	−2	−2	0	2
9	F	−1	−1	−1	−1	−1	0
10	Ne	0	0	0	0	0	0
11	Na	1	1	1	1	1	1
12	Mg	2	2	2	2	2	2
13	Al	3	3	3	3	3	3
17	Cl	−1	−1	−1	−1	5, 7 (0, 1)	3
18	A	0	0	0	0	0	0
19	K	1	1	1	1	1	1
20	Ca	2	2	2	2	2	2
21	Sc	3	3	3	3	3	3
22	Ti	3	3, 4	4	4	4	4
23	V	2	3	4	4, 5	5	5
24	Cr	2	3	3	3	6	4
25	Mn	2	2	2	2	7 (3, 4)	4
26	Fe	0	2	2	3	3	3
27	Co	0	2	2	2	2 (3)	3 (4)
28	Ni	0	2	2	2	2	4 (3)
29	Cu	0	0	2	2	2 (1)	3
30	Zn	2	2	2	2	2	2
31	Ga	0, 3	3	3	3	3	3
33	As	−3, 0	0	3	5	5	5
34	Se	−2	0	0	4	6	6
35	Br	−1	−1	−1	−1, 0	5	5
36	Kr	0	0	0	0	0	0
37	Rb	1	1	1	1	1	1
38	Sr	2	2	2	2	2	2

Z	Element	−0·5V	0	+0·5	+1·0	+1·5	Anhydrous F_2 + KF (or CsF)
39	Y	3	3	3	3	3	3
40	Zr	4	4	4	4	4	4
41	Nb	4	5	5	5	5	5
42	Mo	(3)	(3)	(5)	6	6	6
43	Tc	—	—	4?	—	7	—
44	Ru	0	0, (2)	(3)	(4)	8	5
45	Rh	0	0	0, (3)	(3)	(3)	4
46	Pd	0	0	0, (2)	(2)	(4)	4
47	Ag	0	0	0	1	1	3
48	Cd	0	2	2	2	2	2
49	In	0	3	3	3	3	3
50	Sn	0	2	4	4	4	4
51	Sb	0	0	3	5	5	5
52	Te	0	0	0, 4	4	6 (−2)	6
53	I	−1	−1	−1, 0	0, 5	5, 7 (1, 3)	7
54	Xe	0	0	0	0	0	0
55	Cs	1	1	1	1	1	1
56	Ba	2	2	2	2	2	2
57	La	3	3	3	3	3	3
58	Ce	3	3	3	3	4	4
59	Pr	3	3	3	3	3	4
60	Nd	3	3	3	3	3	3 (4)
63	Eu	2	3	3	3	3	3
64	Gd	3	3	3	3	3	3
70	Yb	3	3	3	3	3 (2)	3
71	Lu	3	3	3	3	3	3
72	Hf	4	4	4	4	4	4
73	Ta	(4)	5	5	5	5	5
74	W	(3)	(5)	6	6	6	6
75	Re	—	(4)	(4), (5)	7	7	6, 7
76	Os	0	0, (3)	(4)	(4)	8	6
77	Ir	0	0	0, (3)	(3), (4)	(4)	6
78	Pt	0	0	0, (2)	(4)	(4)	6
79	Au	0	0	0	0, (3)	(3), (1)	3
80	Hg	0	0	0	2	2 (1)	2
81	Tl	0	1	1	1, 3	3	3
82	Pb	0	2	2	2	4	4
83	Bi	0	0	3	3	3	3
88	Ra	2	2	2	2	2	2
89	Ac	3	3	3	3	3	3
90	Th	4	4	4	4	4	4
91	Pa	4	4, 5	5	5	5	5
92	U	3, 4	4	6	6	6	6
93	Np	3	4	4	5	6	6
94	Pu	3	3	3	3, 4	6	6
95	Am	3	3	3	3	3, 5, 6	4
96	Cm	3	3	3	3	3	4
97	Bk	3	3	3	3	3, 4	4?

oxidation number is the largest obtainable. However, in some cases (Cr, Mn, Ru, Os), oxygen in alkaline solution may produce a higher oxidation number. This is probably caused by the difficulty of co-ordinating more than six fluoride ions to most small central ions.

The stability of the solid, metallic elements introduce rather important variations in Table 26. It is in most cases true that those metals (such as Fe, Co, Ni, Cu, Ru, Rh, Pd, Re, Os, Ir, Pt, Au), forming the strongest complexes, have the highest heats of vaporization, i.e. the strongest bonding by the formation from gaseous atoms. However, this is not quite a general rule; Hg has a lower heat of vaporization than the alkali metals. In the same way as the chelate effect, mentioned p. 255, depends on the dimension in concentration units of the step formation contants, the equilibrium between solid phases, such as metals, and metal complexes in solution depends on the total concentration of the latter. However, since a factor of 10^8 is necessary in the concentration of a monovalent metal for shifting U by 0·5 V, this dependence does not diminish the practical use of Table 26.

The characteristic, flat maximum of the most common oxidation number as a function of Z is very pronounced in Table 26, if for instance the column U $= +0.5$ V is considered, corresponding to a solution which, to most chemists, would neither be very reducing nor very oxidizing:

$$
\left.
\begin{array}{ll}
\text{3d:} & 3, 4, 4, 3, 2, 2, 2, 2, 2, 2 \\
\text{4d:} & 3, 4, 5, (5), (4), (3), (3), (2), 0, 2 \\
\text{5d:} & 3, 4, 5, 6, (4), (4), (3), (2), 0, 0 \\
\text{4f:} & 3, 3, 3, 3, 3, 3, \ldots \\
\text{5f:} & 3, 4, 5, 6, 4, 3, 3, 3, 3, \ldots
\end{array}
\right\}
\qquad (237)
$$

From that point of view, the 5f-group obviously resembles the three ordinary transition groups much more than the lanthanides, at least at the beginning. Hence, the name *uranides* may be quite adequate for the series U, Np, Pu, . . . On the other hand, the spectroscopic evidence from gaseous ions and from complexes, presented above, proves most firmly that it is a 5f-group rather than a 6d-group. Further, interesting chemical differences are observed between the 5f-group on the one hand and the 4d- and 5d-groups on the other: the oxidation numbers $+6$ and $+5$ (except Pa) have the rather unique property of forming dioxo complexes, and $+4$ and $+3$ resemble A elements (eqn. 226), readily forming aquo ions, while the corresponding 4d- and 5d-elements are only known in complexes of halides and other ligands. The latter distinction probably reflects the difference in ionic radii, being much larger in the lanthanides and actinides than in d^n-ions in $+3$ and $+4$. The influence of f-electrons on the chemical properties of substances is rather passive, all the lanthanides thus resembling the closed-shell ions La^{3+} and Lu^{3+}. No clear-cut effect of f-electrons on the chemical properties of the actinides has been demonstrated as yet.

CHAPTER 15

TABLES OF SOME ABSORPTION SPECTRA

TABLES 27–33 are classified according to the electron configuration of the corresponding gaseous ion. As discussed in several of the previous chapters, this is rather a convenient classification than a quite valid approximation to the actual wave function. However, it is a rather unambiguous classification for the complexes compiled here.

TABLE 27. THE ABSORPTION SPECTRA OF d-GROUP COMPLEXES WITH IDENTIFICATION
OF THE EXCITED LEVELS
(The electron transfer spectra of hexahalide complexes, however, are given in Table 32, and the ligand field transitions of octahedral d³, low-spin d⁶ and high-spin d⁸ and of high-spin d⁵ are given in Tables 28–31, "redox" indicates the usual type of electron transfer, "vibr." vibrational structure observed. The values of $S = 0, \frac{1}{2}, 1, \frac{3}{2}, \ldots$ of the excited states are sometimes marked singlets, doublets, triplets, quartets, when no certain Γ_n is known. References followed by * or † indicate assignments by other authors, who either agree (*) or disagree (†) with that proposed here. Shoulders are given in parentheses. Wave numbers are in kK. For abbreviations of ligands, see Table 10, p. 108.)

			Ref.
3d⁰	Ti(IV)		
	TiCl₆²⁻(12 M HCl)	45·0 $\pi \to \gamma_5$	293
	TiCl₄ gaseous	31·4	390
	Ti(O₂)²⁺(?)	23·3 redox	454
	V(V)		
	VO₄³⁻(NaOH)	36·8	81b,* 427
	V(O₂)³⁺(?)	22·2 redox	454
	Cr(VI)		
	CrO₄²⁻	26·8 $\pi \to \gamma_3$, 36·6	81b,* 81d, 620†
	CrO₃(OH)⁻	22, 29, 37	18a, 225
	CrO₃F⁻	22·2, 28·5 vibr., 36 redox	225*
	CrO₂Cl₂	(19·3), 24·0, 33 redox	225*
	Mn(VII)		
	MnO₄⁻	(14·5), (15·9), vibr. 18·6 $\pi \to \gamma_3$,	26,* 81d, 57, 394, 620†
		vibr. 31, 32·3, (44)	
3d¹	Ti(III)		
	Ti(H₂O)₆³⁺	(17·4), 20·3 ²Γ_3, Jahn–Teller	244,* 284
	Ti (urea)₆³⁺	16·4, 18·1 ²Γ_3	207
	TiCl₃, alcohols	(15·2), 16·8 ²Γ_3	207
	Ti(III) 12 M HCl	15·2, 19·1 ²Γ_3	284
	Ti(III)Al₂O₃	17·8 ²Γ_3	402a
	V(IV)		
	VCl₄	9·0 ²Γ_5	439*
	VO²⁺	13·1 ($\epsilon = 16·5$), (16·0, $\epsilon = 7·5$)	163, 284, 494
		d-shell, 41·7 ($\epsilon = 240$, broad)	
	VO enta²⁻	12·8, 17·2, (29·8) d-shell	284
	VO ox₂²⁻	12·6, 16·5, (29·4) d-shell	204, 284
	VO tart²₋	11·0, (17·1), 18·8, 25·3 d-shell	284
	VO SCN⁺	13·9, 17·2 d-shell	163
	VO aca₂	13·3, 17·4, 25 d-shell	145, 284

Table 27—*continued.*

			Ref.
	Cr(V)		
	CrO_4^{3-}	$16 \cdot 0\ \gamma_3 \rightarrow \gamma_5$, $28 \cdot 2\ \pi \rightarrow \gamma_3$	81†
	Mn(VI)		
	MnO_4^{2-}	$16 \cdot 5$, $\gamma_3 \rightarrow \gamma_5$, $22 \cdot 9\ \pi \rightarrow \gamma_3$, $28 \cdot 5$, $33 \cdot 3$	57, 81,† 81b, 375
$3d^2$	V(III)		
	$V(H_2O)_6^{3+}$, alum	$17 \cdot 8\ ^3\Gamma_5$, $25 \cdot 7\ ^3\Gamma_4$	163, 201,† 209, 437*
	V(III) 2 M HCl	$17 \cdot 0\ ^3\Gamma_5$, $25 \cdot 3\ ^3\Gamma_4$	284
	V(III) 12 M HCl	$14 \cdot 8\ ^3\Gamma_5$, $22 \cdot 3\ ^3\Gamma_4$	284
	VF_6^{3-}	$14 \cdot 8\ ^3\Gamma_5$, $23 \cdot 0\ ^3\Gamma_4$	27a*
	V (urea)$_6^{3+}$	$16 \cdot 2\ ^3\Gamma_5$, $24 \cdot 2\ ^3\Gamma_4$, (27)	209*
	V ox$_3^{3-}$	$16 \cdot 45\ ^3\Gamma_5$, $20 \cdot 48\ ^1\Gamma_1$, $23 \cdot 5\ ^3\Gamma_4$	209*, 473c*
	V mal$_3^{3-}$	$16 \cdot 45\ ^3\Gamma_5$, $23 \cdot 7\ ^3\Gamma_4$	209*
	$V(SCN)_6^{3-}$	$16 \cdot 7\ ^3\Gamma_5$, high band at 24	163
	V(III) Al_2O_3	$8 \cdot 8$, $9 \cdot 7\ ^3\Gamma_3$, $^1\Gamma_5$, $17 \cdot 4\ ^3\Gamma_5$, $21 \cdot 02$ $^1\Gamma_1$, $25 \cdot 2\ ^3\Gamma_4$	373,* 402a, 480*
	V enta$^-$	8?, $12 \cdot 5$, $(19 \cdot 4)$, $22 \cdot 6$ d-shell	275
	Mn(V)		
	MnO_4^{3-}	$14 \cdot 8\ \gamma_3 \rightarrow \gamma_5$ $30 \cdot 5\ \pi \rightarrow \gamma_3$	81†
	Fe(VI)		
	FeO_4^{2-}	$12 \cdot 7\ \gamma_3 \rightarrow \gamma_5$ $19 \cdot 6\ \pi \rightarrow \gamma_3$	81†, 81b*
$3d^4$	Cr(II)		
	$Cr(H_2O)_6^{2+}$	$14 \cdot 1\ ^5\Gamma_5$ Jahn–Teller	236,* 467, 513
	Cr en$_2^{2+}$	$18 \cdot 3\ ^5\Gamma_5$ Jahn–Teller	467
	Cr en$_3^{2+}$(?)	$17 \cdot 5$	467
	Cr(II) 12 M HCl	$12 \cdot 6$	513
	Cr dip$_3^{2+}$($^3\Gamma_4$)	$17 \cdot 8$ Cr–dip	637a
	Mn(III)		
	$Mn(H_2O)_6^{3+}$	$21 \cdot 0\ ^5\Gamma_5$ Jahn–Teller	202*
	Mn(III)H_3PO_4	$19 \cdot 6$, redox at $39 \cdot 2$	482
	Mn aca$_3$	10, $17 \cdot 9$ d-shell	82
	Mn aca$_2$($H_2O)_2^+$	$17 \cdot 3$ d-shell	82
	Mn enta($H_2O)^-$	$(21 \cdot 3)$, $22 \cdot 2$	283
	Mn enta(OH)$^{2-}$	20	283
	Mn ox$_3^{3-}$	$(10 \cdot 3$ spin-forbidden), $20 \cdot 1$	279
	$Mn(CN)_6^{3-}$	$30 \cdot 3\ \pi \rightarrow \gamma_5$, $37 \cdot 0$	2, 343
	Mn(III)HCl	$12 \cdot 0$, $17 \cdot 8$, $25 \cdot 2$	293
$3d^5$	Fe(III)		
	$Fe(CN)_6^{3-}$	$24 \cdot 2$, $\pi \rightarrow \gamma_5$, $(30 \cdot 5)$, $33 \cdot 1$, (35), $38 \cdot 5$, $\sim 50\ \pi \rightarrow \gamma_3$	277, 321, 343, 624
	$Fe(SCN)(H_2O)_5^{2+}$	$21 \cdot 8$ redox	469
$3d^6$	Fe(II)		
	$Fe(H_2O)_6^{2+}$	$10 \cdot 4\ ^5\Gamma_3$, $(14 \cdot 4)$, $19 \cdot 8$, $21 \cdot 1$, $22 \cdot 2$, $25 \cdot 9$ triplet states, $(40 \cdot 5)$ $3d^54s$?	98e, 108, 122, 162,* 265, 293
	Fe enta^{2-}	$9 \cdot 7\ ^5\Gamma_3$	275
	Fe en$_3^{2+}$	$11 \cdot 4\ ^5\Gamma_3$, $19 \cdot 5$, $21 \cdot 8$, $24 \cdot 4$ triplets	88c*
	Fe(II)ZnS	$2 \cdot 8$, $3 \cdot 7\ ^5\Gamma_5$ ($+$Co(II) bands)	373a, 605e
	Fe dip$_3^{2+}$	$19 \cdot 1$, $(20 \cdot 2)$ Fe-dip, $(24 \cdot 8)$, $(25 \cdot 8)$, $28 \cdot 5$, $33 \cdot 5$, $(34 \cdot 5)$, $40 \cdot 5$ dip	286, 615*
	Fe phen$_3^{2+}$	$19 \cdot 6$, $(21 \cdot 0)$, $(22 \cdot 8)$ Fe-phen, $31 \cdot 2$, $(34 \cdot 4)$, $37 \cdot 6$, $44 \cdot 4$ phen	54, 286, 615*
	Co(III)		
	CoF_6^{3-}	$11 \cdot 8$, $14 \cdot 4\ ^5\Gamma_3$ Jahn–Teller	98e*
$3d^7$	Co(II)		
	$Co(H_2O)_6^{2+}$	$8 \cdot 2\ ^4\Gamma_5$, $11 \cdot 3\ ^2\Gamma_3$, $16 \cdot 0\ ^4\Gamma_2$ or doublet, $19 \cdot 4$ and $(21 \cdot 55)\ ^4\Gamma_4$	1,* 264, 265, 452*
	$Co(NH_3)_6^{2+}$	$9 \cdot 0\ ^4\Gamma_5$, $18 \cdot 5\ ^4\Gamma_2$, $21 \cdot 1\ ^4\Gamma_4$	22*
	Co en$_3^{2+}$	$\sim 9 \cdot 4\ ^4\Gamma_5$, $18 \cdot 7\ ^4\Gamma_2$, $21 \cdot 7\ ^4\Gamma_4$	22*
	$Co(H_2O)_4Cl_2$	$18 \cdot 4\ ^4\Gamma_4$, $20 \cdot 1$, $20 \cdot 4$ doublets	458,* 404
	Co enta^{2-}	$9 \cdot 1\ ^4\Gamma_5$, $16 \cdot 3\ ^4\Gamma_2$, $19 \cdot 9$, $20 \cdot 6$, $21 \cdot 5\ ^4\Gamma_4$	275
	Co(II)MgO($6O^{2-}$)	$8 \cdot 5\ ^4\Gamma_5$, $17 \cdot 2$, $18 \cdot 7$, $19 \cdot 6\ ^4\Gamma_4$	372,* 459*
	Co(II)MgTiO$_3$($6O^{2-}$)	$13 \cdot 2\ ^4\Gamma_3$, $16 \cdot 8\ ^4\Gamma_4$	516
	Co(II)CaF$_2$($8F^-$)	$3 \cdot 3\ ^4\Gamma_5$, $5 \cdot 9$ and $6 \cdot 6\ ^4\Gamma_4$, $17 \cdot 8$ and $18 \cdot 55\ ^4\Gamma_4$, $19 \cdot 2$, $19 \cdot 6$, $20 \cdot 4$ doublets	551*

TABLE 27—continued.

			Ref.
	$CoCl_4{}^{2-}$	5·7, 6·3 $^4\Gamma_4$, 14·4, 15·2, (15·5), 16·1, (16·4) mixtures of $^4\Gamma_4$, $^2\Gamma_3$, $^2\Gamma_4$, 18·8, (19·8), 22·3, (22·5), 24·2, 24·7, 26·5, 31·2, 33·0, 34·8 doublets, 43·5 redox	22,* 67, 73a, 121, 166a, 235a, 307, 308, 318, 439*
	$CoBr_4{}^{2-}$	4·5, 5·6 $^4\Gamma_4$, 13·9, (14·4), 14·9, (15·5) mixtures of $^4\Gamma_4$ and doublets	68, 98d, 121, 166a, 235a, 283
	$CoI_4{}^{2-}$	4·4, 4·7, 5·4 $^4\Gamma_4$, 12·5, 13·3, 14·0 $^4\Gamma_4$ and doublets	68, 121, 166a, 235a
	$Co(OH)_4{}^{2-}$	16·1, 17·2	539
	$Co(NCS)_4{}^{2-}$	7·8 $^4\Gamma_4$, 16·3 $^4\Gamma_4$, 32·8 thiocyanate	98h,* 121, 320, 539
	$Co(NCS)_4Hg$	8·3 $^4\Gamma_4$, 16·7 $^4\Gamma_4$	98h*
	$Co(H_2O)_5(NCS)^+$	36·7 thiocyanate	533
	$Co(N_3)_4{}^{2-}$	6·6, 7·3 $^4\Gamma_4$, 14·7, 15·3, (16·4) $^4\Gamma_4$ and doublets	98f,* 534
	$Co(II)MgAl_2O_4(4O^{2-})$	7·3 $^4\Gamma_4$, 15·95, 17·05, (18·2) $^4\Gamma_4$ and doublets	459b,* 516*
	$Co(II)Mg_2TiO_4(4O^{2-})$	14·8, 16·0, (17·0), (19·7) $^4\Gamma_4$ and doublets	516*
	$Co(II)ZnO(4O^{2-})$	6·2, 6·6, 7·4 $^4\Gamma_4$, 15·25, 16·3, (16·6), (16·9) $^4\Gamma_4$ and doublets	393,* 459b,* 516
	$Co(II)ZnS(4S^{2-})$	6·0, 7·2 $^4\Gamma_4$, 13·9, (15·0) $^4\Gamma_4$	163a, 555b, 605*
	$Co(II)CdS(4S^{2-})$	5·2, 5·5, 6·2 $^4\Gamma_4$, 13·6, 14·2 $^4\Gamma_4$	459c*
	$Co(CN)_5{}^{3-}$	10·4 ($\epsilon = 3000$), 15·7, 23·2 weak, (28·6), 36·0	183a, 316
3d^8	Ni(II)		
	$NiCl_4{}^{2-}$	7·4 $^3\Gamma_2$, 11·6 $^1\Gamma_{3,\,5}$, 14·2, 15·2, (16·0) $^3\Gamma_4$	58, 73a, 98d, 162b,* 166a, 169b,* 193, 296, 400a
	$NiBr_4{}^{2-}$	7·0 $^3\Gamma_2$, 10·7 $^1\Gamma_{3,\,5}$, 13·3, 14·2 $^3\Gamma_4$	98d, 162b,* 166a, 169b*
	$Ni(II)MgAl_2O_4(4O^{2-})$	(14·2), 15·7, 16·7, (18) $^3\Gamma_4$ and singlets	517
	$Ni(II)Zn_2SiO_4(4O^{2-})$	(14), 15·75, 17·2, (18·5) $^3\Gamma_4$ and singlets	517
	$Ni(II)ZnO(4O^{2-})$	(13·5) $^1\Gamma_3$, $^1\Gamma_5$, 15·1, 16·1, 17·4 $^3\Gamma_4$ and singlets	517
	$Ni(II)CdS(4S^{2-})$	4·2 $^3\Gamma_5$, 8·2 $^3\Gamma_2$, 10·2 $^1\Gamma_{3,\,5}$, 12·5 $^3\Gamma_4$	459c,* 605e*
	Ni temeen$_2{}^{2+}$	23·1 singlet	31
	Ni trien^{2+}, yellow form	22·6 singlet	487
	Ni en$_3{}^{2+}$, yellow form	22·2 singlet	548a
	Ni dtp$_2$(4S)	14·6, 19·1 singlets, 26·2, 31·6, (35·5), 43·9	301c
3d^9	Cu(II)		
	$Cu(H_2O)_6{}^{2+}$	(9·4), 12·6 d-shell, ~50 redox	51,* 159, 236*
	$Cu(NH_3)_4{}^{2+}$	16·9 d-shell	42, 51*
	$Cu(NH_3)_5{}^{2+}$	(11·7), 15·1 d-shell	42, 51*
	Cu en$_2{}^{2+}$	18·2 d-shell	46, 625
	Cu en$_3{}^{2+}$	(11·8), 16·4 d-shell	46
	$CuCl_4{}^{2-}$	~9 d-shell, 26·5, 36·5, (43·5) redox	19, 45, 111, 120, 144,* 293
	$CuBr_4{}^{2-}$	~8 d-shell, 17·0, 19·5 narrow redox 30	19, 120, 277, 333†
	Cu dip$_2(H_2O)_2{}^{2+}$	10·5, 13·9 d-shell	275
	Cu dip$_3{}^{2+}$	~9, 14·7 d-shell	275
	Cu phen$_2(H_2O)_2{}^{2+}$	10·2, 13·3 d-shell	275
	Cu phen$_3{}^{2+}$	~8, 14·7 d-shell	275
	Cu tren^{2+}	(9·6), 11·6, (14·7) d-shell	278
	Cu enta^{2-}	13·6 d-shell, (37·8) redox	275, 278
	Cu ata$^-$	11·4, 12·9 d-shell	278
	Cu ata$_2{}^{4-}$	15·2 d-shell	278
	Cu gly$_2$	15·8	275
	Cu aca$_2$ various solvents		35*
	Cu den (H$_2$O)$^{2+}$	16·3 d-shell	278
	Cu den (NH$_3$)$^{2+}$	17·4 d-shell	278
	Cu den$_2{}^{2+}$	(11·8), 15·9 d-shell	278
3d^{10}	Cu(I)		
	Cu(I) Cl$^-$ solution	36·8 ($\epsilon = 4300$)	157, 159, 394
	Br$^-$ solution	36·3 ($\epsilon = 8500$)	157, 159, 394
	Cu dip$_2{}^+$	23·0 Cu-dip	615*
	Cu phen$_2{}^+$	23·0 Cu-phen	615*

TABLE 27—*continued.*

			Ref.
4d⁰	Nb(V)		
	Nb(V) 10 M HCl	(31·0), 35·5, 41·7 redox	302
	Nb(O₂)³⁺?	27·4 redox	454
	Mo(VI)		
	MoO₄²⁻	44·0 π → γ₃	81
	Mo(VI) HCl	redox above 33	425
	MoF₆ gaseous	54·0 vibr. π → γ₅	579
	Mo(O₂)₂²⁺?	24·1 redox	454
	Tc(VII)		
	TcO₄⁻	34·6 vibr. π → γ₃, 40·5 redox	61, 61a, 81d, 394
	Ru(VIII)		
	RuO₄	26·0 vibr. π → γ₃, 32·3 redox	81d, 93, 394, 606
4d¹	Nb(IV)		
	Nb(IV) 13 M HCl	20·9 (ε ∼ 120)	100
	Nb(IV) 8 M HCl	14·3 (ε ∼ 150)	100
	Mo(V)		
	MoOCl₅²⁻	14·05 and 22·5 d-shell, 28·2 (ε = 500), 32·2 (ε = 4400), 40·7 (ε = 5500) redox	284
	Ru(VII)		
	RuO₄⁻	26·0 π → γ₃, 31 vibr. redox	93, 345a, 567
4d²	Nb(III)		
	Nb(III) 10 M HCl	15·4, 18·9, 22·5 d-shell	100
	Ru(VI)		
	RuO₄²⁻	21·6 γ₃ → γ₅, (26·6) redox	93, 293, 386
	RuO₂Cl₄²⁻	30·5 redox	148, 370a
4d⁴	Ru(IV)		
	Cl₅RuORuCl₅⁴⁻	(13·8) d-shell?, 21·35 (perhaps another complex), 26·0, (27·7) π → γ₅, 39·4 π → γ₃	125, 148, 293, 299 607
4d⁵	Ru(III)		
	Ru(H₂O)₆³⁺	(25) d-shell, 44·5 (ε = 2300) π → γ₅	77
	RuCl(H₂O)₅²⁺	31·8 π → γ₅	77
	RuCl₂(H₂O)₄⁺	(27·8), 33·3 redox	77
	Ru(NH₃)₆³⁺	36·2 (ε = 530) d-shell?	210
	Ru(NH₃)₅Cl²⁺	30·5 redox, 38·5 (ε = 170), 43·5	210
	cis-Ru(NH₃)₄Cl₂⁺	28·4, 32·3 redox, 38·2	210
	Ru(NH₃)₅Br²⁺	25·1 redox, 30·3	210
	Ru(III) 4 M HCl	19·5 γ₃ → γ₃, (26·3), (28·8), 30·3 redox	277, 293
	Ru(CN)₆³⁻	21·8 π → γ₅, 28·2	108a
4d⁸	Pd(II)		
	Pd(H₂O)₄²⁺	26·4 (ε = 86) singlet of d-shell	293
	Pd(OH)₄²⁻?	27·2 (ε = 165)	293
	PdCl₄²⁻	(16·2) (ε = 6) triplet, 21·1 (ε = 166) and (29·8) (ε = 500) two or three singlets of d-shell, 35·8 (ε = 10,500) and 44·9 (ε = 30,000)	90, 123, 293, 498, 572
	PdCl₂(H₂O)₂	(16·85) triplet, 23·9 singlet of d-shell	293
	PdBr₄²⁻	(15·4) triplet, 19·9 and (27·0?) singlets, 30·2, 37·0, and 40·9 redox	14, 293
	PdI₄²⁻	(18·1), 20·5, 24·9 redox	152, 293
	Pd(SCN)₄²⁻	20·0, (24·5) singlets, 32·5 thiocyanate	293
	Pd ox₂²⁻	26·1 singlet	293
	Pd mal₂²⁻	26·9 singlet	293
	Pd gly₂	30·8 (ε = 245) singlet	293
	Pd enta²⁻	29·7 (ε = 1020) singlet	397
	Pd(NH₃)₄²⁺	(27·2) triplet, 33·8 singlet	293
	Pd en₂²⁺	(29·4) triplet, 34·8 singlet	293
	Pd trien²⁺	33·8 singlet	253, 293
	Pd py₄²⁺	∼36·4 singlet	293
	Pd den Cl⁺	29·8 singlet	293
	Pd den Br⁺	29·2 singlet	293
	Pd den (OH)⁺	32·2 singlet	293
	Pd den (NH₃)²⁺	33·8 singlet	293
	Pd den₂²⁺	33·9 singlet	293
	Pd dtp₂(4S)	21·8 singlet, (29·5), 34·0 strong, (37·6), 38·9	301c

TABLE 27—*continued.*

			Ref.
4d¹⁰	Ag(I)		
	Ag(H₂O)₂⁺	(44·7) ($\epsilon = 400$), 47·5 ($\epsilon = 900$), 51·9 ($\epsilon = 1500$)	598
	Ag(NH₃)₂⁺	47, (51)	598
	AgF	44·4, 47·9, 50·3	598
	Ag(I)NaCl sol.	46·3 ($\epsilon = 19{,}000$)	157, 159
	Ag(I)NaBr sol.	(42·0), 44·0 ($\epsilon = 31{,}000$)	157, 159
	Sn(IV)		
	SnCl₆²⁻	44·9 $\pi \rightarrow \gamma_1$	299
	Sb(V)		
	SbCl₆⁻	36·9 $\pi \rightarrow \gamma_1$	111, 299, 424
5d⁰	W(VI)		
	WO₄²⁻	50·3 $\pi \rightarrow \gamma_3$	81
	WF₆ gaseous	57·1 vibr. $\pi \rightarrow \gamma_5$	579
	WCl₆ in CCl₄	(17·1), (19·45), (22·4), 26·4, 29·9 $\pi \rightarrow \gamma_5$	299
	W(O₂)₂²⁺?	33·9	454
	Re(VII)		
	ReO₄⁻	43 vibr. $\pi \rightarrow \gamma_3$, 47	81, 394
	Os(VIII)		
	OsO₄ in H₂O	34·6 vibr., 41·3 vibr. $\pi \rightarrow \gamma_3$	293, 345
	OsO₄, gaseous	31·38, 36·22, 38·10 electronic states with zero vibrational excitation	345
5d¹	Re(VI)		
	ReF₆, gaseous	5·2 vibr. $\Gamma_7(^2\Gamma_5)$, 31·65, 32·36, 33·3 $\Gamma_8(^2\Gamma_3)$ (or ReF₇ ?)	409*
5d²	Os(VI)		
	OsF₆, gaseous	3·9 $\Gamma_1(^3\Gamma_4)$, 4·3 vibr. $\Gamma_4(^3\Gamma_4)$, 8·5 vibr. $^1\Gamma_3$ and $^1\Gamma_5$, 17·3 $^1\Gamma_1$ 35·7 and 40·8 $\pi \rightarrow \gamma_5$	301, 409*†
	OsO₂(OH)₄²⁻	14·7d-shell ?, 29·4 redox	370a
5d⁴	Os(IV)		
	OsCl₆²⁻	10·8, 11·7 $^1\Gamma_3$ and $^1\Gamma_5$, 17·24 vibr. $^1\Gamma_1$, (23·9), 27·0, (29·3), 30·0 $\pi \rightarrow \gamma_5$, (33·2) ($\epsilon = 1000$), (36·0) ($\epsilon = 700$) and 39·2 ($\epsilon = 2000$) $\gamma_5 \rightarrow \gamma_3$?	293, 295
	OsBr₆²⁻	10·6, 11·3 $^1\Gamma_3$ and $^1\Gamma_5$, 16·1 $^1\Gamma_1$, a long series of redox bands (Table 32)	293, 299
	OsF₆²⁻	32·5 $\gamma_5 \rightarrow \gamma_3$, > 50 $\pi \rightarrow \gamma_5$	227
	Pt(VI)		
	PtF₆, gaseous	3·3 vibr. $\Gamma_4(^3\Gamma_4)$, 5·2, 5·8 vibr. Γ_3 and $\Gamma_5(^3\Gamma_4)$, 11·9, 12·6 vibr. $^1\Gamma_3$ and $^1\Gamma_5$, 15·9 vibr. $^1\Gamma_1$, ~25, 32 vibr. $\pi \rightarrow \gamma_5$,	301, 409*†
5d⁵	Ir(IV)		
	IrF₆²⁻	31·6 $\gamma_5 \rightarrow \gamma_3$, 47·0 $\pi \rightarrow \gamma_5$	227
	Ir py Cl₅⁻	16·0, (19·9), 20·5, 24·7 $\pi \rightarrow \gamma_5$, (29·1), $\gamma_5 \rightarrow \gamma_3$?, 37·5 pyridine	285
5d⁶	Pt(IV)		
	Pt(NH₃)₆⁴⁺	(38?) $\gamma_5 \rightarrow \gamma_3$?	194, 277
	Pt(NH₃)₅NH₂³⁺	(36) ($\epsilon = 1800$)	194, 277
	Pt(NH₃)₅Cl³⁺	35·0	194, 277
	Pt(NH₃)₄NH₂Cl²⁺	31, (36)	194, 277
5d⁸	Pt(II)		
	PtCl₄²⁻	(18·2) and 21·15 triplets, 25·7, 30·3 and (37·6) singlets, (43·1), 46·1	15a, 87,* 90, 293, 314c, 466b, 499
	PtBr₄²⁻	(16·5) and 19·7 triplets, 24·1 and (27·8) singlets, 37·3	15a, 293, 314c
	PtI₄²⁻	(16·2) and (18·6) triplets ?, (22·5) singlet ?, 25·7, 30·8, 35·8	293
	Pt(SCN)₄²⁻	(23·8) and (27·0) singlets ?, (37·4) thiocyanate	15a, 293
	Pt(NH₃)₄²⁺	34·7 triplet	87, 293,* 630
	Pt en₂²⁺	35·8 triplet	293, 630
	Pt(CH₃NH₂)₄²⁺	35·0 triplet, (41 ?)	630
	Pt(NH₂OH)₄²⁺	36·7 triplet, (39·3)	630
	Pt(NH₃)Cl₃⁻	(20·7), 24·1 triplets, 28·9, 33·3 singlets, (48) strong	87,* 132

TABLE 27—*continued.*

5d^{10}	Hg(II)		Ref.
	HgCl$_4$$^{2-}$	43·4 (ϵ = 26,000)	158, 167a, 293, 401
	HgBr$_4$$^{2-}$	40·3 (ϵ = 30,000)	158, 401
	HgI$_4$$^{2-}$	30·9 (ϵ = 22,000)	158, 401
	Hg(II) in H$_3$PO$_4$	42·6 (ϵ = 13,900)	73
	Hg enta^{2-}	(39·7)	293
	Pb(IV)		
	PbCl$_6$$^{2-}$	32·6 (ϵ = 9700), 48·1 (ϵ = 24,000)	218

Table 27 contains most of the absorption spectra known of com-complexes, which are not included in other tables. It is arranged according to the three transition groups 3d^q, 4d^q, 5d^q, and within each group according to increasing q, and finally, within the individual electron configurations, according to increasing oxidation number of the metal (or, which is equivalent, to increasing atomic number). There is no definite rule for the order of the ligands for each central ion. Table 27 comprises also the d^0-systems, the corresponding gaseous ion having the inert-gas configuration [A], [Kr] and [Xe]4f^{14} which usually only absorb below 50 kK, if the oxidation number is at least +4. We except here the cases of internal transitions in the ligands. It contains also in principle the d^{10}-systems. However, only a few Cu(I) and Ag(I) complexes are indicated, together with Hg(II) and Pb(IV). The d^{10}-cyanides are mentioned in Table 22. The Laporte-allowed transitions of the square-planar, diamagnetic d^8-tetracyanides are given in Table 22, while the d^8-tetrahalides are given in Table 23, p. 200. Since the latter group also exhibit internal transition ins the partly filled shell, many are given in Table 27 too.

Tables 28–31 indicate the special cases of complexes, having many bands which certainly can be interpreted by the ligand field (or perhaps more exactly, the M.O. theory). Thus, Table 28 lists the octahedral d^3-complexes, of which a large number is known. Table 29 gives the octahedral and tetrahedral complexes of the high-spin d^5-systems Mn(II) and Fe(III) (cf. also Table 15). Some low-spin (and very few high-spin) d^5-systems, of which no ligand field transition, but only electron transfer bands, are known, are given in Tables 27 and 32. Table 30 gives the ligand field transitions of low-spin d^6-systems such as Co(III), Rh(III), Ir(III) and Pt(IV). For the first two elements, the list of complexes is far from complete, and it is advisable to study the literature references for further details. Finally, Table 31 gives the high-spin, octahedral d^8-systems, which in prac-

TABLE 28. ABSORPTION SPECTRA OF OCTAHEDRAL d^3-COMPLEXES
(Notation as in Table 27.)

		$^4\Gamma_5$	$^4\Gamma_4$	$^4\Gamma_4$	$^2\Gamma_3, {}^2\Gamma_4$	$^2\Gamma_5$	Ref.
$3d^3$ V(II)	$V(H_2O)_6^{2+}$	12·35	18·5	27·9	(13·1)	—	21,* 236, 291, 439*
Cr(III)	$Cr(H_2O)_6^{3+}$	17·4	24·6	37·8	(15·0)	(21·0)	264, 315, 439*
	$Cr(III)Al_2O_3$	18·1	24·4	(39·1)	14·45	21·0	112, 147,* 149a, 257, 446, 517a, 551, 555b, 570, 576, 590
	$Cr(III)MgO$	16·2	22·7	(29·7)	14·32	—	371,† 402a
	$Cr_2O_3(6O)$	16·6	21·6	—	13·7, 14·2	19·6	393,* 423a,* 503, 517a, 523
	CrF_6^{3-}	14·9	22·7	34·4	15·7, (16·4)	(22·0)	525*
	$Cr(H_2O)_5F^{2+}$	16·8	24·0	37·6	14·9	—	88b
	$cis\text{-}Cr(H_2O)_4F_2^+$	16·4	23·9	36·8	14·9	—	88b
	$trans\text{-}Cr(H_2O)_4F_2^+$	16·3	22·7, 25·5	37·0	14·8	—	88b
	$CrCl_3(6Cl)$	13·7	18·9	—	(13·15)	—	423a, 525*
	$Cr(H_2O)_5Cl^{2+}$	16·5	23·25	—	—	—	135, 315
	$Cr(H_2O)_4Cl_2^+$	15·75	22·2	36·8	(15)	—	135, 315
	$Cr(III)H_2SO_4$	14·35	21·4	—	14·9, (15·6)	—	525*
	$Cr\ urea_6^{3+}$	16·1	22·6	34·4	14·45, (15·1)	(20·95)	525*
	$Cr\ ox_3^{3-}$	17·5	23·9	—	14·35	—	197a, 265, 473c, 512
	$Cr\ mal_3^{3-}$	17·5	23·9	—	14·45	—	512
	$Cr\ enta\ (H_2O)^-$	18·4	25·6	—	—	—	162a, 275
	$Cr\ enta\ (OH)^{2-}$	17·3	(23·8) 25·3	—	—	—	162a, 275
	$Cr(NH_3)_5Cl^{2+}$	19·4 (22·1)	26·6	—	—	—	362
	$Cr(NH_3)_5Br^{2+}$	19·05 (21·8)	26·5	—	—	—	362
	$Cr(NH_3)_5I^{2+}$	18·5 (21·5)	—	—	—	—	362
	$Cr(NH_3)_5H_2O^{3+}$	20·8	27·8	—	15·15	—	129, 523
	$Cr(NH_3)_5OH^{2+}$	19·7	25·5 (29·8)	—	—	—	523

					References
Cr(NH₃)₆³⁺	21·55	28·5	15·3		52, 362
cis-Cr en₂(H₂O)₂³⁺	20·65	27·25	—	—	514, 619
trans-Cr en₂(H₂O)₂³⁺	19·7 (22·6)	27·7	—	—	619
cis-Cr en₂(OH)₂⁺	19·0	26·55 (29)	—	—	619
trans-Cr en₂(OH)₂⁺	19·9	25·25 (29?)	—	—	619
Cr en₂ox⁺	20·2	27·1	14·5	—	399
Cr en ox₂⁻	18·8	25·4	14·9, 15·5	—	399, 515b
Cr en₃³⁺	21·85	28·5	—	—	514, 619
Cr(III) molybdate	18·5	25·2	—	—	523, 537
Cr(NCS)₆³⁻	17·75	23·8	—	19·05	54, 525, 587
Cr dtp₃(6S)	14·4	18·9	13·15	—	301c, 525
Cr exan₃(6S)	16·0	20·3	—	—	314a
Cr py₃Cl₃	15·9	22·2	13·9	(18·5), (18·8)	343b*
Cr(CN)₆³⁻	26·6	32·2	—	—	343, 525
Mn(IV) MnF₆²⁻	21·75	28·2	16·3	—	292
Mn(IV) molybdate	21·4	—	14·3	—	537
4d³ Mo(III) MoCl₆³⁻	19·15	23·9	9·65	14·8	211,* 284, 293
MoCl₅(H₂O)²⁻	19·75	24·6	9·55	14·75 (15·3)	211,* 284, 293
Tc(IV) TcCl₆²⁻	—	—	—	14·3	107†
TcBr₆²⁻	—	—	9·4	13·5	107†
5d³ Re(IV) ReF₆²⁻	(32)	—	10·8, 11·4	17·7, 18·9 (19·9)	127a,* 301
ReCl₆²⁻	(27·5)	—	9·5	14·2 vibr., 15·7	272, 284, 392
ReBr₆²⁻	—	—	9·3	13·2, 15·1	272
Ir(VI) IrF₆, gaseous	—	—	6·4, 8·5, 9·1	12·6, 15·6	409*

TABLE 29. ABSORPTION SPECTRA OF HIGH-SPIN d^5-COMPLEXES
(Notation as in Table 27.)

	$^4\Gamma_4(G)$	$^4\Gamma_5(G)$	$^4\Gamma_1, {}^4\Gamma_3(G)$	$^4\Gamma_5(D)$	$^4\Gamma_3(D)$	$^4\Gamma_4(P)$	Ref.
$Mn(H_2O)_6^{2+}$	18·9	23·1	24·95 (25·3)	28·0	29·75	32·95	220*, 441*
1·8 M $MnSO_4$	18·8	23·0	25·0 (25·25)	28·1	29·7	32·8	265
2·3 M $MnCl_2$	18·9	23·0	(24·7), 25·0, (25·3)	27·8	29·4	32·0	283, 510
Mn(II) 9 M HCl	19·3	22·7	23·5, 24·3	27·7	28·9	31·3	283
Mn(II) 9 M KSCN	17·4	22·05	24·0	26·9	(28·6)	—	293
$MnO(6O^{2-})$	16·4	20·8	23·8	—	—	—	478a*
$MnF_2(6F^-)$	19·2	23·1	25·3	28·2	30·0	32·8	237, 568*
$MnCl_2(6Cl^-)$	18·5	22·0	23·7	26·9	28·3	—	459*
$MnCl_2(H_2O)_4$	19·3	22·9	24·6	27·6	29·2	31·6	10, 510
$Mnpy_2Cl_2(2N,4Cl)$	16·7, 19·0	20·4	23·7	—	—	—	10, 343b
Mn enta^{2-}	18·8	(21·2)	24·3	27·2	29·2	—	283
Mn en$_3^{2+}$	15·9	20·3	23·7	26·9	—	—	283
Mn Br$_4^{2-}$	21·0	—	22·3	—	—	23·1	166a, 283
$MnS(6S^{2-})$	16·7	19·8	21·6	—	—	—	605e
$Fe(H_2O)_6^{3+}$	12·6	18·5	24·3, 24·6[a]	28·8	30·2	—	265, 321, 486, 509
FeF_6^{3-}	14·2	19·7	25·4 vibr.	—	—	—	295
Fe urea$_6^{3+}$	12·5	17·1	23·1, 23·4	—	—	—	295
Fe ox$_3^{3-}$	10·7	15·2	(22·1)	—	—	—	295
Fe mal$_3^{3-}$	11·0	15·9	22·65, 22·9	(25·8)	—	—	295
$FeCl_4^-$	15·6	16·3	18·8	20·1	22·4[b]	—	156, 321, 396, 402, 421, 633

[a] Redox (electron transfer) band at 42·0.
[b] Redox bands at 27·4, 31·8, 41·1.

TABLE 30. LIGAND FIELD TRANSITIONS IN OCTAHEDRAL, LOW-SPIN d^6-COMPLEXES
(Notation as in Table 27.)

	$^1\Gamma_4$	$^1\Gamma_5$	$^3\Gamma_4, ^3\Gamma_5$	Ref.
$3d^6$ Fe(II) $Fe(CN)_6^{4-}$	31·0	(37·0)	—	91
Co(III) $Co(H_2O)_6^{3+}$	16·6	24·9	—	525*
$Co(CO_3)_3^{3-}$	15·7	22·8	—	275, 410a, 582
$Co\ ox_3^{3-}$	16·5	23·8	—	319, 439,* 473c, 525, 589
Co enta⁻	18·65	26·3	—	275
Co enta$(H_2O)^-$	18·2	26·2	—	275
Co enta$(OH)^{2-}$	17·4	(24·1), 25·9	—	275
$Co(NH_3)_5F^{2+}$	19·45 (21·9)	28·3	10·9 (15·5)	362, 366
$Co(NH_3)_5Cl^{2+}$	18·7 (21·4)	27·5	11·4 (15·2)	31, 63, 178,* 362, 366, 587, 628*
$Co(NH_3)_5Br^{2+}$	18·2 (21·3)	—	10·85 (14·8)	362, 366, 587
$Co(NH_3)_5I^{2+}$	17·2	—	10·7 (13·8)	362, 366
$Co(NH_3)_6^{3+}$	21·0	29·5	13·0 (17·2)	265, 330, 361, 439,* 574*
$Co(NH_3)_5H_2O^{3+}$	20·6	29·4	—	31, 525, 587
$Co(NH_3)_5OH^{2+}$	19·9	27·1	—	31, 525, 587
cis-Co en$_2$Cl$_2^+$	18·9	25·6	—	32, 362
trans-Co en$_2$Cl$_2^+$	16·2, 22·0	25·8	—	24,* 32, 54, 362
cis-Co en$_2$(H$_2$O)$_2^{3+}$	20·2	28·2	—	31, 49
Co en$_3^{3+}$	21·4	29·5	13·7 (17·5)	31, 65, 264, 265, 399
Co den$_2^{3+}$	21·4	29·5	—	65, 609b
trans-Co en$_2$(NH$_3$)$_2^{3+}$	21·4	29·7	—	31, 538
cis-Co en$_2$(NH$_3$)$_2^{3+}$	21·5	29·8	—	31, 538
Co en(NH$_3$)$_4^{3+}$	21·2	29·5	—	538
Co(NH$_3$)$_4$ gly^{2+}	20·3	28·8	—	538
Co(NH$_3$)$_2$ gly$_2^+$	20·2	28·1	—	538
Co(NH$_3$)$_4$ ox$^+$	19·6	28·0	—	587
Co en$_2$ ox$^+$	20·0	28·1	—	587

U

TABLE 30—*continued*.

	$^1\Gamma_4$	$^1\Gamma_5$	$^3\Gamma_4, {}^3\Gamma_5$	Ref.
Co(NH₂OH)₆³⁺	22·8	(30)	—	631
Co (CN)₅H³⁻	32·3	(35·4)	—	183a
Co(CN)₅Cl³⁻	27·0	—	—	4
Co(CN)₅Br³⁻	25·5	—	—	4
Co(CN)₆³⁻	32·4	39·0	—	319, 525
Co(NH₃)₅NO₂²⁺	21·9	—	—	63, 319, 364, 587
cis-Co(NH₃)₄(NO₂)₂⁺	22·4	—	—	31, 587
trans-Co(NH₃)₄(NO₂)₂⁺	22·7	—	—	31
Co cys₃³⁻ (3N, 3S)	17·2	22·6	—	172
Co aes₃(3N, 3S)	17·55	23·05	—	172
Co exan₃(6S)	16·2	20·8	—	314a
Co(NH₃)₅SO₃⁺	21·2	—	—	319
trans-Co(NH₃)₄(SO₃)₂⁻	22·2	—	—	536
cis-Co(NH₃)₄(SO₃)₂⁻	22·0	—	—	536
Co(NH₃)₅NCS²⁺	20·1	—	—	587, 601
Co(NH₃)₅NCSHg	21·0	—	—	524, 601
Co(NH₃)₅CrO₄⁺	18·5	—	—	587
4d⁶ Rh(III) RhBr₆³⁻	18·1	22·2	—	276, 293
RhCl₆³⁻	19·3	24·3	(14·7)	276
Rh(III) 1 M HCl	20·2	25·6	—	276
Rh(H₂O)₆³⁺	25·5	32·8	—	150a, 276
Rh ox₃³⁻	25·1	(30)	(19·2)	276
Rh mal₃³⁻	25·6	30·1	—	293
Rh enta⁻	28·3	—	—	128a, 293
Rh(NH₃)₅Cl²⁺	28·7	36·1	—	276
Rh(NH₃)₅Br²⁺	27·8	—	23·6	276
Rh(NH₃)₅I²⁺	25·8	—	(24·1)	276

Compound				Ref.
$Rh(NH_3)_6^{3+}$	32·7	39·1	—	276
$Rh(NH_3)_5H_2O^{3+}$	31·6	38·1	—	276
$Rh(NH_3)_5OH^{2+}$	31·2	36·0	—	276
cis-$Rh\ en_2Cl_2^+$	28·4	33·9	—	6a
trans-$Rh\ en_2Cl_2^+$	24·6	35·0	—	6a
$Rh\ enCl_4^-$	22·8	27·0	—	6a
$Rh\ en_3^{3+}$	33·2	39·6	—	276
$Rh\ den_2^{3+}$	32·8	(39·2)	—	293
cis-$Rh\ py_2Cl_4^-$	22·4	28·7	—	285
trans-$Rh\ py_2Cl_4^-$	20·0, 23·2	—	—	285
trans-$Rh\ py_4Cl_2^+$	24·5	—	—	295
cis-$Rh\ ox_2Cl_2^{3-}$	23·0	28·4	—	285
trans-$Rh\ ox_2Cl_2^{3-}$	21·3, 25·1	—	—	285
$Rh(SCN)_6^{3-}$	(19·4)	—	—	293
$Rh\ dtp_3(6S)$	21·3	24·4	—	301c
$Rh(OH)_6^{3-}$	23·9	30·1	—	276, 293
Pd(IV) $\ PdCl_6^{2-}$	(20·8)	—	—	90, 292
5d⁶ Ir(III) $\ IrBr_6^{3-}$	22·4	25·8	16·8	276
$IrCl_6^{3-}$	24·1	28·1	(16·3), 17·9	276
$Ir(NH_3)_5Cl^{2+}$	35·0	—	27·8	276
$Ir\ en_3^{3+}$	40·2	—	33·1	276
$Ir\ pyCl_5^{2-}$	23·8	—	18·5	285
cis-$Ir\ py_2Cl_4^-$	—	—	21·6	285
trans-$Ir\ py_2Cl_4^-$	(26·6)	—	19·4	285
$Ir\ oxCl_4^{3-}$	26·2	28·6	18·2	285
$Ir\ dtp_3(6S)$	(23)	—	21·2	301c
Pt(IV) $\ PtBr_6^{2-}$	28·3	—	(19·1)	277, 508
$PtCl_6^{2-}$	31·5	—	22·1	277, 508
PtF_6^{2-}	25	36·4	(22·5), 24·4	508, 610
trans-$Pt(NH_2C_2H_5)_4Cl_2^{2+}$	—	—	22	625

TABLE 31. LIGAND FIELD TRANSITIONS IN OCTAHEDRAL, HIGH-SPIN d^8-COMPLEXES, ALL OF NICKEL (II)

(Notation as in Table 27.)

	$^3\Gamma_5$	$^3\Gamma_4$	$^1\Gamma_3$	$^1\Gamma_5$	$^3\Gamma_4$	Ref.
$Ni(H_2O)_6{}^{2+}$	8·5	13·5	(15·4)	(22·0)	25·3	43, 121, 184, 264, 344, 358, 437b 457, 473b, 493
$NiSO_4, 7H_2O$	8·5	(14·1)	(15·4)	(21·8)	25·7	59, 236*
$(NH_4)_2[Ni(H_2O)_6](SO_4)_2$	8·9	(15·5)	(14·4)	(22·0)	25·8	59
$NiO(6O)$	8·8	13·9	(15·2)	[(21·2)	(23·9)	
$Ni(II) MgO$	8·6	13·8	(14·8)	(21·7)	24·5	120b,* 393*
$Ni(II) MgTiO_3$	—	12·0	(13·3)	(20)	22·2	342, 372,* 459b,* 517
$Ni(II)H_2SO_4$	—	12·2	14·7	—	23·35	517
py $HNiBr_3$(6Br)	—	11·2	(10·3)	17·4	19·9	275
$RbNiBr_3$	—	11·5	(10·4)	17·3	20·2	8,* 59
$NiBr_2$(6Br)	—	12·1	(10·3)	(16·8)	20·7	8,* 59
py $HNiCl_3$(6Cl)	—	11·2	(12·5)	(18·4)	21·3	8,* 59, 632a
$RbNiCl_3$	—	(12·7)	(11·5)	18·6	21·6	8,* 59
$NiCl_2$(6Cl)	—	12·9	(11·6)	19·4	22·1	8,* 59, 632a
$KNiF_3$(6F)	—	12·5	15·3	21·0	23·7	326c,* 525*
Ni ata $(H_2O)_2{}^-$	9·5	16·0	13·3	—	25·6	278
Ni en $(H_2O)_4{}^{2+}$	9·8	15·8	13·6	—	26·8	20,* 278
Ni $gly_2(H_2O)_2$	10·0	16·1	13·3	—	27·0	383
α-Ni enta $(H_2O)^{2-}$	10·1	17·0	12·7	—	26·2	275, 278
β-Ni enta $(H_2O)^-$	9·9	16·9	12·8	—	26·4	278
Ni ata gly^{2-}	9·9	17·1	13·0	—	26·2	278
Ni $ata_2{}^{4-}$	10·4	17·4	13·0	—	27·0	278
Ni enta $(NH_3)^{2-}$	10·2	17·2	12·7	—	26·2	278
Ni $gly_3{}^-$	10·1	16·6	13·1	—	27·6	275, 467a,* 555a
Ni enta $(CN)^{3-}$	10·5	17·5	12·5	—	—	278
Ni en gly_2	10·5	17·3	13·0	—	28·1	278

Ni tren (H$_2$O)$_2$$^{2+}$	10·5	17·8	12·8	—	27·8	278
Ni tren SO$_4$	10·0	17·2	(12·9)	—	26·9	9, 59
Ni tren SO$_4$, 7H$_2$O	10·6	17·9	(12·9)	(23·3)	27·5	9, 59
Ni tren (NCS)$_2$	10·9	17·9	(12·8)	(23·3)	27·8	9, 59, 492
Ni en$_2$ (H$_2$O)$_2$$^{2+}$	10·5	17·9	—	—	28·1	20*
Ni py$_4$ (H$_2$O)$_2$$^{2+}$	10·15	16·5	13·5	—	27·0	59, 278
Ni py$_4$ (NCS)$_2$	10·1	16·7	(12·8)	(22·7)	26·7	9, 59
Ni ata den$^-$	10·9	17·6	12·5	—	28·7	278
Ni trien (H$_2$O)$_2$$^{2+}$	10·4	17·7	(12·5)	—	27·9	253, 287
Ni en$_2$ gly$^+$	10·8	17·9	12·65	—	28·6	278
Ni tren gly$^+$	11·2	18·8	12·6	—	28·2	278
Ni(CH$_3$NH$_2$)$_6$$^{2+}$	10·0	16·7	—	—	27·2	630
Ni(NH$_3$)$_6$$^{2+}$	10·75	17·5	13·15	—	28·2	43, 275, 493
[Ni(NH$_3$)$_6$]Br$_2$	10·8	17·7	—	—	28·2	59, 462
Ni tn$_3$$^{2+}$	10·9	17·8	12·5	—	28·2	278
Ni en$_3$$^{2+}$	11·2	18·35	12·4	(24·0)	29·0	20, 275, 493, 630
Ni tren (NH$_3$)$_2$$^{2+}$	11·0	18·2	12·75	—	28·2	278
Ni$_2$ tren$_3$$^{4+}$	10·7	18·3	12·5	—	28·7	278
Ni tren en^{2+}	11·0	18·7	—	—	29·1	278
Ni den$_2$$^{2+}$	11·5	18·7	(12·4)	—	29·1	253, 278, 609b
Ni den$_2$ (NO$_3$)$_2$	11·2	18·9	—	—	28·6	59
Ni(NH$_2$OH)$_6$$^{2+}$	10·8	18·3	—	—	29·0	630
Ni dip$_3$$^{2+}$	(12·65)	19·2	(11·5)	—	—	275
Ni phen$_3$$^{2+}$	(12·7)	19·3	(11·55)	—	—	275, 493
Ni(II) 9 M KSCN	—	15·3	13·2	—	25·2	275, 293, 319
Ni dtp$_2$py$_2$(4S, 2N)	9·0	14·0	—	—	22·7	—
Ni py$_2$Cl$_2$(4Cl, 2N)	—	14·1	—	23·0	24·1	343b
Ni py$_4$Cl$_2$(2Cl, 4N)	8·5, 10·9	15·9	—	—	25·7	59

TABLE 32. ELECTRON TRANSFER SPECTRA OF d-GROUP HEXAHALIDE COMPLEXES WITH IDENTIFICATION OF THE THREE GROUPS OF EXCITED LEVELS POSSIBLE

(Shoulders are given in parentheses. The wave number unit is kK. Cf. Figs 13–17, pp. 148–152.)

	$\pi \to \gamma_5$	$\pi \to \gamma_3$	$\sigma \to \gamma_3$	Ref.
$4d^0$ MoF$_6$	54·0			579
$5d^0$ WF$_6$	57·1			579
$5d^2$ OsF$_6$	(31), 35·7, 40·8			301, 409
$5d^3$ IrF$_6$	(31), 35·5, 42·0			301, 409
$5d^4$ PtF$_6$	~25, 32·6			301, 409
$5d^5$ IrF$_6^{2-}$	47·0			227
$4d^3$ MoCl$_6^{3-}$	45·3			293
TcCl$_6^{2-}$	29·4, (33)	41·7		74
$4d^4$ RuCl$_6^{2-}$	(17·15), 20·3, (22·9), (24·6)	(36·0), (41·0)		299
$4d^5$ RuCl$_6^{3-}$	(25·6), 28·65, 32·4	43·6		277, 299
RhCl$_6^{2-}$	(10·0), 12·0, 15·6, 16·7	39·2		127b, 301a
$4d^6$ RhCl$_6^{3-}$	—		41·7	299
PdCl$_6^{2-}$	—	29·4		90, 293
$5d^0$ WCl$_6$	(22·4), 26·4, 29·9			293, 299
$5d^3$ ReCl$_6^{2-}$	(31·3), (33·3), 35·6, (39·1)			272, 495
$5d^4$ OsCl$_6^{2-}$	(23·9), 27·0, (29·3), 30·0			299, 346
$5d^5$ OsCl$_6^{3-}$	(32·6), 35·45, 38·2, (39·4)	(47)		299
IrCl$_6^{2-}$	(16·95), 20·5, 23·05, 24·4	43·1		277, 299, 301b
$5d^6$ IrCl$_6^{3-}$	—	48·5		299
PtCl$_6^{2-}$	—	38·2		277, 508
$4d^3$ TcBr$_6^{2-}$	(19·4), 22·5, 26·1	(35·0)		107
$4d^5$ RuBr$_6^{3-}$	(16·4), (19·3), (21·3), 22·5, 23·0, 25·2, (27·2)	(30·1), 33·9		272
$4d^6$ RhBr$_6^{3-}$				299
$5d^3$ ReBr$_6^{2-}$	(20·4), (22·9), (23·8), (26·2), 28·4, 30·8, (33·0)	(38·1)		495, 555
$5d^4$ OsBr$_6^{2-}$	(17·3), (18·95), (19·75), 20·45, 22·6, 23·7, 25·05	(35·7), 40·8		299

Config.	Ion				Refs.
$5d^5$	$OsBr_6^{3-}$	(21·5), (25·6), (27·4), 28·2, 28·8, 31·0, (32·0), (33·6)	31·5, 37·0		299
	$IrBr_6^{2-}$	(11·6), (13·4), (14·3), 14·9, 17·15, 18·35, 19·7	36·8, 41·1		277, 299
$5d^6$	$IrBr_6^{3-}$	—			286, 299
	$PtBr_6^{2-}$		(27·0), 31·8, (33·3)		277, 299, 508
$5d^3$	ReI_6^{2-}	11·6, 14·35, 15·1, (16·7), 17·45, (18·45), 19·35, 20·8, (22·1), 22·55, 23·15, 23·6	25·85, 28·5?	44·2	299, 495
$5d^4$	OsI_6^{2-}	(11·6), 12·3, (14·55), 15·3, 18·2, 18·6	26·8, 30·0, 35·6		
$5d^5$	OsI_6^{3-}	(18·5), 19·1, 20·25, (21·75), 22·65, 25·0, 25·9		44·6	299
$5d^6$	PtI_6^{2-}	—	20·25, (22·75), 29·15	39·8, 43·5	277, 299, 508

TABLE 33. THE ENERGY LEVELS OF THE PARTLY FILLED f-SHELL OF FOUR LANTHANIDES
(Wave numbers are given in kK. Cf. pp. 175–177.)

		Ref.
$4f^2$ Pr(III)	4·3 $^3H_6$, 6·3 $^3F_3$, 6·8 $^3F_4$, 9·8 $^1G_4$, 16·9 $^1D_2$, 20·7 $^3P_0$, 21·3 $^3P_1$, 21·5 $^1I_6$, 22·6 $^3P_2$	119, 134, 168, 223, 271, 280, 550, 560
$4f^3$ Nd(III)	2·0 $^4I_{11/2}$, 4·0 $^4I_{13/2}$, 5·9 $^4I_{15/2}$, 11·5 $^4F_{3/2}$, 12·6 $^4F_{5/2}$, 13·5 $^4F_{9/2}$, 14·7 $^4F_{9/2}$, 15·9 $^2H_{11/2}$, 17·4 $^4G_{7/2}$ + doublets, 19·2 $^4G_{9/2}$?, 20·8–21·7, 23·4 $^2P_{1/2}$, 26·3 (J = $\frac{3}{2}$), 28·2–29·4 ^4D, . . .	79, 115, 262, 269, 280, 502, 560, 620a
$4f^{11}$ Er(III)	6·8 $^4I_{13/2}$, 10·2 $^4I_{11/2}$, 12·5 $^4F_{9/2}$, 15·3 $^4I_{9/2}$, 18·5 $^4S_{3/2}$, 19·2 $^2H_{11/2}$, 20·5 $^4F_{7/2}$, 22·2 $^4F_{5/2}$, 22·6 $^4F_{3/2}$, 24·6 $^2G_{9/2}$, 26·4, ..., 39·2 ^4D, ...	223b, 288, 301f, 562, 620a
$4f^{12}$ Tm(III)	6·0 $^3H_4$, 8·4 $^3H_5$, 12·7 $^3F_4$, 14·5 $^3F_3$, 15·1 $^3F_2$, 21·3 $^1G_4$, 28·0 $^1D_2$, 35·1 $^3P_0$ and $^1I_6$, 36·4 $^3P_1$, 38·2 $^3P_2$	39, 169, 187, 250, 271, 492c, 496a,
$4f^{13}$ Yb(III)	10·3 $^2F_{5/2}$	117, 168, 280, 562

TABLE 34. COMMON CO-ORDINATION NUMBERS IN COMPLEXES WITH RARER CASES
IN PARENTHESES
(Typically electrovalent compounds such as NaCl are not considered. The corresponding electron configurations of the gaseous ions are indicated in the d- and post-d-group cases. Cf. pp. 223 and 232.)

ang = angled, triatomic
arc = Archimedean anti-prism
bip = bipyrimidal
dist. = distorted
lin = linear

oct = octahedral
plan = planar
pyr = pyramidal
te = regular tetrahedral
tri = trigonal (nine-co-ordinated)

	Be(II)	4 te	$3d^0$	Cr(VI)	4 te
	B(III)	4 te, 3 plan	$3d^1$	Cr(V)	4 te
	C(IV)	4 te, (3 plan)	$3d^3$	Cr(III)	6 oct
	N(V)	3 plan	$3d^4$	Cr(II)	4 plan, 6 dist.
	N(III)	2 ang, (3 pyr)	$3d^0$	Mn(VII)	4 te
	N(−III)	3 pyr, 4 te	$3d^1$	Mn(VI)	4 te
	O(II)	2 ang	$3d^2$	Mn(V)	4 te
	O(−II)	4 te, 3 pyr, 2 ang	$3d^3$	Mn(IV)	6 oct
	Mg(II)	6 oct	$3d^4$	Mn(III)	4 plan, 6 dist.
	Al(III)	6 oct, 4 te	$3d^5$	Mn(II)	6 oct, (4 te)
	Si(IV)	4 te, (6 oct)	$3d^6$	Mn(I)	6 oct
	P(V)	4 te, (6 oct), (5 bip)	$3d^2$	Fe(VI)	4 te
	P(III)	3 pyr	$3d^5$	Fe(III)	6 oct, (4 te)
	P(−III)	3 pyr, (4 te)	$3d^6$	Fe(II)	6 oct, (4 te)
	S(VI)	4 te, (6 oct), (3 plan)	$3d^6$	Co(III)	6 oct, ((4 te))
	S(IV)	3 pyr, 2 ang	$3d^7$	Co(II)	6 oct, 4 te
	S(−II)	4 te, 2 ang	$3d^6$	Ni(IV)	6 oct
	Cl(VIII)	4 te	$3d^8$	Ni(II)	6 oct, 4 plan, (4 te), ((5 pyr))
	Cl(V)	3 pyr	$3d^9$	Cu(II)	4 plan, 6 dist. (5 pyr), (4 dist.)
$3d^0$	Ti(IV)	6 oct, 4 te	$3d^{10}$	Cu(I)	4 te, (3 plan), 2 lin
$3d^1$	Ti(III)	6 oct, dist.	$3d^{10}$	Zn(II)	6 oct, 4 te
$3d^0$	V(V)	4 te, dist.	$3d^{10}$	Ga(III)	4 te, 6 oct
$3d^1$	V(IV)	oxygen + dist (5 pyr)		As(V)	4 te, (6 oct)
$3d^2$	V(III)	6 oct, (4 te)		As(III)	3 pyr
$3d^3$	V(II)	6 oct		As(−III)	3 pyr

	Se(VI)	4 te, (6 oct)		5s² Te(IV)	3 pyr, (6 oct)
	Se(IV)	3 pyr		4d¹⁰ I(VII)	4 te
	Se(−II)	2 ang, (4 te)		5s² I(V)	6 oct, dist.
	Br(V)	3 pyr		5p² I(III)	4 plan
	Br(III)	4 plan, 4 dist.		5p⁴ I(I)	2 lin
4d⁰	Zr(IV)	6 oct, 8 dist.		La(III)	9 tri, 12, 8
4d⁰	Nb(V)	6 oct, 4 te		5d⁰ Ta(V)	6 oct, 4 te
4d⁰	Mo(VI)	4 te, 6 oct		5d⁰ W(VI)	4 te, 6 oct
4d¹	Mo(V)	6 dist, 8 dist.		5d⁰ Re(VII)	4 te
4d²	Mo(IV)	8 dist.		5d³ Re(IV)	6 oct, dist.
4d³	Mo(III)	6 oct		5d⁰ Os(VIII)	4 te
4d⁴	Mo(II)	4 plan, dist.		5d² Os(VI)	6 oct, dist.
4d⁰	Tc(VII)	4 te		5d⁴ Os(IV)	6 oct
4d³	Tc(IV)	6 oct		5d⁵ Os(III)	6 oct
4d⁰	Ru(VIII)	4 te		5d⁶ Os(II)	6 oct
4d¹	Ru(VII)	4 te		5d³ Ir(VI)	6 oct
4d²	Ru(VI)	4 te, 6 dist.		5d⁵ Ir(IV)	6 oct
4d³	Ru(V)	6 oct		5d⁶ Ir(III)	6 oct
4d⁴	Ru(IV)	6 oct		5d⁶ Pt(IV)	6 oct
4d⁵	Ru(III)	6 oct		5d⁸ Pt(II)	4 plan
4d⁶	Ru(II)	6 oct, (4)		5d⁸ Au(III)	4 plan
4d⁶	Rh(III)	6 oct		5d¹⁰ Au(I)	2 lin
4d⁶	Pd(IV)	6 oct		5d¹⁰ Hg(II)	2 lin, 4 te, (6 oct)
4d⁸	Pd(II)	4 plan, ((6 dist.))		5d¹⁰ Tl(III)	4 te, 6 oct
4d¹⁰	Ag(I)	2 lin, (3 plan), (4 te), (6 oct)		6s² Tl(I)	6 oct, dist.
4d¹⁰	Cd(II)	6 oct, 4 te		5d¹⁰ Pb(IV)	6 oct
4d¹⁰	In(III)	4 te, 6 oct		6s² Pb(II)	4 dist., various
4d¹⁰	Sn(IV)	4 te, 6 oct		6s² Bi(III)	6 dist., 4 dist.
5s²	Sn(II)	4 dist.		5f⁰ Th(IV)	8 arc, various
4d¹⁰	Sb(V)	4 te, 6 oct		5f⁰ U(VI)	2 oxygens + 3 or 6, 6 oct
5s²	Sb(III)	4 dist.		5f² U(IV)	8 arc, 6 oct
4d¹⁰	Te(VI)	6 oct		5f³ U(III)	9 tri

tice are only represented by Ni(II). It is remarked that high-spin tetrahedral and low-spin square-planar Ni(II), of which the ligand field transitions are much less certainly identified, are relegated to Table 27. In general, no assignment except "d-shell" has been given to transitions in square-planar Ni(II), Cu(II), Pd(II) and Pt(II) complexes, though a discussion of the probable excited states is found on p. 122.

In a few cases, where the assignment of the excited level is uncertain, the molar extinction coefficient ϵ of the maximum is given. "Jahn-Teller" after the energy level of some d^1- and d^4- systems suggest the possible influence of the Jahn–Teller effect, distorting the cubic-octahedral symmetry of six equal ligands.

Table 32 gives a complete list of the electron transfer bands of hexahalide d^q complexes (with $q < 10$) as known at present. Hence, this information is not duplicated in the Tables 27–31. However, many of these hexahalides have ligand field transitions to be found in the appropriate tables.

The gaseous ions with ground configuration s^2 have complexes of which the absorption bands are given in Table 19, p. 186.

Some special types of transitions, such as the Ir–py bands (Table 20) and the thiocyanate bands (Table 21) have been shown in separate lists. The similar transitions from Fe(II) to aromatic heterocyclic compounds, or internal transitions in these ligands, have only been considered (in Table 27) in the notable cases Fe dip_3^{2+} and Fe $phen_3^{2+}$.

The present writer has tried to select a reasonable set of literature references for each complex. This is not meant to be complete, especially in cases like the ruby or $Ni(H_2O)_6^{2+}$, where many different aspects have been studied. It must be emphasized that the identifications of the excited states in many cases have not been performed by the authors cited, but are suggestions by the present writer. There are even cases, where the authors cited have given other assignments. The references by other authors, indicating assignments of excited levels and other theoretical treatment, are marked by an asterisk. In the cases where the present writer proposes another assignment, the reference is marked by a dagger.

Table 33 gives the identification, with the L, S, J symbols of Russell–Saunders coupling in spherical symmetry, of the energy levels of those trivalent lanthanides most investigated, viz. Pr(III), Nd(III), Er(III), Tm(III) and Yb(III). The wave numbers refer in most cases to the aquo ion, either in solution or in solid salts, containing $R(H_2O)_9^{3+}$, but in a few cases to anhydrous fluorides and chlorides. Hence, the nephelauxetic effect is not studied in the table; the reader is advised to see the refs. no. 60, 137, 138, 280 and 298. It is not the intention to tabulate the long lists of lanthanide and actinide energy levels in general; such compilations exist in refs. no. 28, 75, 173, 288, 304, 478, 527, 544, 560, 561 and 562, while the modern theory on multiplet term distribution is treated in refs. no. 133, 259, 269, 282a, 288 and 298. The assignment of each energy level in Table 33 has often a long history of controversial arguments; unfortunately, it would fall outside the limits of this book to tell this story for each individual group of sub-levels.

The inert-gas configurations belonging to the lanthanide and actinide series have been discussed individually in the text (p. 180), such as UF_6 and UO_2^{2+}, and compared to the corresponding 5f- and 5f²-complexes. The cerium (IV) solutions have broad absorption bands in the ultra-violet, but due to the complicated hydrolytic behaviour, it has not been possible to identify species such as a possible aquo ion or hydroxo complexes. In dilute H_2SO_4, several authors (refs. 64, 154, 400) have observed a band maximum of Ce(IV) at 31·4 kK. Analogous to the band at 34·9 kK of UCl_6^{2-} or more exactly, the predominant species of U(IV) in 12 M HCl (refs. 192, 281, and 497a), these bands probably represent electron transfer to the 4f- and 5f-shell, respectively.

Table 34 gives the most usual co-ordination numbers and stereochemical shapes of complexes of central atoms with definite oxidation numbers. It serves to provide an easily accessible survey over the corresponding electron configurations of the gaseous ions. Of course, as discussed on p. 3, it is difficult in some cases to define the oxidation number, e.g. BH_4^- or $B(C_6H_5)_4^-$ can be considered as $B(-V)$ or $B(III)$ compounds, and PCl_3 and PH_3 undoubtedly have resemblances in their sterochemistry. However, the main purpose of Table 34 is to survey the complexes of the metallic elements.

CHAPTER 16

WHAT IS THE USE OF ABSORPTION SPECTRA?

WHEN a new compound is described in chemical literature, it has become more customary to indicate its infra-red spectrum and often its curves of paramagnetic and nuclear-magnetic resonance rather than its absorption spectrum in the visible and near ultra-violet regions. This is rather surprising, since the former information requires apparatus which is more costly and sophisticated than a good spectrophotometer. It is believed that the former type of evidence is much easier to translate into useful knowledge of the molecule, because the theories are available and known by many chemists. However, this belief is not as valid now as it was before 1950, and ordinary spectroscopy will probably be comparatively much more important in the future.

For obtaining a reliable absorption spectrum of a compound, it is often necessary to be more cautious than is the case with most other chemical properties. The usual methods of analysis do not always exactly give the assurance needed for our purpose; the tolerance of elements other than those desired obviously depends on their molar extinction ϵ in the actual case. Hence, the sensitivity is quite specific for some elements, and in this way to some extent resembles reactor material chemistry, where those elements with high cross-sections for slow neutrons are disastrous for the quality. It is often useful to take into consideration the geochemical and other reasons for which some other elements are common impurities in chemicals, even of analytical grade. Thus, Ni and Co have a tendency to occur in each other's salts, and the strongly coloured Ir as traces in Rh. Most chemicals have their ultra-violet absorption influenced by traces of Fe and sometimes by Cu, Pb and Pt (from crucibles). In strongly oxidizing solution, traces of Mn may form MnO_4^-.

Even when the impurities of other elements can be excluded, the percentage analysis of a given compound is not always a good assurance that the light absorption is mainly caused by the species indicated. Here, one problem is the presence of other oxidation numbers (Fe(III) in (II), Ir(IV) in (III), etc.) or of hydrolysis products, having Laporte-allowed electron transfer bands at the places where Laporte-forbidden (and perhaps even spin-forbidden) bands are sought of the main species. One per cent or less of such an impurity, having a percentage composition not very far from the main species, may completely change the spectrum, but have negligible influence on the quantitative analysis or even compensate small errors. It is often recommended to repeat operations (recrystallization, chromatographic separation, etc.) until one further operation does not change the spectrum measurably. When taken *cum grano salis*, this rule can be extremely useful, but it cannot help us to escape the problem discussed in chapter 13 on chemical composition of species in solution, i.e. that the solvent may change the solid, very pure compound into a mixture of species of which the main part is not even represented in the solid. This is not only the case with such ions as $CoCl_4^{2-}$, $Ni(NH_3)_6^{2+}$, and $PdCl_4^{2-}$, but there is a considerable difficulty in obtaining the true spectrum of such ions as $Fe(H_2O)_6^{3+}$ (forming small quantities of brown hydrolysis products) and even $Cr(H_2O)_6^{3+}$ in the ultra-violet, influenced by anion complexes.

When the spectrum of a given species is known with reasonable certainty (and as we saw above, it may be more a question of chemical knowledge and absence of strongly-coloured impurities than one of accurate chemical analysis) it gives the distribution of the energy levels for the internuclear distances of the ground state. These levels can be classified in most cases in M.O. configurations and hence give valuable information about the chemical bonding. Unfortunately, we cannot obtain all chemically pertinent information from the M.O. configurations. The energy quantities of interest to thermodynamicists (free energy of complex formation, etc.) are fairly small, compared to the distribution of excited levels in the visible and the ultra-violet parts of the spectrum, and consequently, the uncertainty of some 1, 5 or 20 kK in different cases of M.O. energy may completely wipe out the interesting chemical differences. Notably, it is very difficult to predict the shape of a given complex, the relative stability of geometrical isomers, and the reaction mechanisms, which all depend on such small quantities. In aqueous and similar solvents, the hydrogen-bonding mechanism and the macroscopic dielectric effects are responsible for similar energy differences between individual complexes, but cannot easily be incorporated into M.O. descriptions.

It might be felt that the amount of absorption spectra and their interpretation accumulated in recent years is so enormous that it is difficult to add substantially new knowledge. However, the present writer does not believe at all in this proposition, one reason being the capability of inorganic chemistry continuously to display new, surprising ligands. Among ligands already known, which will probably be of great interest to the theory of the absorption spectra, are those containing the atoms iodine, sulphur, phosphorus and arsenic. The empty orbitals of the two latter atoms seem to produce "inverted electron transfer" from the partly filled shell in some cases. Another interesting feature is that the π-bonding from the ligands disturb the relative energy of the partly filled shell with the result that the electron transfer spectra of $RuO_2Cl_4^{2-}$ and oxygenated Ru(IV) complexes have a higher wave number and of $MD_2Cl_2^+$ with M = Ru(III), Re(III), Os(III) and D = Nyholm's diarsine C_6H_4 $(As(CH_3)_2)_2$ have lower wave numbers than would be expected from the corresponding hexahalide complexes $RuCl_6$ (extrapolated to have $\pi \rightarrow \gamma_5$ bands at 7 kK), $RuCl_6^{2-}$, $RuCl_6^{3-}$, etc. Other ligand atoms of great probable interest in the future are carbon, boron (which might be reduced to B(I) by accepting two electrons from metals in low oxidation numbers, a hypothetical example being $PtCl_4^{2-}$ adding two BF_3 to form a Pt(IV) (d^6-complex) of Cl^- and BF_3^{2-} with large Δ) and Chatt's hydrogen complexes of Ru(II) and Pt(II). Since H^- does not contain π-, but only a pair of σ-electrons, the electron transfer spectra do not occur at a wave number as low as that expected from the reducing character of the ligand. In this particular case, the distinction from H^+ as ligand (cf. $Co(CO)_4H$ with Co(−I)) can be made by the planar Pt(II) complex, while Pt(0) was expected to form a tetrahedral complex.

As examples of important problems, which have been solved satisfactorily in the period 1956–1959 the following six may be mentioned:

(1) There is no intrinsic difference between the ligand field transitions in 4d- and 5d-complexes, compared to the 3d-complexes. It was often argued from the great differences in magnetochemistry that chemical bonding is essentially more covalent in the former groups. It is now demonstrated with certainty that the numerical value of Δ/B, and to some extent, of ζ_{nd} are the important factors. If the nephelauxetic effect is considered, the chemical bonding in a given oxidation number seems rather more covalent in the 3d-

than in the 4d- and 5d-group. However, this tendency to covalency increases with the oxidation number, with the number of electrons in the partly filled shell, and with the nephelauxetic series of the ligands.

(2) The electron transfer spectra of hexahalide complexes are now well understood. The application of M.O. configurations as classification has been the most conspicuous success for the M.O. theory since the explanation of the paramagnetism of the oxygen molecule.

(3) The high microsymmetry of slightly distorted octahedral complexes (such as Co en_3^{3+}, V ox_3^{3-}, $Rh(NH_3)_5Cl^{2+}$, Co en_2 $(NH_3)_2^{3+}$, Ni gly_3^-, Ni $gly_2(H_2O)_2$, cis-Rh $ox_2Cl_2^{3-}$ and the solid compounds MnO, MnF_2, $KMnF_3$ and $CsNiCl_3$) contrary to the predictions of the strict group theory has been a great surprise, disproving the electrostatic model of the ligand field (as the electrostatic model of the electron transfer spectra has also been disproved). The spectrochemical series has been shown to be mainly a function of the atom directly bound to the metal atom.

(4) The sign of Δ in tetrahedral d^n-complexes, both of halogens and of oxygen, is negative in agreement with the σ-bonding effect on M.O. formation, but disagrees with some previous assumptions for the tetroxo ions.

(5) The orbital energy differences in NpF_6 between the seven 5f-orbitals are now shown to be much smaller than in octahedral d^n-complexes, and comparable to ζ_{5f}. By inference, all energy levels of actinide complexes (excepting the possible cases of the dioxo ions) can be described approximately by a spherical microsymmetry, as is also the case for the lanthanides.

(6) The parameters of interelectronic repulsion are shown to vary strongly with the ionic charge in the lanthanides and the actinides. Hence, the f-shell cannot be qualified as extremely internal in the atom; only for ionic charges $+ 3$ and more, the f-shell obtains a smaller average radius than the atomic core.

Though these six problems are all formulated in chemical spectroscopy, they are all of great consequence to the theory of chemical bonding in the transition groups. As examples of other problems, which have been treated in the period 1956–1959, but which are much less clarified, may be mentioned:

(7) The tetragonality of $d^1(VO^{2+})$ and $d^9(Cu^{2+})$ complexes and in general, the behaviour of Jahn–Teller unstable complexes. While in some cases, such as octahedral Fe(II) and Co(II) complexes, the

theory for O_h is quite satisfactory, deviations already occur for such examples as vanadium (III) and tetrahedral nickel (II), and the highly varying behaviour of Cu(II) is not yet completely explained.

(8) The linear and other complexes of Cu(I), Ag(I) and other d^{10}-systems.

(9) The detailed identification of the energy levels of acetylaceton-ate and ethylenediaminetetra-acetate complexes, where in some cases, a normal "octahedral" spectrum is observed (Cr(III), Co(II), Co(III)) and in other cases, rather anomalous splittings occur (Ti(III), V(III)).

In general, the octahedral complexes are much better understood than the lower symmetries. We may characterize, in 1959, the situation for d^n-complexes:

Symmetry	Ligand field	Electron transfer
octahedral	very good	good
square-planar	good	bad
tetrahedral	good	fair
other symmetries	fair	bad

For giving an impression of how numerical results from the absorption spectra may clarify qualitative discussions of old problems in chemistry, we may mention the concept of *electronegativity*. The present writer has a much higher opinion of this, not too well-defined, notion and Linus Pauling's study of it, than of the hybridization theory, for example. However, this concept has not been mentioned before in this book, because it is difficult to present very clearly. The electronegativity should be a monotonic function of the sum of the electron affinity and the ionization energy of the neutral atom,[*] but is, in practice, determined from chemical argument. It is generally given as a number (3·0 for Cl, 2·6 for I, 1·8 for Cu, 1·2 for Mg, 0·7 for Cs) but for our purpose it is sufficient to consider it as imposing

---

* Mulliken (1949) defined such "absolute electronegativities" from $I(a) - \frac{1}{2}J(aa)$, where $I(a)$ is the ionization energy of the orbital a and $J(aa)$ the Coulomb integral with itself. However, this does not only change the order, in eqn. (238), of H and C and of Al and Mg, when applied to neutral atoms. A more serious consequence is that most monovalent, gaseous ions would have electronega-tivities well above F. The electronegativity is rather a qualitative concept, connected with energy differences such as eqn. (187), p. 240.

an ordered pattern* in a series of elements, arranged in order of electronegativities:

$$F > O > Cl \sim N > Br > I \sim S \sim C > Au > H \sim P$$
$$> As \sim B \sim Hg > Cu \sim Ag \sim Pb > Co \sim Ni > Fe$$
$$> Zn \sim Al > Mg > La \sim Li \sim Ca > Na \sim Ba > K$$
$$\sim Rb > Cs \quad (238)$$

It is argued that the bond between two atoms A and B is mainly covalent if the electronegativities x_A and x_B are nearly equal, while the bond is mainly electrostatic, if x_A and x_B are very different. It is further often assumed that the bond energy is a parabolic function $(x_A - x_B)^2$ for the electrostatic contribution. Actually, most authors try to describe several other properties by the electronegativity; e.g. the difference $x_O > x_S$ is said both to have the effect that S^{2-} is more reducing than O^{2-} and that in SO_4^{2-}, more negative charge is placed on each oxygen than on the sulphur atom. Here, a fundamental difficulty is introduced in the description: the electronegativity must vary with the oxidation number. This is not of much importance in organic compounds, but in inorganic compounds all available evidence would place Mn(VII) near to Au and Mn(II) near to Mg in the list given above.

Ito (1959) argues that since the 3d-metals in average has $x = 1 \cdot 6$, and since the covalent character (cf. Gordy, 1955) of a bond roughly is $1 - 0 \cdot 5 \, |x_A - x_B|$, the mean percentage of covalency in metal–ligand bands is 20 in aquo and 40 in ammonia complexes (assuming $x = 3 \cdot 2$ for O in H_2O and $2 \cdot 8$ for N in NH_3) and that hence aquo complexes have received $1 \cdot 2$ and ammonia complexes $2 \cdot 4$ electrons in the central ion. There is something qualitatively justifiable in these arguments, but a very important fact remains that Mn(IV) is much more covalent than Mn(II).

It is interesting to compare these numerical speculations with the nephelauxetic effect, estimated from the absorption spectra. We saw in chapters 7 and 8 that the spectrochemical series of ligands mainly can be written as

$$I < Br < Cl < S < F < O < N < C \quad (239)$$

* Actually (ref. 301e), the electron transfer spectra, as exemplified in eqn. (187), p. 240, suggest a quantitative measure of a parameter closely related to the electronegativity, one unit of Pauling's scale corresponding to 30 kK. The writer (ref. 301c) attempts to define "differential ionization energies", taking the Madelung energy into account, in order to estimate the actual charges of atoms in complexes.

and the nephelauxetic series of ligands (with some amplification for oxygen complexes)

$$F > O \sim N > Cl \sim C > Br > I \sim S \qquad (240)$$

The latter expression has a certain similarity with the first part of the electronegativity series, eqn. (238).

We saw also that the nephelauxetic effect can be ascribed to a mixture of central field covalency (i.e. the expansion of the radial function, caused by a smaller value Z_{eff} of the effective charge of the central ion than the oxidation number) and of symmetry-restricted covalency (i.e. that the partly filled shell is the anti-bonding linear combination of orbitals in the ligands and in the central ion). Though the two effects are very hard to distinguish experimentally*, we can estimate the upper limit of their sum fairly well. If we compare a certain number of fluoro and chloro complexes (the nephelauxetic effect of most common ligands, such as H_2O, NH_3, and ox^{2-} are confined in the interval between F^- and Cl^-) with isoelectronic gaseous ions (cf. Dunn 1959, the present writer, 1958) we obtain a value of Z_{eff}. The number $z - Z_{eff}$ cannot be considered as the fractional number of electrons present in addition to that expected for the oxidation number z, since as Orgel (1955) pointed out, the effect of widely extended electrons, such as 4s-electrons in gaseous Mn-atoms, is much smaller, some 0·12 per electron. On the other hand, in the symmetry-restricted covalency picture, $1 - \sqrt{\beta}$ would approximately represent the percentage of the partly filled shell M.O. in the ligands. β is the ratio between the interelectronic repulsion parameters in the complex and in the corresponding gaseous ion with the charge z.

It is seen from eqn. (241) that the tendency towards covalent bonding increases strongly with the oxidation number, increases with the number of d-electrons, and decreases, contrary to usual chemical arguments, in the series 3d > 4d > 5d. The nephelauxetic effect is

* One may assume, in analogy with eqn. 84, p. 79, that the interelectronic repulsion parameters J(1, 4), J (3, 4), K (1, 4), K (3, 4) are multiplied by $(1 - \epsilon)$ and J (1, 3) and K (1, 3) by $(1 - \epsilon)^2$ (Koide and Pryce, 1958, and Pappalardo, 1959) compared with the B and C values of J (4, 4), J (4, 5) and K (4, 5) (cf. eqn. 75, p. 73). Hence, it is possible to identify the various values of $\beta_{55} = \beta_{35}$ $(1 - \epsilon)^{-1} = \beta_{33}(1 - \epsilon)^{-2}$ with eqn. 243 interpreting $(1 - \epsilon)$ as the difference between the values of $(1 - X)$ for the σ-antibonding γ_3 and π-antibonding γ_5 subshell, while β_{55} expresses the combined effects of central-field covalency and π-delocalization. In most Cr(III) and Mn(IV) complexes, β_{35} is much smaller than β_{55} (cf. ref. 301d).

remarkably small for fluoro and aquo complexes of 3d-metals with the oxidation number $+2$ (presumably also for PdF_2) and probably due to the central field covalency. However, when Z_{eff} decreases below 1, the symmetry-restricted covalency is probably important. It is interesting that of all the examples given, only $RhCl_6^{3-}$, OsF_6 and IrF_6 have $1 - \sqrt{\beta}$ above 0·30, and none above 0·40. The chemical nature of the three examples might be taken as an argument for strong covalency in the 4d- and 5d-group (and, usually, large electronegativities are assumed for the platinum group metals). However, the reason is simply that the corresponding 3d-compounds $CoCl_6^{3-}$, FeF_6 and CoF_6 (which would be even more covalent) are impossible to prepare, they would spontaneously evolve Cl_2 and F_2.

fluoro and aquo complexes

	Z_{eff}	$z - Z_{eff}$	$1 - \sqrt{\beta}$	X_{corr}
$3d^3$ V(II)H_2O	1·6	0·4	0·04	0·07
$3d^3$ Cr(III)H_2O	1·8	1·2	0·11	0·145
$3d^3$ Mn(IV)F	1·0	3·0	0·25	0·32
$3d^5$ Mn(II)H_2O	1·75	0·25	0·03	0·04
$3d^5$ Fe(III)F	1·75	1·25	0·11	0·15
$3d^8$ Ni(II)H_2O	1·6	0·4	0·05	0·075
$5d^2$ Os(VI)F	1·5	4·5	0·31	0·46
$5d^3$ Re(IV)F	2·2	1·8	0·12	0·19
$5d^3$ Ir(VI)F	0·3	5·7	0·38	0·53
$5d^6$ Pt(IV)F	0·7	3·3	0·27	0·41

$\left. \right\}$ (241)

chloro complexes

	Z_{eff}	$z - Z_{eff}$	$1 - \sqrt{\beta}$	X_{corr}
$3d^3$ Cr(III)	0·4	2·6	0·25	0·36
$3d^5$ Mn(II)	1·5	0·5	0·06	0·08
$3d^5$ Fe(III)	0·75	2·25	0·24	0·31
$3d^8$ Ni(II)	0·7	1·3	0·15	0·19
$4d^3$ Mo(III)	1·4	1·6	0·14	0·225
$4d^6$ Rh(III)	0·2	2·8	0·30	0·46
$5d^3$ Re(IV)	1·5	2·5	0·18	0·31

The percentage of presence in the ligands of the partly filled shell have been estimated somewhat higher by Owen (1955) and Tinkham (1956) from the paramagnetic resonance curves of 3d-aquo and fluoro complexes. As discussed (p. 106), these effects correspond partly to a determination of ζ_{nd} in the complex, partly to an estimate of the influence of the magnetic moments of the nuclei of the central ion and the ligands, allowing a guess to be made of the relative electron densities of the partly filled shell near to these nuclei.

If we consider, as a rough model, the part $(1 - X)$ of the electron density of the partly filled shell to have the radial distribution belonging to the central ion with Z_{eff} smaller than the ionic charge z, and the part X occurring in the ligands, the value of the parameters of interelectronic repulsion such as B would be

$$(1 - X)^2 B(Z_{eff}) + (2X - X^2)B(r_L) \tag{242}$$

with the small correction $B(r_L)$, about 0·2 kK, representing the contribution from the extension* of the partly filled shell in the ligands. If we define β_* as the ratio between ζ_{nd} estimated for the complex and ζ_{nd} in the corresponding gaseous ion, (cf. p. 105) we would have for central-field covalency alone $(X = 0)$:

$$\left.\begin{array}{l} \beta = \dfrac{B(Z_{eff})}{B(z)} = \dfrac{Z_{eff} + Z_c}{z + Z_c} \text{ with } Z_c \sim 0\cdot85 + 0\cdot27q \text{ for } 3d^q \\[3mm] \beta_* = \dfrac{\zeta(Z_{eff})}{\zeta(z)} = \dfrac{(Z_{eff} + Z_n)^2}{(z + Z_n)^2} \text{ with } Z_n \sim 0\cdot4 + 0\cdot5q \text{ for } 3d^q \end{array}\right\} \tag{243}$$

Hence, β^2 has approximately the same value as β_*. This agrees fairly well with the conditions in MO and MF$_2$ of the 3d-group. On the other hand, in Cr(III) complexes, $\beta \sim \beta_*$, and for IrF$_6$, β is slightly smaller than β_*. This can be explained by the introduction of symmetry-restricted covalency $(X > 0)$, giving for ζ_{nd} and presumably also for the proportionality constant of the hyperfine structure from the magnetic moment of the central nucleus

$$\beta_* = (1 - X)\zeta(Z_{eff})/\zeta(z) \tag{244}$$

Hence, in extreme cases of predominant symmetry-restricted effects, $\beta_*^2 \sim \beta \sim (1 - X)^2$. However, we have here neglected the residual

---

* This may possibly be a rather bad approximation, considering the fact that the integrals (eqn. 75, p. 73) for γ_3 and γ_5 electrons, entirely concentrated in the six ligands and having the σ and π wavefunctions from Table 4, p. 58, would have the values J (1, 1) = J (4, 4) = $\frac{1}{4}$J (x, x); J (1, 3) = $\frac{1}{12}$J (x, x); J (4, 5) = $\frac{1}{4}$J (x, x); J (1, 4) = $\frac{1}{12}$J (x, y); and J (3, 4) = $\frac{1}{4}$J (x, y), if only the interaction between p-orbitals x and y on the same ligand atom are considered. In this particular approximation, K (a, b) = J (a, b).

term ~ 0.2 kK in eqn. (242), actually approaching OsF_6 and $RhCl_6{}^{3-}$ to a case with $X \sim 0.5$. In eqn. (241), the value of X_{corr} is that obtained from eqn. (242) with $B(r_L) = 0.2$ kK. In nearly all the examples, X_{corr} is about a third greater than $1 - \sqrt{\beta}$. However, it is a higher limit to the extent of symmetry-restricted covalency, since it has been derived with the assumption $Z_{eff} = z$.

The present writer feels that the nephelauxetic effect is the most informative evidence which has been extracted from absorption spectra for the use in the description of chemical bonding. It gives a quantitative expression for one of the several concepts, which are mixed together in the general notion of electronegativity. It is interesting to note that if Z_{eff} could be found in the halogen ligands (cf. eqn. (151)) it would probably range between, say, -0.9 in fluorine to -0.2 in iodine, probably being a function of the central atom too. There is no reason to believe that the values of Z_{eff} in the central atom and the ligands, when added, would give exactly zero by analogy to the actual electroneutrality of matter. In that way, Berzelius' idea of residual electric charges in radicals (positive in CaO, negative in SO_3, forming the more stable $CaSO_4$) might have an acceptable modern analogy, the bond regions between the atoms being the buffer regions for accumulating the electrons not definitely bound to definite atoms.

One final application of absorption spectra may be mentioned: the study of microsymmetry (e.g. tetrahedral or octahedral co-ordination of Co(II) and Ni(II)) and the relative internuclear distances (Orgel's ruby compression effect, 1957) in solid compounds and minerals. The thorough studies by Schmitz-DuMont and collaborators, Neuhaus, and Pappalardo, have demonstrated that absorption spectra may clarify crystallographic problems in exchange for the large amount of structural data which have been a necessary basis for much of the recent spectroscopic work.

Of course, these remarks do not answer the reader's possible questions: what should be done to identify and interpret the absorption bands of a given set of new species of metal complexes. The present writer is willing, as far as he can, to comment on identifications made by other chemists and physicists and will be glad to see such attempts at an early stage, before they are published in the public literature. Correspondence should be addressed to him at Cyanamid European Research Institute, 91 route de la Capite, Cologny (Geneva), Switzerland.

BIBLIOGRAPHY

1. ABRAGAM, A. and PRYCE, H. M. L. *Proc. Roy. Soc. London* **A206** (1951) 164 and 173.
2. ADAMSON, A. W., WELKER, J. P. and WRIGHT, W. B. *J. Am. Chem. Soc.* **73** (1951) 4786. Mn(III) cyanides.
3. ADAMSON, A. W. *J. Am. Chem. Soc.* **76** (1954) 1578. Chelate effect.
4. ADAMSON, A. W. *J. Am. Chem. Soc.* **78** (1956) 4260. Co(III) cyanides.
4ᵃ. ADAMSON, A. W. *J. Inorg. Nucl. Chem.* **13** (1960) 275. Cr(III) photochemistry.
4ᵇ. ADAMSON, A. W. *Discuss. Faraday Soc.* **29** (1960) 163. Co(III) photochemistry.
4ᶜ. ADDISON, C. C., HATHAWAY, B. J., LOGAN, N., WALKER, A. *J. Chem. Soc.* **1960** 4308. Anhydrous nitrate Cu(II).
5. AHRLAND, S. and GRENTHE, I. *Acta Chem. Scand.* **11** (1957) 1111. Bi(III) complex formation constants.
6. AHRLAND, S., CHATT, J. and DAVIES, N. R. *Quart. Rev.* **12** (1958) 265. A and B elements.
6ᵃ. ANDERSON-JOHNSON, S. and BASOLO, F. *J. Am. Chem. Soc.* **82** (1960) 4423. Rh(III) amine chlorides, and private communication.
7. ARDON, M. and STEIN, G. *J. Chem. Soc.* (1956) 2095. Oxydation products of Cr(II).
8. ASMUSSEN, R. W. and BOSTRUP, O. *Acta Chem. Scand.* **11** (1957) 745. Ni(II) chlorides.
9. ASMUSSEN, R. W. and BOSTRUP, O. *Acta Chem. Scand.* **11** (1957) 1097. Ni(II) complexes.
10. ASMUSSEN, R. W. and SOLING, H. *Acta Chem. Scand.* **11** (1957) 1331. Mn(II) chlorides.
11. ASMUSSEN, R. W., BOSTRUP, O. and JENSEN, J. P. *Acta Chem. Scand.* **12** (1958) 24. Co(III), NO.
12. ASPREY, L. B. and KEENAN, T. K. *J. Inorg. Nucl. Chem.* **7** (1958) 27. Am(III), Am(IV), Cm(III), Cm(IV) fluorides.
12ᵃ. ASPREY, L. B. and CUNNINGHAM, B. B. *Progress Inorg. Chem.* **2** (1960) 267. Unusual oxidation numbers of 4f and 5f elements.
12ᵇ. ASPREY, L. B. and PENNEMAN, R. A. *J. Am. Chem. Soc.* **83** (1961) 2200. Am(IV) fluoride solution.
12ᶜ. AXE, J. D. Electronic structure of octahedrally co-ordinated Protactinium (IV) in Cs_2ZrCl_6. UCRL-9293. Berkeley, July 1960.
13. AYRES, G. H. and QUICK, Q. *Anal. Chem.* **22** (1950) 1403. Ir(IV) in perchloric acid.
14. AYRES, G. H. and TUFFLY, B. L. *Anal. Chem.* **24** (1952) 949. Pd(II) bromides.
15. AYRES, G. H. *Anal. Chem.* **25** (1953) 1622. Rh high oxidation numbers.
15ᵃ. BABAEVA, A. V. *Doklady NAUK Acad. Sci. URSS*, **40** (1943) 61. Pt(II).
16. BAGNALL, K. W., D'EYE, R. W. M. and FREEMAN, J. H. *J. Chem. Soc.* (1955) 2320. Po(II), Po(IV).
17. BAILAR, J. C. *Chemistry of Co-ordination Compounds.* Reinhold, New York, 1956.
18. BAILEY, T. L. *J. Chem. Phys.* **28** (1958) 792. Electron affinity of Fluorine atoms.

18[a]. BAILEY, N., CARRINGTON, A., LOTT, K. A. K. and SYMONS, M. C. R. *J. Chem. Soc.* **1960** 290. Cr(VI), Mn(VII), Re(VII) oxo hydroxo complexes.

18[b]. BALCHAN, A. S. and DRICKAMER, H. G., *J. Chem. Phys.* **35** (1961) 356. Pd(IV), Os(IV), Ir(IV), Pt(IV) hexahalides, pressure effects.

18[c]. BALDWIN, M. E. and TOBE, M. L. *J. Chem. Soc.* **1960** 4275. Co(III) amine thiocyanates.

19. BALLHAUSEN, C. J. *Mat. fys. Medd. Dan. Vid. Selsk.* **29** (1954) No. 4. Electrostatic model Cu(II).

20. BALLHAUSEN, C. J. *Mat. fys. Medd. Dan. Vid. Selsk.* **29** (1955) No. 8. Electrostatic model Ni(II).

21. BALLHAUSEN, C. J. and JØRGENSEN, C. K. *Mat. fys. Medd. Dan. Vid. Selsk.* **29** (1955) No. 14. Electrostatic model V(III), Cr(III), Ni(II).

22. BALLHAUSEN, C. J. and JØRGENSEN, C. K. *Acta Chem. Scand.* **9** (1955) 397. Co(II) chloride and amine complexes.

23. BALLHAUSEN, C. J. *Acta Chem. Scand.* **9** (1955) 821. Band intensities.

24. BALLHAUSEN, C. J. and MOFFITT, W. *J. Inorg. Nucl. Chem.* **3** (1956) 178. Co(III) in tetragonal symmetry, dichroism.

25. BALLHAUSEN, C. J. and ASMUSSEN, R. W. *Acta Chem. Scand.* **11** (1957) 479. Co(III) temperature-independent paramagnetism.

26. BALLHAUSEN, C. J. and LIEHR, A. D. *J. Mol. Spectroscopy* **2** (1958) 342. Intensities of tetrahedral complexes.

27. BALLHAUSEN, C. J. and ANCMON, E. M. *Mat. fys. Medd. Dan. Vid. Selsk.* **31** (1958) No. 9. Table of integrals for the electrostatic model.

27[a]. BALLHAUSEN, C. J. and WINTHER, F. *Acta Chem. Scand.* **13** (1959) 1729. V(III) fluorides.

27[b]. BALLIK, E. A. and RAMSAY, D. A. *J. Chem. Phys.* **31** (1959) 1128. Diatomic carbon molecule.

28. BANKS, C. V. and KLINGMAN, D. W. *Anal. Chim. Acta* **15** (1956) 356. Lanthanides.

29. BANKS, C. V. and BARNUM, D. W. *J. Am. Chem. Soc.* **80** (1958) 4767. Ni(II), Pd(II), Pt(II) interactions in oximes.

29[a]. BANNISTER, E. and COTTON, F. A. *J. Chem. Soc.* **1960** 2276. Co(II) triphenylphosphine oxide complexes.

29[b]. BARRACLOUGH, C. G. and TOBE, M. L. *J. Chem. Soc.* **1961** 1993. Co(III) en-sulphate.

30. BARTLETT, N. and MAITLAND, R. *Acta Cryst.* **11** (1958) 747. Pd(II) fluoride, octahedral micro-symmetry and high-spin.

31. BASOLO, F. *J. Am. Chem. Soc.* **70** (1948) 2634; *Ibid* **72** (1950) 4393; *Ibid.* **75** (1953) 227. Co(III) amine and anion complexes.

32. BASOLO, F., BALLHAUSEN, C. J. and BJERRUM, J. *Acta Chem. Scand.* **9** (1955) 810. Co(III).

33. BASOLO, F. and PEARSON, R. G. *The Kinetics of Inorganic Reactions.* John Wiley, New York, 1958.

33[a]. BASU, G. and BASU, S. *Z. physik. Chem.* (*Leipzig*) **215** (1960) 309. Cu(II) Schiff' base complexes.

34. BAYLISS, N. S. and REES, A. L. G. *J. Chem. Phys.* **8** (1940) 377; *Trans. Faraday Soc.* **35** (1939) 792. Halogen molecules.

34[a]. BECQUEREL, J. and OPECHOWSKI, W. *Physica* **6** (1939) 1039. Ni(II) theory.

35. BELFORD, R. L., CALVIN, M. and BELFORD, G. *J. Chem. Phys.* **26** (1957) 1164. Cu(II) acetylacetonates.

35[a]. BELFORD, R. L. and BELFORD, G. *J. Chem. Phys.* **34** (1961) 1330. U(VI) wave functions.

35ᵇ. BELLUGUE, J. and DAUDEL, R. *Revue Scient. (Paris)* **84** (1946) 541. Electronegativity.

36. BERNAL, I., EBSWORTH, E. A. V. and WEIL, J. A. *Proc. Chem. Soc.* (1957) 57. Co(III, IV) paramagnetic resonance.

36ᵃ. BERNAL, I. and HARRISON, S. E. *J. Chem. Phys.* **34** (1961) 102. Cr(I) nitrosyl pentacyanide.

37. BERRY, A. J. and LOWRY, T. M. *J. Chem. Soc.* (1928) 1748. Tl(III) halides.

37ᵃ. BERRY, R. S. *J. Chem. Phys.* **27** (1957) 1288. Gaseous alkali halide molecules.

38. BETHE, H. *Ann. Physik* [5] **3** (1929) 133. Foundation of ligand field theory.

39. BETHE, H. and SPEDDING, F. H. *Phys. Rev.* **52** (1937) 454. Tm(III).

40. BEVAN, H., DAWES, S. V. and FORD, R. A. *Spectrochim. Acta* **13** (1958) 43. Ti(IV) oxide.

41. BINGEL, W. *Z. Naturforsch.* **9a** (1954) 675. F^k integrals for hydrogenic radial functions.

42. BJERRUM, J. *Mat. fys. Medd. Dan. Vid. Selsk.* **11** (1932) No. 5; *Ibid.* **11** (1932) No. 10; *Ibid.* **12** (1934) No. 15. Cu(II) ammonia complexes.

43. BJERRUM, J. *Metal Ammine Formation in Aqueous Solution.* Haase and Son, Copenhagen, 1941 (reprinted 1957).

44. BJERRUM, J. *Mat. fys. Medd. Dan. Vid. Selsk.* **21** (1944) No. 4. Principle of corresponding solutions.

45. BJERRUM, J. *Mat. fys. Medd. Dan. Vid. Selsk.* **22** (1946) No. 18. Cu(II) chlorides.

46. BJERRUM, J. and NIELSEN, E. J. *Acta Chem. Scand.* **2** (1948) 297. Cu(II) ethylenediamine complexes.

47. BJERRUM, J. and LAMM, C. G. *Acta Chem. Scand.* **4** (1950) 997. Cu(II) amine complexes.

48. BJERRUM, J. *Chem. Rev.* **46** (1950) 381. Complex formation constants.

49. BJERRUM, J. and RASMUSSEN, S. E. *Acta Chem. Scand.* **6** (1952) 1265. Co(III) ethylenediamine complexes.

50. BJERRUM, J. and JØRGENSEN, C. K. *Acta Chem. Scand.* **7** (1953) 951. Nd(III) in aqueous alcohol.

51. BJERRUM, J., BALLHAUSEN, C. J. and JØRGENSEN, C. K. *Acta Chem. Scand.* **8** (1954) 1275. Electrostatic model, Cu(II) aquo and amine complexes.

52. BJERRUM, J. and LAMM, C. G. *Acta Chem. Scand.* **9** (1955) 216. Cr(III) ammonia complexes.

53. BJERRUM, J. and JØRGENSEN, C. K. *Rec. trav. chim. Pays-Bas* **75** (1956) 658. Ligand field stabilization.

54. BJERRUM, J., ADAMSON, A. W. and BOSTRUP, O. *Acta Chem. Scand.* **10** (1956) 329. Solvation effects. Fe(II), Cr(III), Co(III) complexes.

55. BJERRUM, J., SCHWARZENBACH, G. and SILLÉN, L. G. *Stability Constants of Metal-ion Complexes, with Solubility Products of Inorganic Substances.* Part I. Organic Ligands (Special Publication No. 6, 1957) and Part II. Inorganic Ligands (Special Publication No. 7, 1958), Chemical Society, London.

56. BJERRUM, N. *Z. anorg. Chem.* **63** (1909) 140. Cr(III), dependence of the absorption spectrum on first co-ordination sphere only.

56ᵃ. BLASIUS, E., PREETZ, W. and SCHMITT, R. *J. Inorg. Nucl. Chem.* **19** (1961) 115. Ir(IV), Pt(IV) chlorides.

57. BOEF, G. DEN, BECK, H. J. VAN DER and BRAAF, T. *Rec. trav. chim. Pays-Bas* **77** (1958) 1064. Mn(VI), Mn(VII) tetroxo ions.

58. BOSTON, C. R. and SMITH, G. P. *J. Phys. Chem.* **62** (1958) 409. Ni(II) chloride melts.

59. Bostrup, O. and Jørgensen, C. K. *Acta Chem. Scand.* **11** (1957) 1223. Ni(II) amine and anion complexes in solid state and solution.

60. Boulanger, F. *Ann. Chim.* [12] **7** (1952) 732. Pr(III), Nd(III) solid compounds.

61. Boyd, G. E., Cobble, J. W., Nelson, C. M. and Smith, W. T. *J. Am. Chem. Soc.* **74** (1952) 556. Tc(VII) tetroxo ions.

61ᵃ. Boyd, G. E. and Larson, Q. V. *J. Phys. Chem.* **64** (1960) 988. Tc(VII).

62. Brandt, Dwyer, F. P. and Gyarfas, E. C. *Chem. Rev.* **54** (1954) 959. Dipyridyl and phenanthroline complexes.

63. Brasted, R. C. and Hirayama, C. *J. Phys. Chem.* **63** (1959) 780. Co(III) amine and chloro complexes.

64. Brickler, C. E. and Sweetser, P. B. *Anal. Chem.* **25** (1953) 764. Ce(IV).

65. Brigando, J. *Bull. Soc. Chim. France* (1957) 211. Co(III) amine complexes.

66. Brigando, J. *Bull. Soc. Chim. France* (1957) 503. Ni(II), Cu(I), Au(I) cyanides.

67. Brode, W. R. *Proc. Roy. Soc. London* A**118** (1928) 286. Co(II) chlorides.

68. Brode, W. R. and Morton, R. A. *Proc. Roy. Soc. London* A**120** (1928) 21. Co(II) bromides and iodides.

69. Broer, L. J. F., Gorter, C. J. and Hoogschaagen, J. *Physica* **11** (1945) 231. Band intensities in lanthanides.

70. Brosset, C. *Arkiv Kemi, Mineral., Geol.* A**25** (1948) No. 19. Crystallography of Pt(II, IV) salts.

71. Brown, D. A. *J. Chem. Phys.* **28** (1958) 67. F^k-integrals in the 3d-group.

72. Brunetti, R. *Nuovo Cimento* **5** (1928) 391; *Ibid.* **6** (1929) 347. Ligand fields in Pr(III).

73. Buck, R. P., Singhadeja, S. and Rogers, L. B. *Anal. Chem.* **26** (1954) 1240. Cu(I) and Ni(II) cyanides, nitrate and several other anions.

73ᵃ. Buffagni, S. and Dunn, T. M. *Nature* **188** (1960) 937. Mn(II), Co(II), Ni(II) tetrachlorides.

73ᵇ. Bullen, G. J., Mason, R. and Pauling, P. *Nature* **189** (1961) 291. Crystal structure of Ni_3aca_6.

74. Busey, R. H. and Larson, Q. V. *One Hundred and Thirty-third American Chemical Society Meeting, San Francisco* 1958. Tc(IV), Re(II), Re(III).

75. Butement, F. D. S. *Trans. Faraday Soc.* **44** (1948) 617. Sm(II), Eu(II), Yb(II).

76. Böke, K. *Z. physik. Chem.* **10** (1957) 45 and 59; *Ibid.* **11** (1957) 326. X-ray absorption edges.

76ᵃ. Cabezas, A. Y. and Lindgren, J. *Phys. Rev.* **120** (1960) 920. Tm(III).

77. Cady, H. H. and Connick, R. E. *J. Am. Chem. Soc.* **80** (1958) 2646. Ru(III) aquo and chloride complexes.

78. Candela, G. A., Hutchinson, C. A. and Lewis, W. B. *J. Chem. Phys.* **30** (1959) 246. U(IV), Pu(IV) magnetism and ligand fields.

79. Carlson, E. and Dieke, G. H. *J. Chem. Phys.* **29** (1958) 229; *Ibid.* **34** (1961) 1602. Nd(III) fluorescence.

80. Carlson, W. T., Fields, P. R., Stewart, D. C. and Keenan, T. K. *J. Inorg. Nucl. Chem.* **6** (1958) 213. Cm(III) aquo ion.

80ᵃ. Carnall, W. T. and Fields, P. R. *J. Am. Chem. Soc.* **81** (1959) 4445. Cm(III).

81. Carrington, A., Schonland, D. and Symons, M. C. R. *J. Chem. Soc.* **1956** 4710; *Ibid.* **1957** 659. Tetroxo ions (cf. ref. 81ᵇ).

81ª. CARRINGTON, A., INGRAM, D. J. E., LOTT, K. A. K., SCHONLAND, D. and SYMONS, M. C. R. *Proc. Roy. Soc. London* A254 (1959) 101. Paramagnetic resonance of tetroxo ions.

81ᵇ. CARRINGTON, A. and SYMONS, M. C. R. *J. Chem. Soc.* 1960 889. V(V), Cr(VI), Mn(VI), Mn(VII), Fe(VI) tetroxo ions.

81ᶜ. CARRINGTON, A. and SCHONLAND, D. S. *Mol. Phys.* 3 (1960) 331. Tetroxo ions.

81ᵈ. CARRINGTON, A. and JØRGENSEN, C. K. *Mol. Phys.* in press. Tetroxo ions.

82. CARTLEDGE, G. H. *J. Am. Chem. Soc.* 74 (1952) 6015. Mn(III) oxalates and acetylacetonates.

83. CATALAN, M. A., RÖHRLICH, F. and SHENSTONE, A. G. *Proc. Roy. Soc. London* A221 (1954) 421. F^k-integrals in the 3d-group.

83ª. CAUCHOIS, Y. *J. Phys. Radium* 13 (1952) 113; *Ibid.* 16 (1955) 253. X-ray absorption edges.

83ᵇ. CHAKRAVORTY, A. *Naturwiss.* 48 (1961) 375. Au(III) chloride.

83ᶜ. CHAKRAVORTY, A. and BASU, S. *J. Inorg. Nucl. Chem.* 17 (1961) 55. Ni(II), Cu(II) biguanide crystals.

84. CHARLOT, G. *Théorie et méthode nouvelle d'analyse qualitative* (2nd Ed.). Masson, Paris, 1946.

85. CHARLOT, G. and GAUGUIN, R. *Les méthodes d'analyse des réactions en solution.* Masson, Paris, 1951.

86. CHARLOT, G. and GAUGUIN, R. *Dosages Colorimétriques.* Masson, Paris, 1952.

87. CHATT, J., GAMLEN, G. A. and ORGEL, L. E. *J. Chem. Soc.* (1958) 486. Pt(II) chloride and amine complexes, ligand field theory.

88. CHATT, J., GAMLEN, G. A. and ORGEL, L. E. *J. Chem. Soc.* (1959) 1047. Pt(II). Spectrochemical series Te, Se, S, As, N, P.

88ª. CHATT, J. and HAYTER, R. G. *J. Chem. Soc.* 1961 772. Ru(II) phosphine halide hydrides. Spectrochemical position of H^-, CH_3^- and $C_6H_5^-$.

88ᵇ. CHIA, YUAN-TSAN and KING, E. L. *Discuss. Faraday Soc.* 29 (1960) 109 and private communication. Cr(III) fluorides.

88ᶜ. CIAMPOLINI, M., PAOLETTI, P. and SACCONI, L. *J. Chem. Soc.* 1960 4553. Fe(II), Zn(II) ethylenediamine complexes.

89. CLAUSS, D. and LISSNER, A. *Z. anorg. Chem.* 297 (1958) 300. Mn(I), Re(I) cyanides.

90. COHEN, A. J. and DAVIDSON, N. *J. Am. Chem. Soc.* 73 (1951) 1955. Pd(II), Pd(IV), Pt(II) chlorides.

91. COHEN, A. J. and PLANE, R. A. *J. Phys. Chem.* 61 (1957) 1096. Fe(II) cyanides.

91ª. COHEN, D. and CARNALL, W. T. *J. Phys. Chem.* 64 (1960) 1933. U(III), U(IV).

91ᵇ. COLLET, V. *Thesis* Paris 1959 (Gauthier-Villars). X-ray absorption edges.

91ᶜ. COLLIN, J. *J. Chimie Physique* 57 (1960) 424. C(IV) oxide and sulphide.

91ᵈ. COLLMAN, J. P., MOSS, R. A., MALTZ, H. and HEINDEL, C. C. *J. Am. Chem. Soc.* 83 (1961) 531. Halogenated Cr(III), Co(III) acetylacetonates.

91ᵉ. COLTON, R., LEVITIUS, R. and WILKINSON, G. *J. Chem. Soc.* 1960 5275. Tetrahedral Re(III).

92. CONDON, E. U. and SHORTLEY, G. H. *Theory of Atomic Spectra* (2nd Ed.). University Press, Cambridge, 1953.

93. CONNICK, R. E. and HURLEY, C. R. *J. Am. Chem. Soc.* 74 (1952) 5012. Ru(VI), Ru(VII), Ru(VIII).

94. CONWAY, J. G., WALLMANN, J. C., CUNNINGHAM, B. B. and SHALIMOFF, G. V. *J. Chem. Phys.* 27 (1957) 1416. U(III), Np(III), Cm(III).

95. CONWAY, J. G. UCRL-8613; *J. Chem. Phys.* **31** (1959) 1002. U(IV) in fluorite.
95ᵃ. CONWAY, J. G. and GRUBER, J. B. *J. Chem. Phys.* **32** (1960) 1586. Pm(III) chloride.
96. COOK, S. P. and DIEKE, G. H. *J. Chem. Phys.* **27** (1957) 1213. Gd(III).
97. COTTON, F. A. and BALLHAUSEN, C. J. *J. Chem. Phys.* **25** (1956) 617. X-ray absorption edges.
98. COTTON, F. A. and HANSON, F. A. *J. Chem. Phys.* **28** (1958) 83. X-ray absorption edges.
98ᵃ. COTTON, F. A., MONCHAMP, R. R., HENRY, R. J. M. and YOUNG, R. C. *J. Inorg. Nucl. Chem.* **10** (1959) 28. Mn(I), Fe(II) nitrosyl cyanides.
98ᵇ. COTTON, F. A. and GOODGAME, D. M. L. *J. Am. Chem. Soc.* **82** (1960) 2967 and 5771. Ni(II) phosphine and phosphine oxide halides.
98ᶜ. COTTON, F. A. and HOLM, R. H. *J. Am. Chem. Soc.* **82** (1960) 2979 and 2983. Co(II) acetylacetonates and phosphine oxide complexes.
98ᵈ. COTTON, F. A. and FRANCIS, R. *J. Am. Chem. Soc.* **82** (1960) 2986 and *J. Inorg. Nucl. Chem.* **17** (1961) 62. Co(II), Ni(II) halides in dimethyl-sulphoxide.
98ᵉ. COTTON, F. A. and MEYERS, M. D. *J. Am. Chem. Soc.* **82** (1960) 5023. Fe(II) aquo, high-spin Co(III) fluoride.
98ᶠ. COTTON, F. A. and GOODGAME, M. *J. Am. Chem. Soc.* **83** (1961) 1777. Co(II) cyanates, azides.
98ᵍ. COTTON, F. A., FAUT, O. D., GOODGAME, D. M. L. and HOLM, R. H. *J. Am. Chem. Soc.* **83** (1961) 1780. Co(II) phosphine halides.
98ʰ. COTTON, F. A., GOODGAME, D. M. L., GOODGAME, M. and SACCO, A. Co(II) phosphine thiocyanates, Hg(II), Co(II) thiocyanate.
99. COULSON, C. A., CRAIG, D. P. and JACOBS, J. *Proc. Roy. Soc. London* A**206** (1951) 297. M.O. configuration intermixing.
100. COZZI, D. and VIVARELLI, S. *Z. anorg. Chem.* **279** (1955) 165. Nb(II), Nb(III), Nb(IV).
101. COZZI, D. and PANTANI, F. *J. Inorg. Nucl. Chem.* **8** (1958) 385. Rh(III) chloride complex formation constants.
102. CRAIG, D. P., MACCOLL, A., NYHOLM, R. S., ORGEL, L. E. and SUTTON, L. E. *J. Chem. Soc.* (1954) 332 and 354. Overlap integrals with hydrogenic radial functions.
103. CRAIG, D. P. and MAGNUSSON, E. A. *J. Chem. Soc.* (1956) 4895; *Discuss. Faraday Soc.* **26** (1958) 116 (cf. p. 179). Influence of V_0 on radial functions.
103ᵃ. CRAYTON, P. H. and MATTERN, J. A. *J. Inorg. Nucl. Chem.* **13** (1960) 248. Co(III) amine complexes.
103ᵇ. CROSBY, G. A., WHAN, R. E. and ALIRE, R. M. *J. Chem. Phys.* **34** (1961) 743. Lanthanide acetylacetonates, energy transfer to triplet states of the ligands.
104. CROUTHAMEL, C. E. and JOHNSON, C. E. *Anal. Chem.* **26** (1954) 1284. Mo(V), W(V) thiocyanates.
105. CROUTHAMEL, C. E. *Anal. Chem.* **29** (1957) 1756. Tc(V) thiocyanates.
106. CUNNINGHAM, B. B., GRUEN, D. M., CONWAY, J. G. and McLAUGHLIN, R. D. *J. Chem. Phys.* **24** (1956) 1275. Pu(III) fluorescence.
106ᵃ. CUNNINGHAM, B. B. Comparative Chemistry of the Actinide and Lanthanide Elements. *17th Internat. Congress Pure Appl. Chem.*, Munich, p. 64, 1959.
106ᵇ. CURTIS, N. F. *J. Chem. Soc.* **1960** 4409. Ni(II) Schiff' base complex from acetone.
107. DALZIEL, J., GILL, N. S., NYHOLM, R. S. and PEACOCK, R. D. *J. Chem. Soc.* (1958) 4012. Tc(IV), Re(IV).

108. DAINTON, F. S. *J. Chem. Soc.* (1952) 1533. V(II), Cr(II), Fe(II).
108ᵃ. DEFORD, D. D. and DAVIDSON, A. W. *J. Am. Chem. Soc.* **73** (1951) 1469. Ru(II), Ru(III) cyanides.
109. DELÉPINE, M. *Compt. rend.* **236** (1953) 559; 1115. Alcohol catalysis of Rh(III) reactions.
110. DELÉPINE, M. *Compt. rend.* **248** (1959) 2682; *Ann Chim.* [13] **4** (1959) 1115. Ir(III, IV) nitrogen sulphates.
110ᵃ. DELÉPINE, M. *Compt. rend.* **251** (1960) 2633. Ir(III, IV) nitride sulphate pyridine complexes.
111. DESESA, M. A. and ROGERS, L. B. *Anal. Chim. Acta* **6** (1952) 534. Fe(III), Cu(II), Sb(V) in hydrochloric acid.
111ᵃ. DESKIN, W. A. *J. Am. Chem. Soc.* **80** (1958) 5680. Ni(II) dithiomalonates.
112. DEUTSCHBEIN, O. *Ann. Physik* [5] **14** (1932) 753; *Ibid.* **20** (1934) 828. Ruby lines of Cr(III).
113. DEXTER, D. L. *Phys. Rev.* **108** (1957) 707. Exciton model in NaCl.
113ᵃ. DEXTER, D. L. *J. Phys. Chem. Solids* **8** (1959) 473. Photoconductivity, RbI, AgCl, CdS, TlCl, PbSe.
113ᵇ. DHARMATTI, S. S. and KANEKAR, C. R. *J. Chem. Phys.* **31** (1959) 1436. Co(III) nuclear magnetic resonance chemical shifts.
114. DICKENS, P. G. and LINNETT, J. W. *Quart. Rev.* **11** (1957) 291. Correlation effects.
115. DIEKE, G. H. and HEROUX, L. *Phys. Rev.* **103** (1956) 1227. Nd(III).
116. DIEKE, G. H. and SINGH, S. *J. Opt. Soc. Am.* **46** (1956) 495. Dy(III) (cf. ref. 288).
117. DIEKE, G. H. and CROSSWHITE, H. M. *J. Opt. Soc. Am.* **46** (1956) 885. Yb(III).
118. DIEKE, G. H. and LEOPOLD, L. *J. Opt. Soc. Am.* **47** (1957) 944. Gd(III).
119. DIEKE, G. H. and SARUP, R. *J. Chem. Phys.* **29** (1958) 741. Pr(III).
119ᵃ. DIEKE, G. H., CROSSWHITE, H. M. and DUNN, B. *J. Opt. Soc. Am.* **51** (1961) 820. Gaseous 4f ions.
120. DOEHLEMANN, E. and FROMHERZ, H. *Z. physik. Chem.* **A171** (1934) 371. Cu(II), Cd(II).
120ᵃ. DOYLE, W. P. *J. Phys. Chem. Solids* **4** (1958) 144. As(III), Sb(III), Bi(III) oxides.
120ᵇ. DOYLE, W. P. and LONERGAN, G. A. *Discuss. Faraday Soc.* **26** (1958) 27. Co(II), Ni(II) oxide.
121. DREISCH, T. and TROMMER, W. *Z. physik. Chem.* **B37** (1937) 37. Co(II), Ni(II), anion complexes.
122. DREISCH, T. and KALLSCHEUER, O. *Z. physik. Chem.* **B45** (1939) 19. Cr(II), Fe(II), U(IV).
123. DROLL, H. A., BLOCK, B. P. and FERNELIUS, W. C. *J. Phys. Chem.* **61** (1957) 1000. Pd(II).
123ᵃ. DRUDING, L. F. and CORBETT, J. D. *J. Am. Chem. Soc.* **81** (1959) 5512; *Ibid.* **83** (1961) 2462. Nd(II) chloride and iodide.
124. DUNINA, A. A., MORGENSTERN, Z. L. and SHAMOSKY, L. M. *Optika y Spektroskopia* **4** (1958) 105. In(I) in alkali halide crystals.
125. DUNITZ, J. D. and ORGEL, L. E. *J. Chem. Soc.* (1953) 2594. M.O. treatment of dimeric Ru(IV).
126. DUNITZ, J. D. and ORGEL, L. E. *J. Phys. Chem. Solids* **3** (1957) 20 and 318. Jahn–Teller effect and values of Δ for oxides of the 3d-group.
126ᵃ. DUNITZ, J. D. and ORGEL, L. E. *Adv. Inorg. Chem. Radiochem.* **2** (1960) 1. Stereochemistry of ionic solids.

127. DUNN, T. M. *J. Chem. Soc.* (1959) 623. Nephelauxetic effect.
127ᵃ. DUNN, T. M. and PEACOCK, R. D. Private communication.
127ᵇ. DWYER, F. P. and NYHOLM, R. S. *Nature* 160 (1947) 502. Rh(IV) chloride.
128. DWYER, F. P. and GYARFAS, E. C. *J. Am. Chem. Soc.* 73 (1951) 2322. Os(II), (III) dipyridyl complexes.
128ᵃ. DWYER, F. P. and GARVAN, F. L. *J. Am. Chem. Soc.* 82 (1960) 4823. Rh(III) enta complex.
128ᵇ. EARWICKER, G. A. *J. Chem. Soc.* 1960 2620. Pd(II) sulphites.
128ᶜ. EBY, J. E., TEEGARDEN, K. J. and DUTTON, D. B. *Phys. Rev.* 116 (1959) 1099. Alkali metal halide crystals.
129. EDELSON, M. R. and PLANE, R. A. *J. Phys. Chem.* 63 (1959) 327. Cr(III) aquo and ammonia complexes, photochemistry.
129ᵃ. EDWARDS, A. L., SLYKHOUSE, T. E. and DRICKAMER, H. G. *J. Phys. Chem. Solids* 11 (1959) 140. Pressure experiments with GaP, GaAs, GaSb, ZnO, ZnS, ZnSe, ZnTe.
130. EISENSTEIN, J. C. and PRYCE, M. H. L. *Proc. Roy. Soc. Lond* A229 (1955) 20; A238 (1957) 31. Np(VI), Np(V) and Pu(VI) with assumption of great λ-splitting.
131. EISENSTEIN, J. C. *J. Chem. Phys.* 25 (1956) 142. s-, p-, d- and f-electrons in many low symmetries.
131ᵃ. EISENSTEIN, J. C. *J. Chem. Phys.* 34 (1961) 1628. Re (IV) chloride, d^3 intermediate coupling determinants in octahedral symmetry.
132. ELLEMAN, T. S., REISHUS, J. W., MARTIN, D. S. *J. Am. Chem. Soc.* 80 (1958) 536. Pt(II).
133. ELLIOTT, J. P., JUDD, B. R. and RUNCIMAN, W. A. *Proc. Roy. Soc. London* A240 (1957) 509. Calculation of f^n-multiplet term energies.
134. ELLIS, C. B. *Phys. Rev.* 49 (1936) 875; *Ibid.* 55 (1939) 1114. Pr(III).
135. ELVING, P. J. and ZEMEL, B. *J. Am. Chem. Soc.* 79 (1957) 1281. Cr(III) aquo and chloride complexes.
135ᵃ. EMERSON, K. and GRAVEN, W. M. *J. Inorg. Nucl. Chem.* 11 (1959) 309. Cr(III) hydrolysis.
136. EMSCHWILLER, G. Suppl. *Ricerca Scientifica* 27 (1957) 164. Reaction mechanisms of Fe(II) cyanides.
136ᵃ. ENGLMAN, R. *Mol. Phys.* 3 (1960) 23 and 48. Band shape and intensity of Laporte-forbidden transitions in aquo ions.
137. EPHRAIM, F. and BLOCH, R. *Ber.* 59 (1926) 2692; *Ibid.* 61 (1928) 65 and 72. Pr(III), Nd(III), Sm(III) solid compounds.
138. EPHRAIM, F. and RÂY, P. *Ber.* 62 (1929) 1509, 1520 and 1639. Pr(III), Nd(III), Sm(III) solid compounds.
139. EPHRAIM, F. and MEZENER, M. *Helv. Chim. Acta* 16 (1933) 1260. U(IV) halides.
139ᵃ. EPPLER, K. A. and DRICKAMER, H. G. *J. Phys. Chem. Solids* 6 (1958) 180. Pressure effects on Tl(I).
140. EVANS, M. G., GEORGE, P. and URI, N. *Trans. Faraday Soc.* 45 (1949) 230. Fe(III).
140ᵃ. EVANS, M. G. and NANCOLLAS, G. H. *Trans. Faraday Soc.* 49 (1953) 363. Co(III) ion-pairs.
141. EYRING, H., WALTER, J. and KIMBALL, G. E. *Quantum Chemistry.* John Wiley, New York, 1944.
141ᵃ. FACKLER, J. P. and COTTON, F. A. *J. Am. Chem. Soc.* 82 (1960) 5005; *Ibid.* 83 (1961), 2818 and 3775. Low-spin monomeric and high-spin trimeric Ni(II) acetylacetonate.
142. FAJANS, K. *Naturwiss.* 11 (1923) 165. Polarization.

142ᵃ. FEHRMANN, K. R. A. and GARNER, C. S. *J. Am. Chem. Soc.* **82** (1960) 6294. Cr(III) amine fluorides.

143. FEIGL, F. *Specific, Selective, Sensitive Reactions.* Academic Press, New York, 1949.

144. FELSENFELD, G. *Proc. Roy. Soc. London* A**236** (1956) 506. Cu(II) chlorides, electrostatic model.

145. FELTHAM, R. D. and CALVIN, M. Private communication. V(IV) acetyl-acetonates.

145ᵃ. FELTHAM, R. D. *J. Inorg. Nucl. Chem.* **16** (1961) 197. Cr(I) benzene cation.

145ᵇ. FERGUSON, J. *J. Chem. Phys.* **32** (1960) 528 and 533. Co(II) pyridine chlorides.

145ᶜ. FERGUSON, J. *Spectrochim. Acta* **17** (1961) 316. Ni(II), Cu(II) Schiff' base complexes.

146. FIGGIS, B. N., LEWIS, J., NYHOLM, R. S. and PEACOCK, R. D. *Discuss. Faraday Soc.* **26** (1958) 103. Magnetism of Ru(III), Ru(IV), Re(IV), Os(V), Ir(V).

147. FINKELSTEIN, R. and VAN VLECK, J. H. *J. Chem. Phys.* **8** (1940) 790. Ruby lines of Cr(III), ligand field theory.

148. FLETCHER, J. M., WOODHEAD, J. L., GREENFIELD, B. F., GILL, N. and HARDY, C. J. *Symposium on Co-ordination Compounds, London* 1959, and *J. Chem. Soc.* **1961** 2000. Ru(III, IV) polymeric complexes.

149. FLODMARK, S. *Arkiv. Fysik* **14** (1959) 513. M.O. theory, metal borides.

149ᵃ. FORD, R. A. and HILL, O. F. *Spectrochim. Acta* **16** (1960) 493. Cr(III) ruby.

149ᵇ. FORD, R. A. and WILLIAMS, M. M. R. *Spectrochim. Acta* **16** (1960) 721. Dy(III) in ZnS.

150. FORMÁNEK, J. *Die qualitative Spektralanalyse.* Rudolf Mückenberger, Berlin, 1905.

150ᵃ. FORRESTER, J. S. and AYRES, G. H. *J. Phys. Chem.* **63** (1959) 1979. Rh(III) hexaaquo ion.

151. FRANCK, J. and SCHEIBE, G. *Z. physik. Chem.* A**139** (1928) 22. Electron transfer spectra of halide ions.

152. FRASER, J. G., BEAMISH, F. E. and MCBRYDE, W. A. E. *Anal. Chem.* **26** (1954) 495. Pd(II) iodide.

153. FREED, S. *Phys. Rev.* **38** (1931) 2122. Ce(III) aquo ion.

154. FREEDMAN, A. J. and HUME, D. N. *Anal. Chem.* **22** (1950) 932. Ce(IV).

154ᵃ. FREEMAN, A. J. and WATSON, R. E. *Phys. Rev.* **118** (1960) 1168. Expanded 3d-radial function studied by neutron diffraction.

155. FRIED, S. and HINDMAN, J. C. *J. Am. Chem. Soc.* **76** (1954) 4863. Pa(IV).

156. FRIEDMAN, H. L. *J. Am. Chem. Soc.* **74** (1952) 5. Fe(III) chlorides.

157. FROMHERZ, H. and MENSCHICK, W. *Z. physik. Chem.* B**3** (1929) 1. Cu(I), Ag(I) chlorides.

158. FROMHERZ, H. *Z. Elektrochem.* **37** (1931) 553. Hg(II), Pb(II) aquo and chloride complexes.

159. FROMHERZ, H. and KUN-HOU LIH. *Z. physik. Chem.* A**153** (1931) 321. Cu(II), Ag(I), Cd(II), Tl(I), Pb(II) aquo and halide complexes.

160. FROMHERZ, H. and WALLS, H. J. *Z. physik. Chem.* A**178** (1936) 29. Sn(II).

160ᵃ. FRONÆUS, S. *Thesis*, Lund 1948. Cu(II) carboxylates.

161. FRÖMAN, A. *Phys. Rev.* **112** (1958) 870. Correlation effects of He- and Ne-like systems.

161ᵃ. FRÖMAN, A. *Relativistic Corrections in Many-Electron Systems.* Preprint No. 38. Quantum Chemical Group, Uppsala, 1960.

162. FURLANI, C. and SARTORI, G. *Gazz. chim. ital.* **87** (1957) 371 and 380. Fe(II) complexes and multiplet terms of Fe^{2+}.

162ª. FURLANI, C., MORPURGO, G. and SARTORI, G. *Z. anorg. Chem.* **303** (1960) 1. Cr(III) enta complexes.

162ᵇ. FURLANI, C. and MORPURGO, G. *Z. physik Chem.* **28** (1961) 93. Ni(II) tetrachloride and tetrabromide.

162ᶜ. FURLANI, C. and FURLANI, A. *J. Inorg. Nucl. Chem.* **19** (1961) 51. Mn(II) tetrachloride and bromide.

163. FURMAN, S. C. and GARNER, C. S. *J. Am. Chem. Soc.* **72** (1950) 1785; *Ibid.* **73** (1951) 4528. V(III), V(IV) aquo and thiocyanate complexes.

163ª. GARLICK, G. F. *J. J. Phys. Chem. Solids* **8** (1959) 449. Infra-red emission of semiconductors. Co(II) in ZnS.

164. GEISLER, H. F. and HELLWEGE, K. H. *Z. Physik* **136** (1953) 293. Tb(III).

164ª. GEORGE, P. and McCLURE, D. S. *Progress Inorg. Chem.* **1** (1959) 382. Ligand field stabilization.

164ᵇ. GIACOMETTI, G. and TURCO, A. *J. Inorg. Nucl. Chem.* **15** (1960) 242. Ni(II) phosphine chlorides.

165. GIELESSEN, J. *Ann. Physik* [5] **22** (1935) 537. Narrow bands of cooled salts of Mn(II), Co(II), Ni(II).

165ª. GIESBRECHT, E. *J. Inorg. Nucl. Chem.* **15** (1960) 265. Ce(III) triphosphate.

166. GILL, N. S., NYHOLM, R. S. and PAULING, P. *Nature* **182** (1958) 168. Mn(II), Fe(II), Co(II), Ni(II), Cu(II) tetrachloro complexes.

166ª. GILL, N. S. and NYHOLM, R. S. *J. Chem. Soc.* **1959** 3997. Mn(II), Co(II), Ni(II), Cu(II) tetrahalides.

167. GILLESPIE, R. J. and NYHOLM, R. S. *Quart. Rev.* **11** (1957) 339. Stereochemical effects of lone-pairs.

167ª. GLASNER, A. and REISFELD, R. *J. Chem. Phys.* **32** (1960) 956. Sb(III), Hg(II), Bi(III) halides.

168. GOBRECHT, H. *Ann. Physik* [5] **28** (1937) 673; *Ibid.* **31** (1938) 181 and 755. Multiplet splitting in the lanthanides.

169. GOBRECHT, H. *Ann. Physik* [6] **7** (1950) 88. Tm(III)

169ª. GOODGAME, D. M. L. and COTTON, F. A. *J. Am. Chem. Soc.* **82** (1960) 5774. Ni(II) triphenylarsenic oxide halides.

169ᵇ. GOODGAME, D. M. L., GOODGAME, M. and COTTON, F. A. Private communication. Tetrahedral Mn(II) and Ni(II) chlorides, bromides, iodides.

170. GOODMAN, G. L. and FRED, M. *J. Chem. Phys.* **30** (1959) 849. Np(VI). Ligand field theory, assuming large γ_n-splitting of 5f in the hexafluoride.

171. GORDY, W. *Discuss. Faraday Soc.* **19** (1955) 14. Microwave spectroscopy, electronegativity.

172. GORIN, G., SPESSARD, J. E., WESSLER, G. A. and OLIVER, J. P. *J. Am. Chem. Soc.* **81** (1959) 3193. Co(III) cysteine and mercapto aminoethanol complexes.

173. GORTER, C. J. and HOOGSCHAGEN, J. *Physica* **14** (1948) 197. Review of the lanthanide absorption spectra.

173ª. GOUTERMAN, M. *J. Mol. Spectr.* **6** (1961) 138. Porphyrines.

173ᵇ. GRAHAM, J. *J. Phys. Chem. Solids* **17** (1960) 18. Cr(III) in Al_2O_3.

174. GRANT, M. I. *Trans. Faraday Soc.* **31** (1935) 433. Sn(IV) iodide.

175. GRANTHAM, L. R. F., ELLEMAN, T. S. and MARTIN, D. S. *J. Am. Chem. Soc.* **77** (1955) 2965. Pt(II).

175ª. GREEN, M. and LINNETT, J. W. *J. Chem. Soc.* **1960** 4959. Simple molecules with an odd number of electrons.

176. GRIFFITH, J. S. *J. Inorg. Nucl. Chem.* **2** (1956) 1 and 229. Spin-pairing energies in octahedral d-complexes.

177. GRIFFITH, J. S. *J. Inorg. Nucl. Chem.* **3** (1956) 15. Energy of the metallic elements.

178. GRIFFITH, J. S. and ORGEL, L. E. *J. Chem. Soc.* (1956) 4981. M.O. theory Co(III) halide ammonia complexes.
179. GRIFFITH, J. S. and ORGEL, L. E. *Trans. Faraday Soc.* **53** (1957) 601. Co(III) temperature-independent paramagnetism.
180. GRIFFITH, J. S. and ORGEL, L. E. *Quart. Rev.* **11** (1957) 381. Review of ligand field theory.
181. GRIFFITH, J. S. *Trans. Faraday Soc.* **54** (1958) 1109; *Ibid.* **56** (1960) 193. Ligand field theory, magnetism of octahedral d-complexes.
182. GRIFFITH, J. S. *Discuss. Faraday Soc.* **26** (1958) 81. M.O. theory of phthalocyanines.
182ᵃ. GRIFFITH, J. S. *The Theory of Transition-Metal Ions*. Cambridge University Press, 1961.
182ᵇ. GRIFFITH, J. S. *Mol. Phys.* **3** (1960) 477. Intensity of Laporte-forbidden bands.
183. GRIFFITH, W. P., LEWIS, J. and WILKINSON, G. *J. Chem. Soc.* (1958) 3993; *Ibid.* (1959) 872; *J. Inorg. Nucl. Chem.* **7** (1958) 38. NO-complexes.
183ᵃ. GRIFFITH, W. P. and WILKINSON, G. *J. Chem. Soc.* **1959** 2757. Co(III), Rh(III) hydride pentacyanides.
184. GRIFFITHS, J. H. E. and OWEN, J. *Proc. Roy. Soc. London* A213 (1952) 459. Ni(II) aquo ion.
185. GRIFFITHS, J. H. E., OWEN, J. and WARD, I. M. *Proc. Roy. Soc. London* A219 (1953) 526.
185ᵃ. GRIFFITHS, T. R. and SYMONS, M. C. R. *Mol. Phys.* **3** (1960) 90; and *Trans. Faraday Soc.* **56** (1960) 1125. Ion-pairs, iodide solutions.
186. GROSS, E. F. Suppl. *Nuovo Cimento* **3** (1956) 672; and *J. Phys. Chem. Solids* **8** (1959) 172. Narrow lines of Cu(I) oxide, exciton model.
187. GRUBER, J. B. and CONWAY, J. G. UCRL-8839; *J. Chem. Phys.* **32** (1960) 1178. Tm(III).
187ᵃ. GRUBER, J. B. and CONWAY, J. G. *J. Inorg. Nucl. Chem.* **14** (1960) 303. Pm(III).
188. GRUEN, D. M. *J. Chem. Phys.* **20** (1952) 1818. Np(V) and Pu(VI) dioxo ions.
189. GRUEN, D. M. and FRED, M. *J. Am. Chem. Soc.* **76** (1954) 2117 and 3850. U(IV) oxides and fluorides.
190. GRUEN, D. M. *J. Inorg. Nucl. Chem.* **4** (1957) 74. Co(II), Ni(II), Cu(II), Pr(III), Nd(III), U(VI) in molten salts.
191. GRUEN, D. M., GRAF, P. and FRIED, S. *Sixteenth International Chemical Congress, Paris*, 1957, p. 319. Butterworth, London, 1958. U(IV), Np(IV), Pu(III), Pu(IV) in molten salts.
192. GRUEN, D. M. and MCBETH, R. L. *J. Inorg. Nucl. Chem.* **9** (1959) 290. U(III), U(IV), U(VI) in chloride melts.
193. GRUEN, D. M. and MCBETH, R. L. *J. Phys. Chem.* **63** (1959) 393. Ni(II) in chloride melts.
194. GRÜNBERG, A. A. and FAERMAN, G. P. *Z. anorg. Chem.* **193** (1930) 193. Pt(IV) amide and ammonia complexes.
194ᵃ. GUILLAUMONT, R., MUXART, R., BOUÍSSIÈRES, G. and HAÏSSINSKY, M. *J. Chimie Physique* **57** (1960) 1019. Pa(IV) and Pa(V).
194ᵇ. HAIM, A. and WILMARTH, W. K. *J. Am. Chem. Soc.* **83** (1961) 509. Fe(II), Fe(III), Co(III) cyanides.
195. HAÏSSINSKY, M. *J. Chem. Soc.* (1949) S241; *J. Chim. Phys.* **47** (1950) 15; *Experientia* **9** (1953) 117, Uranides.
195ᵃ. HAÏSSINSKY, M., *J. Chimie Physique* **46** (1949) 298. Electronegativity.

195ᵇ. HAÏSSINSKY, M. *17th Int. Congress Pure Appl. Chem.*, Munich, p. 185, 1959. Comparison of uranide and lanthanide chemistry.

196. HAKEN, H. *Z. Physik.* **155** (1959) 223. Exciton theory.

197. HALL, G. G. *Reports Progress Physics* **22** (1959) 1. Review of quantum mechanics applied to theoretical chemistry.

197ᵃ. HALPERN, J. and HARKNESS, A. C. *J. Chem. Phys.* **31** (1959) 1147. Cr(III), Mn(VII), Fe(III), Co(II), Ni(II), Cu(II) deuterium oxide effect.

197ᵇ. HALPERN, J. *Quart. Rev.* **15** (1961) 207. Reaction mechanism of electron transfer between complexes in solution.

198. HANSON, M. W., BRADBURY, W. C. and CARLTON, J. K. *Anal. Chem.* **29** (1957) 490. Te(IV).

199. HARBOTTLE, G. and DODSON, R. W. *J. Am. Chem. Soc.* **73** (1951) 2442. Tl(III).

200. HARNED, H. S. and OWEN, B. B. *The Physical Chemistry of Electrolytic Solutions* (2nd Ed.). ACS Monograph No. 95, New York, 1950.

200ᵃ. HARRIS, C. M., NYHOLM, R. S. and PHILLIPS, D. J. *J. Chem. Soc.* **1960** 4379. Ni(II) diarsine halides.

201. HARTMANN, H. and ILSE, F. E. *Z. Naturforsch.* **6a** (1951) 751. Electrostatic model of V(III).

202. HARTMANN, H. and SCHLÄFER, H. L. *Z. Naturforsch.* **6a** (1951) 754 and 760. Ti(III), V(III), Cr(III), Mn(III) aquo ions.

203. HARTMANN, H. *Theorie der Chemischen Bindung auf Quantentheoretischer Grundlage.* Julius Springer, Göttingen, 1954.

204. HARTMANN, H. and SCHLÄFER, H. L. *Angew. Chem.* **66** (1954) 768. Review.

205. HARTMANN, H. and FISCHER-WASELS, H. *Z. physik. Chem.* **4** (1955) 297. Ni(II), electrostatic model for tetrahedral and square-planar complexes.

206. HARTMANN, H. and KRUSE, H. H. *Z. physik. Chem.* **5** (1955) 9. Electrostatic model of Cr(III).

207. HARTMANN, H., SCHLÄFER, H. L. and HANSEN, K. H. *Z. anorg. Chem.* **284** Ti(III), V(III), Cr(III), Mn(III) aquo ions.

208. HARTMANN, H., FURLANI, C. and BÜRGER, A. *Z. physik. Chem.* **9** (1956) 6 V(III) oxalates.

209. HARTMANN, H. and FURLANI, C. *Z. physik. Chem.* **9** (1956) 162. V(III) urea and malonate complexes.

210. HARTMANN, H. and BUSCHBECK, C. *Z. physik. Chem.* **11** (1957) 120. Ru(III) chloride and ammonia complexes.

211. HARTMANN, H. and SCHMIDT, H. J. *Z. physik. Chem.* **11** (1957) 234. Mo(III).

212. HARTMANN, H. *Z. Elektrochem.* **61** (1957) 908. Review.

213. HARTMANN, H. and SCHLÄFER, H. L. *Angew. Chem.* **70** (1958) 155. Review.

214. HARTREE, R. D. *J. Opt. Soc. Am.* **46** (1956) 350. Radial functions in the 3d-group.

215. HARTREE, R. D. *The Calculation of Atomic Structures.* John Wiley, New York, 1957.

215ᵃ. HASTINGS, J. M., ELLIOTT, N. and CORLISS, L. M. *Phys. Rev.* **115** (1959) 13. Neutron diffraction of Mn(II).

216. HAYES, W. *Discuss. Faraday Soc.* **26** (1958) 58. Paramagnetic resonance of Cr(I), Mn(II), Fe(I), Co(I), Ni(I) in sodium fluoride crystals.

217. HAWKINS, G. L. and GARNER, C. S. *J. Am. Chem. Soc.* **80** (1958) 2946. W(III).

218. HEAL, H. G. and MAY, J. *J. Am. Chem. Soc.* **80** (1958) 2374. Pb(IV) chlorides.

Y

219. HEIDT, L. J. and BERESTECKI, J. *J. Am. Chem. Soc.* **77** (1955) 2049. Ce(III) in perchlorate solutions.
220. HEIDT, L. J., KOSTER, G. F. and JOHNSON, A. M. *J. Am. Chem. Soc.* **80** (1958) 6471. Mn(II) aquo ion, ligand field theory.
221. HELLWEGE, K. H. *Ann. Physik* [6] **4** (1948) 95, 127, 136, 143, 150 and 357. Ligand fields of every symmetry possible.
222. HELLWEGE, K. H. and KAHLE, H. G. *Z. Physik* **129** (1951) 62. Eu(III).
223. HELLWEGE, A. M. and HELLWEGE, K. H. *Z. Physik* **130** (1951) 549; *Ibid.* **135** (1953) 92. Pr(III).
223ᵃ. HELLWEGE, K. H., HESS, G. and KAHLE, H. G. *Z. Physik.* **159** (1960) 333. Pr(III).
223ᵇ. HELLWEGE, K. H., HÜFNER, S. and KAHLE, H. G. *Z. Physik.* **160** (1960) 149 and 162. Er(III).
224. HELMHOLZ, L. *J. Am. Chem. Soc.* **61** (1939) 1544. Nd(III) aquo ion nine-co-ordination.
225. HELMHOLZ, L., BRENNAN, H. and WOLFSBERG, M. *J. Phys. Chem.* **23** (1955) 853. Cr(VI) oxo- and chloro-complexes, M.O. theory.
226. HENRI, V. and BIELECKI, J. *Physik. Z.* **14** (1913) 516. Gaussian shape of absorption bands.
227. HEPWORTH, M. A., ROBINSON, P. L. and WESTLAND, G. J. *J. Chem. Soc.* (1958) 611. Os(IV), Ir(IV) fluorides.
228. HERBERT, R. H. and IRVINE, J. W. *J. Am. Chem. Soc.* **78** (1956) 905. Formation of monochloro complex of Ni(II).
228ᵃ. HERMAN, F. and MCCLURE, D. S. *Bull. Am. Phys. Soc.* **5** (1960) 48. Cu(I) halides.
229. HERSHENSON, H. M. *Ultraviolet and Visible Absorption Spectra Index* 1930–54. Academic Press, New York, 1956.
230. HERZBERG, G. *Molecular Spectra and Molecular Structure* Vol. I. *Spectra of Diatomic Molecules* (2nd Ed.). D. Van Nostrand, New York, 1950.
231. HERZFELD, C. M. *Phys. Rev.* **107** (1957) 1239. Nitrogen atoms, trapped in cool materials.
231ᵃ. HERZFELD, C. M. and MEIJER, P. H. E. *Solid State Phys.* in press. Group theory and ligand field theory.
232. HERZOG, S. and SCHÖN, W. *Z. anorg. Chem.* **297** (1958) 323. Preparation of Cr(0) dipyridyl.
232ᵃ. HIDAKA, J., FUJITA, J., SHIMURA, Y. and TSUCHIDA, R. *Bull. Chem. Soc. Japan* **32** (1959) 1317. Co(III) amine thiophosphates and thiosulphates.
232ᵇ. HIDAKA, J., SHIMURA, Y. and TSUCHIDA, R. *Bull. Chem. Soc. Japan* **33** (1960) 847. Co(III) enta complexes.
233. HINDMAN, J. C. and WEHNER, P. *J. Am. Chem. Soc.* **75** (1953) 2869. Re(IV).
234. HOLLECK, L. and HARTINGER, L. *Angew. Chem.* **67** (1955) 648. Lanthanides. acetonates.
234ᵃ. HOLLOWAY, F. *J. Am. Chem. Soc.* **74** (1952) 224. Cr(VI) phosphates.
235. HOLM, R. H. and COTTON, F. A. *J. Am. Chem. Soc.* **80** (1958) 5658. Acetyl-acetonates.
235ᵃ. HOLM, R. H. and COTTON, F. A. *J. Chem. Phys.* **31** (1959) 788. Co(II) tetrahalides.
235ᵇ. HOLM, R. H. *J. Am. Chem. Soc.* **82** (1960) 5632. Ni(II), Cu(II) Schiff' base complexes.
236. HOLMES, O. G. and MCCLURE, D. S. *J. Chem. Phys.* **26** (1957) 1686. V(III), Cr(II), Cr(III), Co(II), Ni(II), Cu(II) aquo ions, ligand field theory.
236ᵃ. HONDA, M. and SCHWARZENBACH, G. *Helv. Chim. Acta* **40** (1957) 27. Co(III), Ni(II), Cu(II) Schiff' base complexes.

236^b. HORAK, Z. *Czech. J. Phys.* B10 (1960) 405. Mn(II), Fe(III) X-ray emission.

237. HROSTOWSKI, H. J. and KAISER, R. H. *Bull. Am. Phys. Soc.* 4 (1959) 167. Mn(II) fluorides.

238. HUND, F. *Z. Physik* 73 (1931) 1 and 565; *Ibid.* 74 (1932) 429. Molecular orbital theory.

239. HUNT, J. P. and TAUBE, H. *J. Chem. Phys.* 19 (1951) 602. Cr(III) kinetics.

240. HUSH, N. S. and PRYCE, H. M. L. *J. Chem. Phys.* 26 (1957) 143; *Ibid.* 28 (1958) 244. Variation of ionic radii, caused by ligand field stabilization.

241. HUSH, N. S. *Discuss. Faraday Soc.* 26 (1958) 145. Ligand field stabilization.

242. HUTCHINSON, C. A. and CANDELA, G. A. *J. Chem. Phys.* 27 (1957) 707. Magnetism of U(IV) chlorides.

242^a. HUTCHINSON, C. A. and WEINSTOCK, B. *J. Chem. Phys.* 32 (1960) 56. Np(VI) fluoride.

243. IGUCHI, K. *J. Chem. Phys.* 23 (1955) 1983. M.O. configuration intermixing.

244. ILSE, F. E. and HARTMANN, H. *Z. physik. Chem.* 197 (1951) 239. Electrostatic model Ti(III).

245. INGOLD, C. K., NYHOLM, R. S. and TOBE, M. L. *J. Chem. Soc.* (1956) 1691 and 1707. Co(III) thiocyanate hydroxo, ammonia complexes.

245^a. INSKEEP, R. G. and BJERRUM, J. *Acta Chem. Scand.* 15 (1961) 62. Cr(III) dip and phen complexes.

246. IRVING, H. and WILLIAMS, R. J. P. *J. Chem. Soc.* (1953) 3192. The order of complexity constants for a given ligand with 3d-ions.

247. ITO, K. and KURODA, Y. *J. Chem. Soc. Japan* 76 (1955) 545, 766 and 934. M.O. theory, Wolfsberg–Helmholz model for Co(III).

248. ITO, K. *Naturwiss.* 46 (1959) 445. Electronegativity and covalent bonding in complexes.

249. JAHN, H. A. and TELLER, E. *Proc. Roy. Soc. London* A161 (1937) 220. Jahn–Teller effect.

249^a. JAHODA, F. C. *Phys. Rev.* 107 (1957) 1261. Ba(II) oxide, theory of reflection.

249^b. JAKUSZEWSKI, B. *J. Chem. Phys.* 31 (1959) 846. Absolute potential of hydrogen electrode.

250. JOHNSEN, U. *Z. Physik* 152 (1958) 454. Tm(III) aquo ions.

251. JOHNSON, F. D. and WILLIAMS, F. E. *J. Chem. Phys.* 20 (1952) 124. Tl(I) in alkali halide crystals.

252. JONASSEN, H. B. and DOUGLAS, B. E. *J. Am. Chem. Soc.* 71 (1949) 4094 Ni(II) amine complexes.

253. JONASSEN, H. B. and CULL, N. L. *J. Am. Chem. Soc.* 71 (1949) 4097. Pd(II) amine complexes.

254. JONASSEN, H. B., REEVES, R. E. and SEGAL, L. *J. Am. Chem. Soc.* 77 (1955) 2667 and 2748. Cu(II) amine complexes.

255. JONES, H. C., ANDERSON, J. A. and STRONG, W. W. *Carnegie Inst. Publ.* No. 110, 130 and 160. Washington 1909–11. Co(II), Ni(II), Pr(III), Nd(III), Sm(III), U(IV), U(VI) in different solvents.

256. JOOS, G. *Ann. Physik* 81 (1926) 1076; *Ibid.* 85 (1928) 641. General remarks on transition group colours.

257. JOOS, G. and SCHNETZLER, K. *Z. physik. Chem.* B20 (1933) 1. Cr(III).

258. JORDAHL, O. M. *Phys. Rev.* 45 (1934) 87. Magnetism and ligand field theory for Cu(II).

258^a. JORTNER, J., RAZ, B. and STEIN, G. *Trans. Faraday Soc.* 56 (1960) 1273. Iodide.

258^b. JORTNER, J., RAZ, B. and STEIN, G. *J. Chem. Phys.* 34 (1961) 1455. Hydroxide.

259. JUDD, B. R. *Proc. Roy. Soc. London* A**228** (1955) 120. Eu(III), Gd(III) theory.

260. JUDD, B. R. *Proc. Phys. Soc.* A**69** (1956) 157. Multiplet splitting in the lanthanides.

261. JUDD, B. R. *Proc. Roy. Soc. London* A**232** (1955) 458; *Ibid.* A**241** (1957) 122. Lanthanide double nitrates and icosahedral ligand fields.

261ᵃ. JUDD, B. R. *Proc. Roy. Soc. London* A**241** (1957) 414. Pr(III).

262. JUDD, B. R. *Proc. Roy. Soc. London* A**251** (1959) 134. Nd(III) theory.

263. JØRGENSEN, C. K. *Acta Chem. Scand.* **8** (1954) 175. Cr(III), Co(II), Ni(II), Cu(II) in aqueous alcohol (cf. ref. 310).

264. JØRGENSEN, C. K. *Acta Chem. Scand.* **8** (1954) 1495. Cr(III), Ni(II) aquo, Co(III) amine complexes. Gaussian shape of absorption bands.

265. JØRGENSEN, C. K. *Acta Chem. Scand.* **8** (1954) 1502. Mn(II), Fe(II), Co(II) aquo, Cr(III) oxalate, Ni(II) aquo and amine complexes. Spin-forbidden bands.

266. JØRGENSEN, C. K. *Acta Chem. Scand.* **9** (1955) 116. Theory for d³, d⁴, d⁵ in octahedral complexes.

267. JØRGENSEN, C. K. and BJERRUM, J. *Acta Chem. Scand.* **9** (1955) 180. Ligand field stabilization.

268. JØRGENSEN, C. K. *Mat. fys. Medd. Dan. Vid. Selsk.* **29** (1955) No. 7. U(IV) aquo ion, theory (cf. ref. 298).

269. JØRGENSEN, C. K. *Mat. fys. Medd. Dan. Vid. Selsk.* **29** (1955) No. 11. Nd(III), Pm(III), Sm(III), Gd(III), actinides, theory.

270. JØRGENSEN, C. K. *Acta Chem. Scand.* **9** (1955) 405. Band intensities.

271. JØRGENSEN, C. K. *Acta Chem. Scand.* **9** (1955) 540. Pr(III), Tm(III).

272. JØRGENSEN, C. K. *Acta Chem. Scand.* **9** (1955) 710. Re(IV) chlorides and bromides. Vibrational structure of narrow, spin-forbidden bands.

273. JØRGENSEN, C. K. *Acta Chem. Scand.* **9** (1955) 717. The possible influence of 4s on the 3d-energy levels.

274. JØRGENSEN, C. K. *J. Inorg. Nucl. Chem.* **1** (1955) 301. Variation of Landé parameters in gaseous transition group ions.

275. JØRGENSEN, C. K. *Acta Chem. Scand.* **9** (1955) 1362. Aquo, enta, aca, dip, phen complexes of V(III), Cr(III), Co(II), Ni(II), Cu(II). Intermixing of S in Ni(II).

276. JØRGENSEN, C. K. *Acta Chem. Scand.* **10** (1956) 500. Halide oxalate and amine complexes of Rh(III), Ir(III).

277. JØRGENSEN, C. K. *Acta Chem. Scand.* **10** (1956) 518. Halide and amine complexes of Ru(II), Ru(III), Ir(IV), Pt(IV).

278. JØRGENSEN, C. K. *Acta Chem. Scand.* **10** (1956) 887. Amine and amino-acid complexes of Ni(II), Cu(II). Stereochemical effects, entropy effect.

279. JØRGENSEN, C. K. (1956) *Report to the Tenth Solvay Council. Quelques Problèmes de Chimie Minérale*, p. 355. R. Stoops, Bruxelles, 1956. Review, discussion of Δ.

280. JØRGENSEN, C. K. *Mat. fys. Medd. Dan. Vid. Selsk.* **30** (1956) No. 22. Nephelauxetic effect Ce(III), Pr(III), Nd(III), Sm(III), Gd(III), Yb(III).

281. JØRGENSEN, C. K. *Acta Chem. Scand.* **10** (1956) 1503. U(III) chloro complexes.

282. JØRGENSEN, C. K. *Acta Chem. Scand.* **10** (1956) 1505. Spin-pairing energy, normal oxidation potentials.

282ᵃ. JØRGENSEN, C. K. *Energy Levels of Complexes and Gaseous Ions.* Gjellerup, Copenhagen, 1957.

283. JØRGENSEN, C. K. *Acta Chem. Scand.* **11** (1957) 53. Mn(II), nephelauxetic effect.

284. JØRGENSEN, C. K. *Acta Chem. Scand.* **11** (1957) 73. Anion complexes of Ti(III), Ti(III, IV), V(IV), Mo(III), Mo(V). Constitution of vanadyl ion.
285. JØRGENSEN, C. K. *Acta Chem. Scand.* **11** (1957) 151. Oxalate, halide, pyridine complexes of Rh(III), Ir(III), prepared by M. Delépine.
286. JØRGENSEN, C. K. *Acta Chem. Scand.* **11** (1957) 166. Ir(III), U(VI), electron transfer bands, Fe(II) dip and phen complexes, the Ir–py band.
287. JØRGENSEN, C. K. *Acta Chem. Scand.* **11** (1957) 1399. Ni(II) amine complexes in strong salt solutions.
288. JØRGENSEN, C. K. *Acta Chem. Scand.* **11** (1957) 1981. Dy(III), Ho(III), Er(III) aquo ions. Interelectronic repulsion parameters.
289. JØRGENSEN, C. K. *J. Inorg. Nucl. Chem.* **4** (1957) 369. Variation of term distances with ionic charge in gaseous ions.
290. JØRGENSEN, C. K. *Acta Chem. Scand.* **12** (1958) 903. Diagonal elements of interelectronic repulsion in general M.O. theory of octahedral complexes.
291. JØRGENSEN, C. K. *Acta Chem. Scand.* **12** (1958) 1537. V(II) aquo ion.
292. JØRGENSEN, C. K. *Acta Chem. Scand.* **12** (1958) 1539. Mn(IV) hexafluoro complex. Prediction of band positions of hexafluorides.
293. JØRGENSEN, C. K. Contract no. DA-91-508-EUC-247 with European Research Office, U.S. Department of the Army, Frankfurt am Main. Absorption Spectra of Complexes of Heavy Metals. Chemical and spectrochemical studies, e.g. of Ru(III), Ru(IV), Rh(III), Pd(II), Pd(IV), Sn(II), Sb(III), W(VI), Re(IV), Os(III), Os(IV), Ir(IV), Pt(II), Pt(IV), Au(III), Hg(II), Pb(II), Bi(III). ASTIA document no. 157158, September 1958.
294. JØRGENSEN, C. K. 16. *Congrès International de chimie pure et appliquée, Paris,* 1957, p. 313. Butterworth, London, 1958. 5f- and 6d-orbitals.
295. JØRGENSEN, C. K. *Discuss. Faraday Soc.* **26** (1958) 110. Nephelauxetic effect, central-field and symmetry-restricted covalency.
296. JØRGENSEN, C. K. *Mol. Phys.* **1** (1958) 410. Ni(II) tetrachloro complexes in salt melts.
297. JØRGENSEN, C. K. *Acta Chem. Scand.* **13** (1959) 196. Ir(III, IV) nitrogen sulphates.
298. JØRGENSEN, C. K. *Mol. Phys.* **2** (1959) 96. U(IV), Cm(III) and other actinides.
299. JØRGENSEN, C. K. *Mol. Phys.* **2** (1959) 309. Electron transfer spectra of hexahalide complexes of Ru(III), Ru(IV), Rh(III), Pd(IV), Sn(IV), Sb(V), Re(IV), Os(III), Os(IV), Ir(III), Ir(IV), Pt(IV), Pb(IV). Effects of intermediate coupling.
300. JØRGENSEN, C. K. *J. Chimie Physique* **56** (1959) 889. Review of spectra of complexes.
301. JØRGENSEN, C. K. *Mol. Phys.* **3** (1960) 201. Re(VI), Os(VI), Ir(VI), Pt(VI). Np(VI) hexafluorides.
301[a]. JØRGENSEN, C. K. *Mol. Phys.* **4** (1961) 231. Rh(IV) hexachloride.
301[b]. JØRGENSEN, C. K. *Mol. Phys.* **4** (1961) 235. Ir(IV), Ag(I), Tl(I) chlorides.
301[c]. JØRGENSEN, C. K. *Adv. Chem. Phys.* In press. Review.
301[d]. JØRGENSEN, C. K. *Progress Inorg. Chem.* In press. Nephelauxetic series.
301[e]. JØRGENSEN, C. K. *Solid State Phys.* In press. Review.
301[f]. KAHLE, H. G. *Z. Physik.* **161** (1961) 486. Er(III).
301[g]. KAMIMURA, H., KOIDE, S., SEKIYAMA, H. and SUGANO, S. *J. Phys. Soc. Japan* **15** (1960) 1264. Magnetochemistry of 4d and 5d complexes.
302. KANZELMEYER, J. H., RYAN, J. and FREUND, H. *J. Am. Chem. Soc.* **78** (1956) 3020. Nb(V) in hydrochloric acid.
302[a]. KAPLAN, H. *J. Chem. Phys.* **26** (1957) 1704. Ammonia molecule.

303. KATO, S. *Sci. Papers Inst. Phys. Chem. Research (Tokyo)* **13** (1930) 49. Various transition group aquo ions.

303ª. KATO, M. *Z. physik. Chem.* **23** (1960) 375 and 391. Ni(II), Cu(II) biuret complexes.

303ᵇ. KATO, M. *Bull. Aichi Gaguku Univ.* **9** (March 1960) 95. Cu(II).

304. KATZ, J. J. and SEABORG, G. T. *The Chemistry of the Actinide Elements.* Methuen, London, 1957.

305. KATZ, M. L. and NIKOLSKY, V. K. *Optika y Spektroskopia* **4** (1958) 354. In(I) in alkali halide crystals.

306. KATZIN, L. I. and GEBERT, E. *J. Am. Chem. Soc.* **72** (1950) 5457. Co(II) in organic solvents.

307. KATZIN, L. I. *J. Chem. Phys.* **20** (1952) 1165. Co(II), Ni(II) halides and thiocyanates.

308. KATZIN, L. I. *J. Am. Chem. Soc.* **76** (1954) 3089. Co(II) chloride.

309. KATZIN, L. I. *J. Chem. Phys.* **23** (1955) 2055. Electron transfer spectra as shifted halide bands.

310. KATZIN, L. I., JØRGENSEN, C. K. and BJERRUM, J. *Nature* **175** (1955) 425 and 426. Solvation and anion complexes in alcohol.

311. KATZIN, L. I. *Nature* **182** (1958) 1013. Tetrahedral Ni(II).

312. KAUER, *Z. physik. Chem.* **6** (1956) 105. X-ray absorption edges.

313. KAUZMANN, W. *Quantum Chemistry.* Academic Press, New York, 1957.

313ª. KEATING, K. B. and DRICKAMER, H. G. *J. Chem. Phys.* **34** (1961) 140 and 143. Pressure effects on Pr(III), Nd(III), Er(III), U(III), U(IV).

314. KEYES, R. W. *J. Chem. Phys.* **29** (1958) 523. M.O. description of diatomic molecules.

314ª. KIDA, S. and YONEDA, H. *J. Chem. Soc. Japan* **76** (1955) 1059. Cr(III), Co(III), Ni(II) xanthates and other sulphur-containing complexes.

314ᵇ. KIDA, S. *Bull. Chem. Soc. Japan* **29** (1956) 805. Cu(II).

314ᶜ. KIDA, S. *Bull. Chem. Soc. Japan* **33** (1960) 587. Pt(II) tetrachloride and tetrabromide.

314ᵈ. KIDA, S. *Bull. Chem. Soc. Japan* **33** (1960) 1204. Ni(II), Pt(II) dithiooxalates.

315. KING, E. L., WOODS, J. W. and GATES, H. S. *J. Am. Chem. Soc.* **80** (1958) 5015. Cr(III) chloro complexes, weak band of Ce(III).

315ª. KING, E. L., ESPENSON, J. H. and VISCO, R. E. *J. Phys. Chem.* **63** (1959) 755. Co(III) ion-pairs.

316. KING, N. K. and WINFIELD, M. E. *J. Am. Chem. Soc.* **80** (1958) 2060. Co(II) cyanides.

317. KIMBALL, G. E. *J. Chem. Phys.* **8** (1940) 188. Group theory and valence bonds.

318. KISS, A. and GERANDAS, M. *Z. physik. Chem.* **A180** (1937) 117. Co(II) chlorides.

319. KISS, A. and CZEGLEDY, D. *Z. anorg. Chem.* **235** (1938) 407; *Ibid.* **239** (1938) 27. Co(III) amine and anion complexes; *Ibid.* **245** (1941) 355. Ni(II).

320. KISS, A. and CZOKAN, P. *Z. physik. Chem.* **A186** (1940) 239; **A188** (1941) 27. Co(II) chloride and thiocyanate complexes.

321. KISS, A., ABRAHAM, J. and HEGEDUS, I. *Z. anorg. Chem.* **244** (1940) 98. Fe(III) aquo and anion complexes.

322. KISS, A. and SZABO, R. *Z. anorg. Chem.* **252** (1943) 172. Ni(II) dioximes.

323. KLEINER, W. H. *J. Chem. Phys.* **20** (1952) 1784. Attempts to calculate Δ.

324. KLEMM, W. *Angew. Chemie* **66** (1954) 461. Review of fluorides.

325. KLINKENBERG, P. F. A. *Physica* **16** (1950) 618. Gaseous Th²⁺.

326. KLÄNING, U. *Acta Chem. Scand.* **11** (1957) 1313. Cr(VI) in alcohols.

326ᵃ. KNOX, R. S. and DEXTER, D. L. *Phys. Rev.* **104** (1956) 1245. In(I), Tl(I).
326ᵇ. KNOX, R. S. and INCHAUSPÉ, N. *Phys. Rev.* **116** (1959) 1093. Alkali metal halide crystals, theory.
326ᶜ. KNOX, K., SHULMAN, R. G. and SUGANO, S. *Bull. Am. Phys. Soc.* **5** (1960) 415. Ni(II) fluorides.
327. KOIDE, S. and PRYCE, M. H. L. *Phil. Mag.* **3** (1958) 607. Mn(II), band intensities. Covalency parameters.
328. KOIDE, S. *Phil. Mag.* **4** (1959) 243. Co(II), band intensities.
329. KONDO, Y. and NAKAHARA, K. *J. Chem. Soc. Japan* **75** (1954) 17. Temperature effects on absorption spectra.
330. KONDO, Y. *Bull. Chem. Soc. Japan* **28** (1955) 497. Solid salts of Co(III) hexammonia ion.
331. KORTÜM, G. *Kolorimetrie, Photometrie und Spektrophotometrie* (3rd Ed.) Springer-Verlag 1955.
331ᵃ. KOSTER, G. F. Quart. Progress Rep. January 1960. *Solid-state and Molecular Theory Group, M.I.T.* Interelectronic repulsion parameters in general M.O. theory.
332. KOSOWER, E. M. *J. Am. Chem. Soc.* **78** (1956) 5700. Electron transfer bands of pyridinium iodide.
333. KOSOWER, E. M., MARTIN, R. L. and MELOCHE, V. W. *J. Am. Chem. Soc.* **79** (1957) 1509. Cu(II) bromides.
333ᵃ. KOSOWER, E. M., MARTIN, R. L. and MELOCHE, V. W. *J. Chem. Phys.* **26** (1957) 1353. Iodide.
333ᵇ. KOSOWER, E. M., SKORCZ, J. A., SCHWARZ, W. M. and PATTON, J. W. *J. Am. Chem. Soc.* **82** (1960) 2188. Double band of pyridinium iodides.
334. KOTANI, M. *J. Phys. Soc. Japan* **4** (1949) 293. Paramagnetism as function of Landé parameters and temperature.
335. KOTANI, M., MIZUNO, Y., KAYAWA, K. and ISHIGURO, E. *J. Phys. Soc. Japan* **12** (1957) 707. Oxygen molecule.
336. KRAUS, K. A. and NELSON, F. *J. Am. Chem. Soc.* **72** (1950) 3901; *Ibid.* **77** (1955) 3721. U(IV), Pu(IV).
337. KRAUS, K. A. and NELSON, F. *Proc. Geneva Conf.* **7** (1955) 113, 131 and 245. Anion complexes and hydrolysis products of various elements.
338. KRUMHOLZ, P. *J. Am. Chem. Soc.* **75** (1953) 2163. Fe(II) complexes of aliphatic diimines.
339. KRÖGER, F. A. *Physica* **6** (1939) 369. Mn(II) in zinc sulphide.
340. KRÖGER, F. A. and BAKKER, J. *Physica* **8** (1941) 628. Ce(III) fluorescence.
341. KRÖGER, F. A. *Some Aspects of the Luminiscence of Solids.* Elsevier, Amsterdam, 1948. Cr(III), Mn(II), Mn(IV).
342. KRÖGER, F. A., VINCK, H. J. and BOOMGAARD, J. VAN DEN *Physica* **18** (1952) 77. Ni(II) in magnesium oxide.
342ᵃ. KRÖGER, F. A. *Ergeb. Exakt. Naturwiss.* **29** (1956) 61. Phosporescence.
343. KUROYA, H. and TSUCHIDA, R. *J. Chem. Soc. Japan* **61** (1940) 597. Cr(III), Fe(III), Co(III), Ni(II) cyanides.
343ᵃ. KÖHLER, H. and SCHEIBE, G. *Z. anorg. Chem.* **285** (1956) 221. Aminobenzenes.
343ᵇ. KÖNIG, E. and SCHLÄFER, H. L. *Z. physik. Chem.* **26** (1960) 371. Cr(III), Mn(II), Co(II), Ni(II), Cu(II), Mo(III), Ir(III), Pt(II), Tl(III) halide pyridine complexes.
344. LACROIX, R. *Arch. Sci. (Genève)* **8** (1955) 317 and 321. Ni(II) and V(III) aquo ions.

345. LANGSETH, A. and QVILLER, B. Z. physik. Chem. B27 (1934) 79. Os(VIII) tetroxide.

345ᵃ. LARSEN, R. P. and ROSS, L. E. Analyt. Chem. 31 (1959) 176. Ru(VII) tetroxo ion.

346. LARSON, L. L. and GARNER, C. S. J. Am. Chem. Soc. 76 (1954) 2180. Os(IV) hexachloro complex.

347. LASWICK, J. A. and PLANE, R. A. J. Am. Chem. Soc. 81 (1959) 3564. Cr(III) hydrolysis products.

348. LATIMER, W. M. The Oxidation States of the Elements and their Potentials in Aqueous Solution (2nd Ed.). Prentice-Hall, New York, 1952.

349. LATIMER, W. M. J. Chem. Phys. 23 (1955) 90. Heats of hydration of gaseous ions and the crystallographic radii.

349ᵃ. LAYZER, D. Annals of Physics 8 (1959) 271. Near orbital degeneracy in atomic spectroscopy.

350. LÉDEN, I. and CHATT, J. J. Chem. Soc. (1955) 2936. Pt(II) complex formation constants.

351. LEDERER, M. J. Chromatogr. 1 (1958) 279. Rh(III), separation of complexes.

352. LENNARD-JONES, J. Proc. Roy. Soc. London A198 (1949) 1 and 14. Equivalent orbitals.

353. LEUSSING, D. L. J. Am. Chem. Soc. 80 (1958) 4180. Co(II), Ni(II) thioglycollates.

354. LEUSSING, D. L. J. Am. Chem. Soc. 81 (1959) 4208. Ni(II) dimercaptopropanol complexes.

354ᵃ. LEUSSING, D. L. and ALBERTS, G. S. J. Am. Chem. Soc. 82 (1960) 4458. Ni(II) ethanedithiol complexes.

354ᵇ. LEUSSING, D. L. and MISLAN, J. P. J. Phys. Chem. 64 (1960) 1908. Fe(III) dimercaptopropanol complexes.

355. LEVER, F. M. and POWELL, A. R. Conference on Co-ordination Compounds, London, 1959. Preparation of Ru(II) and Ru(III) hexammonia complexes.

356. LEWIS, J., IRVING, R. J. and WILKINSON, G. J. Inorg. Nucl. Chem. 7 (1958) 32. Infra-red spectra of nitrosyl complexes.

356ᵃ. LEWIS, J. and WILKINS, R. G. Modern Coordination Chemistry, Interscience, New York, 1960 (the chapter by T. M. DUNN, p. 229, on visible and ultraviolet spectra is particularly important for our purpose).

357. LIEHR, A. D. and BALLHAUSEN, C. J. Phys. Rev. 106 (1957) 1161. Model of band intensities from hydrogenic 4p-radial functions.

358. LIEHR, A. D. and BALLHAUSEN, C. J. Ann. Physics (New York) 6 (1959) 134. Intermediate coupling in V(III) and Ni(II) complexes.

358ᵃ. LIESER, K. H. Z. anorg. Chem. 305 (1960) 255. Te(—II) octa (Ag(I)) complex.

359. LIFSCHITZ, J. and ROSENBOHM, E. Z. physik. Chem. 97 (1920) 1. Various transition group complexes.

359ᵃ. LINDERBERG, J. and SHULL, H. J. Mol. Spectr. 5 (1960) 1. Near orbital degeneracy of 2s and 2p.

359ᵇ. LINDERBERG, J. Phys. Rev. 121 (1961) 816. Expansion in Taylor series of Z of Hartree–Fock energies.

359ᶜ. LINDQVIST, I. Nova Acta Reg. Soc. Scient. Upsal. IV 17 (1960) No. 11. Bond length variations.

360. LINGANE, J. J. and SMALL, L. A. J. Am. Chem. Soc. 71 (1949) 973. W(III).

361. LINHARD, M. Z. Elektrochem. 50 (1944) 224. Co(III) hexammonia, influence of ion-pair formation.

362. LINHARD, M. and WEIGEL, M. Z. anorg. Chem. 264 (1951) 321; Ibid. 266 (1951) 49. Co(III) carboxylate and halide-pentammonia complexes.

362ᵃ. LINHARD, M. and WEIGEL, M. *Z. anorg. Chem.* **266** (1951) 73. Co(III) amine fluorides.

363. LINHARD, M. and WEIGEL, M. *Z. anorg. Chem.* **271** (1952) 101. Co(III) electron transfer.

364. LINHARD, M., SIEBERT, H. and WEIGEL, M. *Z. anorg. Chem.* **278** (1955) 287. Cr(III), Co(III) nitrite and thiocyanate complexes.

365. LINHARD, M. and WEIGEL, M. *Z. physik. Chem.* **5** (1955) 20. Cr(III) chloro, amine complexes.

366. LINHARD, M. and WEIGEL, M. *Z. physik. Chem.* **11** (1957) 308. Co(III) halide-pentammonia, relative intensities of the spin-forbidden bands.

367. LINNETT, J. W. *J. Chem. Soc.* (1956) 275. M.O. description of diatomic molecules.

368. LINNETT, J. W. *Discuss. Faraday Soc.* **26** (1958) 7. Review on transition groups.

369. LONGUET-HIGGINS, H. C., ÖPIK, U., PRYCE, M. H. L. and SACK, R. A. *Proc. Roy. Soc. London* A**244** (1958) 1. Jahn–Teller effect.

370. LOTT, K. A. K. and SYMONS, M. C. R. *J. Chem. Soc.* (1959) 829. Mn(V). Order of M.O. energies in tetroxo complexes.

370ᵃ. LOTT, K. A. K. and SYMONS, M. C. R. *J. Chem. Soc.* **1960** 973. Ru(VI), Os(VI) oxo chlorides.

371. LOW, W. *Phys. Rev.* **105** (1957) 801. Cr(III) in magnesium oxide.

372. LOW, W. *Phys. Rev.* **109** (1959) 247 and 256. Ni(II) and Co(II) in magnesium oxide.

373. LOW, W. *Z. phys. Chem.* **13** (1957) 107. V(III) in aluminium oxide.

373ᵃ. LOW, W. and WEGER, M. *Phys. Rev.* **118** (1960) 1130. Fe(II), Co(II) in ZnS.

374. LOW, W. *Survey of Paramagnetic Resonance in Solids.* Supplement *Solid State Physics.* Academic Press, New York, 1960.

375. LUX, H. and NIEDERMANN, T. *Z. anorg. Chem.* **285** (1956) 246. Mn(VI), Mn(VII) in hydroxide melts.

376. LÖWDIN, P. O. Suppl. *Phil. Mag.* **5** (1956) No. 17. Quantum theory of solids.

377. LÖWDIN, P. O and YOSHIZUMI, H. *Correlation Problem in Many-Electron Quantum Mechanics.* Kvantkemiska Gruppen, Uppsala, 1957, and *Adv. Chem. Phys.* **2** (1959) 207 and 323.

378. LÖWDIN, P. O. and SHULL, H. *Phys. Rev.* **101** (1956) 1730. Correlation effects in the helium atom.

379. LÖWDIN, P. O. *J. Phys. Chem.* **61** (1957) 55. Review of quantum chemistry.

380. MACBETH, A. K. and MAXWELL, N. I. *J. Chem. Soc.* (1923) 370. Bi(III).

381. MAGNUSSON, E. A. *Rev. Pure Appl. Chem.* **7** (1957) 195. Chemical bonding in complexes.

381ᵃ. MAIR, G. A., POWELL, H. M. and HENN, D. E. *Proc. Chem. Soc.* **1960** 415. Crystal structure Ni(II) triarsine bromide.

382. MAKI, A. H. and McCARVEY, B. R. *J. Chem. Phys.* **29** (1958) 31 and 35. Paramagnetic resonance of Cu(II).

383. MAKI, G. *J. Chem. Phys.* **28** (1958) 651; *Ibid.* **29** (1959) 162; *Ibid.* **29** (1958) 1129. High-spin and low-spin Ni(II) complexes, electrostatic model.

384. MALATESTA, L. and VATTARINO, L. *J. Chem. Soc.* (1956) 1867. Colours of Rh(I) isonitrile complexes.

384ᵃ. MALATESTA, L. *Progress Inorg. Chem.* **1** (1959) 284. Isonitrile complexes.

385. MALM, J. G., WEINSTOCK, B. and WEAVER, E. E. *J. Phys. Chem.* **62** (1958) 1506. Np(VI), Pu(VI) hexafluorides.

385ᵃ. MANDEL, G., BAUMAN, R. P. and BANKO, E. *J. Chem. Phys.* **33** (1960) 192. Ce(III) infra-red.

334 BIBLIOGRAPHY

386. MARSHALL, E. D. and RICKARD, R. R. *Anal. Chem.* **22** (1950) 795. Ru(VI) tetroxo ion.
387. MARTELL, A. E. and CALVIN, M. *Chemistry of the Metal Chelate Compounds.* Prentice-Hall, New York, 1952.
388. MARTIENSSEN, W. *J. Phys. Chem. Solids* **2** (1957) 257. Analogy between iodide and xenon transitions.
389. MARTIN, B. and WAIND, G. M. *J. Chem. Soc.* (1958) 4284. Co(III), Rh(III), Ir(III) dipyridyl complexes.
390. MASON, D. M. and VANGO, S. P. *J. Phys. Chem.* **60** (1956) 622. Ti(IV) chloride gas.
391. MATHIEU, J. P. *Bull. Soc. Chim., France* **3** (1935) 463 and 476; *Ibid.* **4** (1937) 687; *Ibid.* **5** (1938) 105. Co(III) amine and anion complexes; optical rotation.
392. MAUN, E. K. and DAVIDSON, N. *J. Am. Chem. Soc.* **72** (1950) 2254. Re(IV).
392ª. McCARTHY, M. and ROBINSON, G. W. *Mol. Phys.* **2** (1959) 415. Na and Hg atoms in solid A, Kr, Xe.
393. McCLURE, D. S. *J. Phys. Chem. Solids* **3** (1957) 311. Co(II), Ni(II) in zinc oxide, and other oxides, Cr(III).
394. McCLURE, D. S. *Solid State Physics* **9** (1959) 399. Review of d^n, f^n, sp, and charge transfer spectra, especially in crystals.
395. McCONNELL, H. and DAVIDSON, N. *J. Am. Chem. Soc.* **72** (1950) 3168. Cu(I, II) chlorides.
395ª. McCONNELL, H. M. *J. Chem. Phys.* **20** (1952) 700. C(IV), N(V) nonbonding $M.O. \rightarrow \pi^*$.
396. McCUSHER, P. A. and KENNARD, S. M. S. *J. Am. Chem. Soc.* **81** (1959) 2976. Fe(II), chlorides.
396ª. McGLYNN, S. P. and KASHA, M. *J. Chem. Phys.* **24** (1956) 480. Electron transfer in oxo complexes.
397. McNEVIN, W. M. and KRIEGE, O. H. *J. Am. Chem. Soc.* **77** (1955) 6149, Pd(II) enta complexes.
398. McWEENY, R. *Electronic Structure of Molecules. Some Recent Developments.* Technical Report No. 7, Solid-State and Molecular Theory Group. Massachusetts Institute of Technology, 1955.
399. MEAD, A. *Trans. Faraday Soc.* **30** (1934) 1052. Cr(III), Co(III) oxalate and amine complexes.
400. MEDALIA, A. I. and BURME, B. J. *Anal. Chem.* **23** (1951) 453. Ce(IV).
400ª. MEEK, D. W., STRAUB, D. K. and DRAGO, R. S. *J. Am. Chem. Soc.* **82** (1960) 6013. Mn(II), Co(II), Ni(II), Cu(II) halides in dimethylsulphoxide.
401. MERRITT, C., HERSHENSON, H. M. and ROGERS, L. B. *Anal. Chem.* **25** (1953) 572. Hg(II), Tl(I), Pb(II), Bi(III) chlorides, bromides, iodides.
402. METZLER, D. E. and MYERS, R. J. *J. Am. Chem. Soc.* **72** (1950) 3776. Fe(III) chlorides.
402ª. MINOMURA, S. and DRICKAMER, H. G. *J. Chem. Phys.* **35** (1961) 903. Pressure effects on Ti(III), V(III), Cr(III), Co(II), Ni(II).
403. MITCHELL, G. and BEEMAN, W. W. *J. Chem. Phys.* **20** (1952) 1298. X-ray absorption edges.
404. MIZUNO, J., UKEI, S. and SUGAWARA, T. *J. Phys. Soc. Japan* **14** (1959) 383. Crystallography of $CoCl_2$, $6H_2O$.
405. MOELLER, T. and ULRICH, W. F. *J. Inorg. Nucl. Chem.* **2** (1956) 164. Lanthanide acetylacetonates.
406. MOFFITT, W. *Proc. Roy. Soc. London* **A210** (1951) 224 and 245. M.O. theory (atoms in molecules), oxygen molecule.

407. MOFFITT, W. and BALLHAUSEN, C. J. *Ann. Rev. Phys. Chem.* **7** (1956) 107. Review of ligand field theory.
408. MOFFITT, W. *J. Chem. Phys.* **25** (1956) 467 and 1189. Optical rotation.
409. MOFFITT, W., GOODMAN, G. L., FRED, M. and WEINSTOCK, B. *Mol. Phys.* **2** (1959) 109. Re(VI), Os(VI), Ir(VI), Pt(VI) fluorides.
410. MOORE, C. E. *Atomic Energy Levels.* Nat. Bur. Stand. Circ. No. 467. Vol. I (H to V), II (Cr to Nb), III (Mo to Ac except Ce–Lu). Washington 1949, 1952, 1958.
410ᵃ MOOSER, E. and PEARSON, W. B. *Acta Cryst.* **12** (1959) 1015. Ionic radius ratio.
410ᵇ. MORI, M., SHIBATA, M., KYUNO, E. and ADACHI, T. *Bull. Chem. Soc. Japan* **29** (1950) 883. Co(III) carbonate.
410ᶜ. MORI, M., SHIBATA, M., KYUNO, E. and HOSHIYAMA, K. *Bull. Chem. Soc. Japan* **31** (1958) 291. Co(III) amine carbonates.
410ᵈ. MORI, M., SHIBATA, M., KYUNO, E. and OKUBO, Y. *Bull. Chem. Soc. Japan* **31** (1958) 940. Co(III) ata complexes.
410ᵉ. MOSS, T. S. *Photoconductivity in the Elements.* Butterworths, London, 1952.
411. MULLIKEN, R. S. *Phys. Rev.* **40** (1932) 55; *Ibid.* **41** (1932) 49 and 751; *Ibid.* **43** (1933) 279. M.O. theory.
412. MULLIKEN, R. S. *Rev. Mod. Phys.* **4** (1932) 1. M.O. delocalization.
413. MULLIKEN, R. S. *J. Chem. Phys.* **1** (1933) 492; *Ibid.* **2** (1934) 782. Valence states, electro-negativity.
414. MULLIKEN, R. S. *J. Chem. Phys.* **3** (1935) 375, 506, 517 and 586. M.O. theory.
415. MULLIKEN, R. S. *J. Chem. Phys.* **8** (1940) 234 and 382. Halide and halogen molecules.
416. MULLIKEN, R. S. *Phys. Rev.* **57** (1940) 500; *Ibid.* **61** (1942) 277. Halides.
417. MULLIKEN, R. S. and RIECKE, C. A. *Rep. Progress Phys.* **8** (1941) 231. Electron transfer spectra.
418. MULLIKEN, R. S. *J. Chim. Physique* **46** (1949) 497. M.O. theory, review.
419. MURMANN, R. K., TAUBE H. and POSEY, F. A. *J. Am. Chem. Soc.* **79** (1957) 262. Co(III), Cr(II) kinetics.
420. MURRELL, J. N. *Mol. Phys.* **1** (1958) 384. Pyridine and other heterocyclic compounds.
420ᵃ. MURRELL, J. N. *Quart. Rev.* **15** (1961) 191. Charge-transfer spectra of organic molecular compounds.
420ᵇ. MUTO, Y. *Bull. Chem. Soc. Japan* **33** (1960) 604. Cu(II) Schiff' base complexes.
421. NACHTRIEB, H. N. and CONWAY, J. C. *J. Am. Chem. Soc.* **70** (1948) 3547. Fe(III) chlorides.
421ᵃ. NAKAHARA, A. *Bull. Chem. Soc. Japan* **32** (1959) 1195. Cu(II) amino-acid complexes.
422. NAKAMOTO, K., FUJITA, J., KOBAYASHI, M. and TSUCHIDA, R. *J. Chem. Phys.* **27** (1957) 439. Wolfsberg-Helmholz model of Co(III) chloro amine complexes.
422ᵃ. NAKAMURA, D., KURITA, Y., ITO, K. and KUBO, M. *J. Am. Chem. Soc.* **82** (1960) 5783. Nuclear quadrupole resonance of Pt(IV) halides.
422ᵇ. NESBET, R. K. *Phys. Rev.* **119** (1960) 658. Antiferromagnetism Mn(II), Fe(II), Co(II), Ni(II) oxides.
422ᶜ. NESBET, R. K. *Bull. Am. Phys. Soc.* **5** (1960) 460. Nitrogen molecule and antiferromagnetism at large internuclear distances.
423. NEUHAUS, A. and SCHILLY, W. *Fortschr. Mineralogie* **36** (1958) 64. Cr(III) in minerals.

423ᵃ. NEUHAUS, A. *Z. Kristallog.* **113** (1960) 195. Cr(III) minerals.
424. NEUMANN, H. M. *J. Am. Chem. Soc.* **76** (1954) 2611. Sb(V) chlorides.
425. NEUMANN, H. M. and COOK, N. C. *J. Am. Chem. Soc.* **79** (1959) 3026. Mo(VI) in hydrochloric acid.
426. NEWMAN, L. and HUME, D. N. *J. Am. Chem. Soc.* **79** (1959) 4576 and 4581. Bi(III) chlorides and bromides.
427. NEWMAN, L., LA FLEUR, W., BROUSAIDES, F. J. and ROOS, A. M. *J. Am. Chem. Soc.* **80** (1958) 4491. V(V) in alkaline solution.
428. NEWMAN, L. and QUINLAN, K. P. *J. Am. Chem. Soc.* **81** (1959) 547. V(V) in acid solution.
428ᵃ. NEWMAN, R. and CHENKO, R. M. *Phys. Rev.* **115** (1959) 1147. *L, S* coupling Co(II) infra-red.
428ᵇ. NIGHTINGALE, E. R. *Analyt. Chem.* **31** (1959) 146 and 958. Mn(III) tri-ethanolamine.
429. NIKITINE, S. *Phil. Mag.* **4** (1959) 1. Narrow lines of cooled Cu(I) oxides and halides and of Hg(II) and Pb(II) iodides, exciton theory.
429ᵃ. NIKITINE, S. *J. Phys. Chem. Solids* **8** (1959) 190. Cu(I), Ag(I), Tl(I), Pb(II) exciton series.
430. NYHOLM, R. S. *Chem. Rev.* **53** (1953) 263. Chemical bonding, mainly hybridization language.
431. NYHOLM, R. S. *Quart. Rev.* **7** (1953) 377. Magneto-chemistry.
432. NYHOLM, R. S. *Report to the Tenth Solvay Council. Quelques Problèmes de Chimie Minérale*, p. 225. Stoops, Bruxelles, 1956. Magneto-chemistry.
433. NYHOLM, R. S. and SUTTON, G. J. *J. Chem. Soc.* (1958) 560, 564, 567 and 572. Preparation and colours of Cr(III), Mn(II), Mn(III), Ru(II), Ru(III), Os(II), Os(III), Os(IV) diarsine complexes.
433ᵃ. NYHOLM, R. S. *Proc. Chem. Soc.* **1961** 273. Electron coufigurations and electronegativity.
433ᵇ. NYMAN, C. J. and PLANE, R. A. *J. Am. Chem. Soc.* **82** (1960) 5787. Pt(IV) tris(ethylenediamine) ion-pairs.
434. OKAMOTO, Y. *Nachr. Akad. Wiss. Göttingen* **2a** No. 14 (1956) 275. AgCl, AgBr crystals.
435. ONSTOTT, E. I. and BROWN, C. J. *Anal. Chem.* **30** (1958) 172. Tb(III).
436. ÖPIK, U. and PRYCE, M. H. L. *Proc. Roy. Soc. London* **A238** (1957) 425. Jahn–Teller effect, e.g. in Cu(II).
437. ORGEL, L. E. *J. Chem. Soc.* (1952) 4756. Ligand field stabilization, absorption spectra.
438. ORGEL, L. E. *Quart. Rev.* **8** (1954) 422. Electron transfer spectra.
439. ORGEL, L. E. *J. Chem. Phys.* **23** (1955) 1004. Ligand field theory, V(IV) chloride, V(II), V(III), Cr(III), Fe(II), Co(II), Ni(II), Cu(II) aquo ions, Co(III) amine complexes.
440. ORGEL, L. E. *J. Chem. Phys.* **23** (1955) 1819. Effects of partly covalent bonding.
441. ORGEL, L. E. *J. Chem. Phys.* **23** (1955) 1824. Mn(II) aquo ion, band widths.
442. ORGEL, L. E. *J. Chem. Phys.* **23** (1955) 1958. Mn(II), fluorescence.
443. ORGEL, L. E. *Reports to the Tenth Solvay Council. Quelques Problèmes de Chimie Minérale*, p. 289. Stoops, Bruxelles, 1956. Review of ligand field theory.
444. ORGEL, L. E. and MULLIKEN, R. S. *J. Am. Chem. Soc.* **79** (1957) 4839. Charge-transfer spectra of molecular compounds in solution.
445. ORGEL, L. E. and DUNITZ, J. D. *Nature* **179** (1957) 462. Cu(II) stereo-chemistry.
446. ORGEL, L. E. *Nature* **179** (1957) 1348. Cr(III) in ruby. Ion compression.

447. ORGEL, L. E. *J. Chem. Soc.* (1958) 4186. Cu(I), Ag(I), Au(I), Hg(II) stereo-chemistry.
448. ORGEL, L. E. *Mol. Phys.* 1 (1958) 322. Pb(II) chemical shifts of nuclear magnetic moments.
449. ORGEL, L. E. *Discuss. Faraday Soc.* 26 (1958) 138. Ferroelectricity and stereochemistry of oxides.
449ª. ORGEL, L. E. *J. Chem. Soc.* 1959 3815. Pb(II) stereochemistry.
449ᵇ. ORGEL, L. E. *An Introduction to Transition-Metal Chemistry.* Methuen, London, 1960.
449ᶜ. ORGEL, L. E. *Nature* 187 (1960) 504. Oxygen trimetallic acetates.
449ᵈ. ORGEL, L. E. and JØRGENSEN, C. K. *Mol. Phys.* 4 (1961) 215. Ir(III, IV) and Ru(III, IV) polynuclear complexes.
450. OVERHAUSER, A. W. *Phys. Rev.* 101 (1956) 1702. Group theory applied to alkali halide crystals.
451. OWEN, J. and STEVENS, K. W. H. *Nature* 171 (1953) 836. Ir(IV) hyperfine-structure of paramagnetic resonance curves.
451ª. OWEN, J. *Discuss. Faraday Soc.* 19 (1955) 127. Paramagnetic resonance and evidence for partly covalent bonding.
452. OWEN, J. *Proc. Roy Soc. London* A227 (1955) 183. Ligand field theory, gyromagnetic factors, V(III), Cr(III), Fe(II), Co(II), Ni(II), Cu(II) aquo ions.
453. OWEN, J. *Discuss. Faraday Soc.* 26 (1958) 53. Ir(IV) co-operative effects on magnetism.
454. PALILLA, F. C., ADLER, N. and HISKEY, C. F. *Anal. Chem.* 25 (1953) 926. Ti(IV), V(V), Zr(IV), Nb(V), Mo(VI), Ta(V), W(VI) peroxo complexes.
455. PALMER, W. G. *Valency.* Cambridge University Press, 1948.
456. PAPPALARDO, R. *Phil. Mag.* 2 (1957) 1397. Mn(II) hexaquo ion.
457. PAPPALARDO, R. *Nuovo Cimento* Series 10, 6 (1957) 392. Ni(II) hexaquo ion.
458. PAPPALARDO, R. *Phil. Mag.* 4 (1959) 219. Co(II) chloride, bromide hexa-hydrates (cf. ref. 404).
459. PAPPALARDO, R. *J. Chem. Phys.* 31 (1959) 1050; *Ibid.* 33 (1960) 612. Mn(II) chloride and bromide.
459ª. PAPPALARDO, R. *J. Mol. Spectr.* 6 (1961) 554. Cu(II) in ZnO and other oxides.
459ᵇ. PAPPALARDO, R., WOOD, D. L. and LINARES, R. C. Private communication. Co(II), Ni(II) in oxides.
459ᶜ. PAPPALARDO, R. and DIETZ, R. E. *Phys. Rev.* 123 (1961) 1188. Fe(II), Co(II), Ni(II), Cu(II) in CdS.
459ᵈ. PARIS, J. P. and BRANDT, W. W. *J. Am. Chem. Soc.* 81(1959) 5001. Ru(II), dip luminescence.
460. PARISER, R. *J. Chem. Phys.* 24 (1956) 250. Aromatic hydrocarbons, con-figuration intermixing.
460ª. PARKER, C. A. *Proc. Roy. Soc. London* A220 (1953) 104. Fe(III) oxalates.
461. PARR, R. G. and JOY, H. W. *J. Chem. Phys.* 26 (1957) 424. Non-integral exponents in radial functions.
462. PARSONS, R. W. and DRICKAMER, H. G. *J. Chem. Phys.* 29 (1958) 930. Influence of high pressure on Cr(III) and Ni(II) ligand field spectra.
463. PAULING, L. *J. Am. Chem. Soc.* 53 (1931) 1367. Hybridization theory.
464. PAULING, L. *The Nature of the Chemical Bond.* Cornell University Press, Ithaca, 1944.
465. PAULING, L. *J. Chem. Soc.* (1948) 1461. The magnetochemical criterion cannot distinguish between covalent and electrovalent bonding.

466. PAUNCZ, R. *Acta Vålådalensia.* Kvantkemiska Gruppen, Uppsala, 1958.
466ᵃ. PEACOCK, R. D. *Progress Inorg. Chem.* 2 (1960) 193. Review of fluorides.
466ᵇ. PEARSON, G. P., GRAY, H. B. and BASOLO, F. *J. Am. Chem. Soc.* 82 (1960) 787. Pt(II) chloride solvent effects.
467. PECSOK, R. L. and BJERRUM, J. *Acta Chem. Scand.* 11 (1957) 1419. Cr(II) aquo and amine complexes.
467ᵃ. PELLETIER, S. *J. Chimie Physique* 57 (1960) 287. Ni(II) amino-acid complexes.
468. PENNEY, W. G. *Trans. Faraday Soc.* 36 (1940) 627. Ligand field stabilization.
469. PERRIN, D. D. *J. Am. Chem. Soc.* 80 (1958) 3852. Fe(III) thiocyanates.
469ᵃ. PERRIN, D. D. *Analyt. Chem.* 31 (1959) 1181. Fe(III) acetate.
470. PESTEMER, M. and ALSLEV-KLINKER, A. *Z. Elektrochem.* 53 (1949) 387. Ni(II) carboxylate complexes.
471. PETERSEN, R. *J. Phys. Chem. Solids* 1 (1957) 284. Analogy between iodide and xenon energy levels.
472. PFLAUM, R. T. and POPOV, A. I. *Anal. Chim. Acta* 13 (1955) 165. Different Cr(III), Co(II), Ni(II), Cu(II) salts in dimethylformamide.
473. PHILIPP, H. R. and TAFT, E. A. *J. Phys. Chem. Solids* 1 (1956) 159. Crystalline alkali iodides.
473ᵃ. PHILIPP, H. R. and TAFT, E. A. *Phys. Rev.* 120 (1960) 37. Silicon absorption bands.
473ᵇ. PIPER, T. S. and KOERTGE, N. *J. Chem. Phys.* 32 (1960) 559. Ni(II) co-excitation of water vibrations.
473ᶜ. PIPER, T. S. and CARLIN, R. L. *J. Chem. Phys.* 33 (1960) 608. V(III), Cr(III) Co(III) oxalates, dichroism.
473ᵈ. PIPER, W. W., JOHNSON, P. D. and MARPLE, D. T. F. *J. Phys. Chem. Solids* 8 (1959) 457. Zn(II) sulphide.
473ᵉ. POË, A. J. and VAIDYA, M. S. *J. Chem. Soc.* 1961 1023. Criticism of A and B character of elements.
474. POLDER, D. *Physica* 9 (1942) 709. Cu(II), ligand field theory.
475. POSEY, F. A. and TAUBE, H. *J. Am. Chem. Soc.* 78 (1956) 15. Ion-pairs Co(III) hexammonia sulphate.
476. POULET, H. *J. Chimie Physique* 54 (1957) 258. Ligand fields and chemical bonding.
477. POULSEN, I. and BJERRUM, J. *Acta Chem. Scand.* 9 (1955) 1407. Ni(II), Cu(II), heats of formation of amine complexes.
477ᵃ. POULSEN, K. G. and GARNER, C. S. *J. Am. Chem. Soc.* 81 (1959) 2615. Cr(III) binuclear complexes.
478. PRANDTL, W. and SCHEINER, K. *Z. anorg. Chem.* 220 (1934) 107. Lanthanide aquo ions.
478ᵃ. PRATT, G. W. and COELHO, R. *Phys. Rev.* 116 (1959) 281. Mn(II), Co(II) oxide.
478ᵇ. PRENER, J. S., HANSON, R. E. and WILLIAMS, F. E. *J. Chem. Phys.* 21 (1953) 759. Hg atoms in zeolite.
478ᶜ. PROLL, P. J., SUTCLIFFE, L. H. and WALKLEY, J. *J. Phys. Chem.* 65 (1961) 455. Co(II) acetates.
479. PRYCE, M. H. L. *Discuss. Faraday Soc.* 26 (1958) 21. Ligand field theory.
480. PRYCE, M. H. L. and RUNCIMAN, W. A. *Discuss. Faraday Soc.* 26 (1958) 34. V(III) in aluminium oxide.
481. PRZIBRAM, K. *Z. Physik* 102 (1936) 331; *Ibid.* 107 (1937) 709. Sm(II), Eu(II) fluorescence.
482. PURDY, W. C. and HUME, D. N. *Anal. Chem.* 27 (1955) 256. Mn(III) phosphoric, sulphuric acids.

483. QUILL, L. L., SELWOOD, P. W. and HOPKINS, B. S. *J. Am. Chem. Soc.* **50** (1928) 2929. Nd(III) in concentrated acids.
484. RABINOWITCH, E. and THILO, E. *Periodisches System, Geschichte und Theorie.* Ferdinand Enke, Stuttgart, 1930.
485. RABINOWITCH, E. *Rev. Mod. Phys.* **14** (1942) 112. Electron transfer spectra, mainly of crystalline and gaseous alkali halides.
486. RABINOWITCH, E. and STOCKMAYER, W. H. *J. Am. Chem. Soc.* **64** (1942) 335. Fe(III) aquo and halide complexes.
487. RACAH, G. *Phys. Rev.* **62** (1942) 438; *Ibid.* **63** (1942) 367. d^n-systems in spherical symmetry.
488. RACAH, G. *Phys. Rev.* **76** (1949) 1352. Theory of interelectronic repulsion in f^n-systems.
489. RACAH. G. *Physica* **16** (1950) 651. Theory of gaseous Th^{2+}.
490. RACAH, G. *Proceedings of the Rydberg Centennial Conference on Atomic Spectroscopy* (2nd Ed.) **50**, 21, p. 31. Lunds Universitets Årsskrifts, 1955.
490ᵃ. RACAH, G. and SHADMI, Y. *Phys. Rev.* **119** (1960) 156. Racah–Trees and seniority number corrections in gaseous 3d ions.
490ᵇ. RANDLES, J. E. B. *Ann. Rep. Chem. 1959 London* **56** (1960) 33. Hydration energies.
491. RASMUSSEN, S. E. *Acta Chem. Scand.* **10** (1956) 1279. Chelate effect.
492. RASMUSSEN, S. E. *J. Inorg. Nucl. Chem.* **8** (1958) 441. Crystal structure of Ni tren(NCS)$_2$.
492ᵃ. RICH, R. L. and TAUBE, H. *J. Am. Chem. Soc.* **76** (1954) 2608. Pt(IV) reaction kinetics.
492ᵇ. RICHARDS, D. H. and SYKES, K. W. *J. Chem. Soc.* **1960** 3626. Fe(III) hydrolysis.
492ᶜ. RIDLEY, E. C. *Proc. Cambridge Phil. Soc.* **56** (1960) 41. Hartree SCF, Pr(III), Tm(III).
493. ROBERTS, G. L. and FIELD, F. H. *J. Am. Chem. Soc.* **72** (1950) 4232. Co(II), Ni(II) aquo, oxalate, amine complexes.
493ᵃ. ROBINSON, G. W. *J. Mol. Spectr.* **6** (1961) 58. Spectra and energy transfer between molecules in crystalline inert gases.
494. ROSOTTI, F. J. C. and ROSOTTI, H. S. *Acta Chem. Scand.* **9** (1955) 1177. Vanadyl (IV) ion.
494ᵃ. ROYER, D. J. *J. Inorg. Nucl. Chem.* **11** (1959) 151. Cu(II) picoline complexes.
495. RULFS, C. and MEYER, R. *J. Am. Chem. Soc.* **77** (1955) 4505. Re(IV) chlorides, bromides and iodides.
496. RUNCIMAN, W. A. *Reports Progress Phys.* **21** (1958) 30. Spectra of crystals.
496ᵃ. RUNCIMAN, W. A. and WYBOURNE, B. G. *J. Chem. Phys.* **31** (1959) 1149. Pr(III), Tm(III).
497. RUSSELL, C. D., COOPER, G. R. and VOSBURGH, W. C. *J. Am. Chem. Soc.* **65** (1943) 1301. Connections between magnetic moment, spectra and stability of Ni(II) complexes.
497ᵃ. RYAN, J. L. *J. Phys. Chem.* **65** (1961) 1856. U(IV), Np(IV), Pu(IV) hexachlorides.
498. SAMUEL, R. and DÉSPANDE, A. R. R. *Z. Physik.* **80** (1933) 395. Pd(II) and various other transition group complexes.
499. SAMUEL, R. and UDDIN, M. *Trans. Faraday Soc.* **31** (1935) 423. Co(III) amine, Rh(III), Pt(II) halide complexes.
500. SANDELL, E. B. *Colorimetric Determination of Traces of Metals* (3rd Ed.). Interscience Publishers, New York, 1959.
501. SANTEN, J. H. VAN and WIERINGEN, J. S. VAN, *Rec. trav. chim. Pays-Bas* **71** (1952) 420. Variation of ionic radii with sub-shell configuration.

502. SATTEN, R. A. *J. Chem. Phys.* **21** (1953) 637. Nd,(III) identification of quartet levels.

502[a]. SATTEN, R. A. and MARGOLIS, J. S. *J. Chem. Phys.* **32** (1960) 573. f² in octahedral symmetry.

502[b]. SATTEN, R. A., YOUNG, D. and GRUEN, D. M. *J. Chem. Phys.* **33** (1960) 1140. U(IV) chlorides.

503. SAUER, H. *Ann. Physik* **87** (1928) 197. Cr(III) oxide.

504. SAUM, G. A. and HENSLEY, E. B. *Phys. Rev.* **113** (1959) 1019. Mg(II), Ca(II), Sr(II), Ba(II) oxides, sulphides, selenides and tellurides.

505. SCAIFE, D. E. *Conference on Co-ordination Compounds, London,* 1959. V(III) tetrachloro and bromo complexes.

505[a]. SCHEFFER, E. R. and HAMMAKER, E. M. *J. Am. Chem. Soc.* **72** (1950) 2575. Cr(III), Mn(III) fluorides.

506. SCHEIBE, G. *Z. physik. Chem.* B5 (1929) 355. Electron transfer bands of halide ions.

506[a]. SCHILT, A. A. *J. Am. Chem. Soc.* **82** (1960) 3000 and 5779. Fe(II), Fe(III) dip, phen cyanides.

507. SCHLAPP, R. and PENNEY, W. G. *Phys. Rev.* **42** (1932) 666. Ligand field theory applied to magnetochemistry.

508. SCHLESINGER, H. and TAPLEY, M. *J. Am. Chem. Soc.* **46** (1924) 276. Pt(IV) halides.

509. SCHLÄFER, H. L. *Z. physik. Chem.* **4** (1955) 116. Mn(II), Fe(III) (cf. ref. 283).

510. SCHLÄFER, H. L. *Z. physik. Chem.* **6** (1956) 201. Mn(II) chlorides.

511. SCHLÄFER, H. L. *Z. physik. Chem.* **8** (1956) 373. Model of phenanthroline complexes.

512. SCHLÄFER, H. L. *Z. physik. Chem.* **11** (1957) 65. Cr(III) ruby lines, photochemistry.

513. SCHLÄFER, H. L. and SKOLUDEK, H. *Z. physik. Chem.* **11** (1957) 277. Cr(II) aquo and chloro complexes.

514. SCHLÄFER, H. L. and KOLLRACK, R. *Z. physik. Chem.* **18** (1958) 348. Cr(III) amine complexes.

515. SCHLÄFER, H. L. and KÖNIG, E. *Z. physik. Chem.* **19** (1959) 265. Electrostatic model of pyridine complexes.

515[a]. SCHLÄFER, H. L. and KLING, O. *Z. anorg. Chem.* **302** (1959) 1. Cr(III) amine complexes.

515[b]. SCHLÄFER, H. L., KLING, O., MÄHLER, L. and OPITZ, H. P. *Z. physik. Chem.* **24** (1960) 307. Cr(III) amine oxalates.

515[c]. SCHLÄFER, H. L. and OPITZ, H. P. *Z. Elektrochem.* **65** (1961) 372. Cr(III), Mn(II), Fe(III), Co(II), Ni(II), Cu(II) in dimethylsulphoxide.

515[d]. SCHMIDT, G. and HERR, W. *Z. Naturforsch.* **16a** (1961) 748. Re(IV), Os(IV), Ir(IV), Pt(IV) bromides.

516. SCHMITZ-DUMONT, O., BROKOPF, H. and BURKHARDT, K. *Z. anorg. Chem.* **295** (1958) 7. Co(II) in mixed oxides.

517. SCHMITZ-DUMONT, O., GÖSSLING, H. and BROKOPF, H. *Z. anorg. Chem.* **300** (1959) 159. Ni(II) in mixed oxides.

517[a]. SCHMITZ-DUMONT, O. and REINEN, D. *Z. Elektrochem.* **63** (1959) 978. Cr(III) in mixed oxides.

517[b]. SCHNEPP, O. and DRESSLER, K. *J. Chem. Phys.* **33** (1960) 49. Solid Kr and Xe.

517[c]. SCHNEPP, O. *J. Phys. Chem. Solids* **17** (1961) 188. Mg and Mn atoms in solid inert gases.

517[d]. SCHONLAND, D. S. *Proc. Roy. Soc. London* A254 (1960) 111. Paramagnetic resonance in Mn(VI) tetroxo ion.

518. SCHWARZENBACH, G. *Helv. Chim. Acta* **35** (1952) 2344. Chelate effect.
519. SCHWARZENBACH, G. and GUT, R. *Helv. Chim. Acta* **39** (1956) 1589. Half-filled shell effect in formation constants of lanthanide complexes.
520. SCHWARZENBACH, G. *Experientia* Suppl. **5** (1956) 162. Review of organic ligands.
521. SCHWARZENBACH, G. *Die Komplexometrische Titration* (2nd Ed.). Ferdinand Enke, Stuttgart, 1956.
522. SCHÄFFER, C. E. and JØRGENSEN, C. K. *J. Inorg. Nucl. Chem.* **8** (1958) 143. Nephelauxetic series.
523. SCHÄFFER, C. E. *J. Inorg. Nucl. Chem.* **8** (1958) 149. Cr(III) amine and oxide complexes.
524. SCHÄFFER, C. E. *Conference on Co-ordination Compounds, London*, 1959. Cr(III), Co(III) thiocyanate + Ag(I), Hg(II).
525. SCHÄFFER, C. E. Private communication.
526. SEABORG, G. T., KATZ, J. J. and MANNING, W. M. *The Transuranium Elements*. Nat. Nucl. Energy Ser., IV, Vol. 14B, 1949.
527. SEABORG, G. T. and KATZ, J. J. *The Actinide Elements*. Nat. Nucl. Energy Ser., IV, Vol. 14A, 1954.
528. SEARCY, A. W. *J. Chem. Phys.* **28** (1958) 1237; *Ibid.* **31** (1959) 1. Angles and lone-pairs in complexes.
529. SEITZ, F. *J. Chem. Phys.* **6** (1938) 150. Tl(I) in alkali halide crystals.
530. SEITZ, F. *The Modern Theory of Solids* (1st Ed.). McGraw-Hill, New York and London, 1940.
531. SEITZ, F. *Rev. Mod. Phys.* **18** (1946) 384; *Ibid.* **26** (1954) 7. Colour centres in alkali halide crystals.
531ᵃ. SELBIN, J. *J. Inorg. Nucl. Chem.* **17** (1961) 84. Co(III) tetraethylene pentamine nitrite.
532. SELWOOD, P. W. *J. Am. Chem. Soc.* **52** (1930) 3112 and 4308. Nd(III).
533. SENISE, P. and PERRIER, M. *J. Am. Chem. Soc.* **80** (1958) 4194. Co(II) mono-complex of thiocyanate.
534. SENISE, P. *J. Am. Chem. Soc.* **81** (1959) 4196. Co(II) azide complexes.
534ᵃ. SHELDON, J. C. *Nature* **184** (1959) 1210; *Ibid. J. Chem. Soc.* **1960** 1007. Mo(II) chlorides.
535. SHEREMETIEV, G. D. *Optika y Spektroskopia* **2** (1957) 99. U(VI) hexafluoride.
535ᵃ. SHERIF, F. G. and ORABY, W. M. *J. Inorg. Nucl. Chem.* **17** (1961) 152. Cr(III) azides.
536. SHIMURA, Y. *Bull. Chem. Soc. Japan* **25** (1952) 49; *Ibid.* **29** (1956) 311. Co(III) spectrochemical series.
537. SHIMURA, Y., ITO, H. and TSUCHIDA, R. *J. Chem. Soc. Japan* **75** (1954) 560. Mn(IV) heteropoly molybdates.
538. SHIMURA, Y. *Bull. Chem. Soc. Japan* **28** (1955) 572; *Ibid.* **31** (1958) 173, 311 and 315. Co(III) oxalate, glycinate, other amino-acid and amine complexes; inclination parameter, optical rotation.
539. SHIMURA, Y. and TSUCHIDA, R. *Bull. Chem. Soc. Japan* **30** (1957) 502. Co(II), Co(III) (tetrahedral), Mn(IV) hetero wolframates.
539ᵃ. SHIVELY, R. R. and WEYL, W. A. *J. Phys. Chem.* **55** (1951) 512. Fe(II, III) colours.
540. SHUKLA, S. K. *J. Chromatogr.* **1** (1958) 457. Separation of Rh(III) anion complexes.
541. SIDGWICK, N. V. *The Chemical Elements and their Compounds*. Oxford University Press, 1950.
542. SILLÉN, L. G. *Quart. Rev.* **13** (1959) 146. Hydrolysis of metal ions in constant salt mediae.

Z

542ᵃ. SILVERMAN, J. N., PLATAS, O. and MATSEN, F. A. *J. Chem. Phys.* **32** (1960) 1402. Two-electron atoms.

543. SIMPSON, E. A. and WAIND, G. M. *J. Chem. Soc.* (1958) 1746. Cu(I) cyanides.

544. SJOBLOM, R. and HINDMAN, J. C. *J. Am. Chem. Soc.* **73** (1951) 1744. Np(III), Np(IV), Np(V), Np(VI).

544ᵃ. SLATEN, L. E. and GARNER, C. S. *J. Phys. Chem.* **63** (1959) 1214. Cr(III) amine chlorides.

545. SLATER, J. C. *Phys. Rev.* **36** (1930) 57. Hydrogenic radial functions.

546. SLATER, J. C. *Electronic Structure of Atoms and Molecules.* Technical Report No. 3, Solid-State and Molecular Theory Group, Massachusetts Institute of Technology, 1953.

546ᵃ. SLATER, J. C. *Quantum Theory of Atomic Structure, Vol. I and II.* McGraw-Hill, New York, 1960.

546ᵇ. SLYKHOUSE, T. E. and DRICKAMER, H. G. *J. Phys. Chem. Solids* **7** (1958) 207. Pressure effects on Cu(I), Ag(I) halides.

547. SMITH, G. S. and HOARD, J. L. *J. Am. Chem. Soc.* **81** (1959) 549 and 556. Crystallographic study of Co(III) and Ni(II) enta-complexes.

548. SONE, K. *J. Am. Chem. Soc.* **75** (1953) 5207. Acetylacetonates, hydroxy-quinolinates and complexes of other organic ligands.

548ᵃ. SONE, K. and KATO, M. *Z. anorg. Chem.* **301** (1959) 277. Yellow Ni(II) amine complexes in hot alcohol.

549. SPEAKMAN, J. C. *An Introduction to the Electronic Theory of Valency.* (3Ed.) Edward Arnold, London, 1955.

550. SPEDDING, F. H. *Phys. Rev.* **58** (1940) 255. Pr(III).

550ᵃ. SPICER, W. E. and SOMMER, A. H. *J. Phys. Chem. Solids* **8** (1959). 437. Au(—I) Cs(I) compound.

551. STAHL-BRADA, R. and LOW, W. *Phys. Rev.* **113** (1959) 775. Co(II), cubic co-ordination in fluorite, and Cr(III), ruby.

551ᵃ. STAPLES, P. J. and TOBE, M. L. *J. Chem. Soc.* **1960** 4803 and 4812. Co(III) amine azides.

552. STEIN, G. and TREININ, A. *Trans. Faraday Soc.* **55** (1959) 1086. Iodide and other anions.

553. STELLING, O. *Z. Physik* **50** (1928) 506; *Z. physik. Chem.* **B7** (1930) 210; *Ibid.* **B24** (1934) 282. X-ray absorption edges of phosphorus, sulphur compounds.

554. STELLING, O. *Z. physik. Chem.* **B16** (1932) 303. X-ray absorption edges of chloro complexes.

555. STEPHENS, D. R. and DRICKAMER, H. G. *J. Chem. Phys.* **30** (1959) 1518. Pressure dependence of electron transfer spectra. Fe(III), Co(III), Re(IV).

555ᵃ. STEPHENS, D. R. and DRICKAMER, H. G. *J. Chem. Phys.* **34** (1961) 937. Pressure effects on octahedral Ni(II).

555ᵇ. STEPHENS, D. R. and DRICKAMER, H. G. *J. Chem. Phys.* **35** (1961) 427 and 429. Pressure effects on Cr(III) ruby and on tetrahedral Co(II) and Ni(II).

556. STERN, F. *Phys. Rev.* **104** (1956) 684. F^k integrals of gaseous Fe, Ni.

557. STERN, A. and DEZELIC, M. *Z. physik. Chem.* **A180** (1937) 131. Metal porphyrines.

558. STEUNENBERG, R. K. and VOGEL, R. C. *J. Am. Chem. Soc.* **78** (1956) 901. Halogen molecules.

559. STEVENS, K. W. H. *Proc. Roy. Soc. London* **A219** (1953) 542. Ligand field theory.

560. STEWART, D. C. *Light Absorption . . . I.* AECD-2389, Berkeley, 1948. Ce(III), Pr(III), Nd(III), Sm(III), Eu(III), Gd(III).

561. STEWART, D. C. *Light Absorption . . . II.* ANL-4812, Argonne, 1952. Pm(III), U(III), N(IV), Np(III), Pu(III), Am(III).
562. STEWART, D. C. *Light Absorption . . . III.* ANL-5624, Argonne, 1956. Tb(III), Dy(III), Ho(III), Er(III), Tm(III), Yb(III).
563. STEWART, D. C. and KATO, D. *Anal. Chem.* **30** (1958) 164. Tb(III).
564. STOKES, S. and DUNCAN, A. B. F. *J. Am. Chem. Soc.* **80** (1958) 6177. Fluromethanes.
565. STOLARCZYK, L. *Roczniki Chem.* **31** (1957) 1273. Cu(II) in alcohols, electrostatic model.
566. STOLARCZYK, L. *Acta Chem. Scand.* **12** (1958) 1885. Conditions for Gaussian shape of absorption bands.
567. STONER, G. A. *Anal. Chem.* **27** (1955) 1186. Ru(VII) tetroxo ion.
568. STOUT, J. W. *J. Chem. Phys.* **31** (1959) 709; *Ibid.* **33** (1960) 303. Mn(II) fluoride.
569. STRICKLAND, J. D. H. *J. Am. Chem. Soc.* **74** (1952) 862. Mo(V, VI) complexes.
569ª. STRICKLER, S. J. and KASHA, M. *J. Chem. Phys.* **34** (1961) 1077. Solvent effects on halide ions.
569ᵇ. STROHMEIER, W. and GERLACH, K. *Z. physik. Chem.* **27** (1961) 439. Cr(O), Mo(O), W(O) pyridine and piperidine carbonyls.
570. SUGANO, S. and TANABE, Y. *Discuss. Faraday Soc.* **26** (1958) 43. Cr(III) in ruby.
570ª. SUGANO, S. *Progr. Theor. Phys. Suppl.* No. 14 (1960) 66. Narrow bands of octahedral complexes.
571. SULZER, P. and WIELAND, K. *Helv. Phys. Acta* **25** (1952) 653. Shape of absorption bands as function of temperature.
572. SUNDRAM, A. K. and SANDELL, E. B. *J. Am. Chem. Soc.* **77** (1955) 855. Pd(II) chloro and nitrate complexes.
573. TAFT, E. A. and PHILIPP, H. R. *J. Phys. Chem. Solid* **3** (1957) 1. Crystalline alkali halides.
574. TANABE, Y. and SUGANO, S. *J. Phys. Soc. Japan* **9** (1954) 753. Ligand field theory for octahedral complexes.
575. TANABE, Y. and SUGANO, S. *J. Phys. Soc. Japan* **9** (1954) 766. Tanabe–Sugano diagrams. Discussion of V(III), Cr(II), Cr(III), Fe(II), Co(II), Co(III), Ni(II).
576. TANABE, Y. and SUGANO, S. *J. Phys. Soc. Japan* **12** (1957) 556. Cr(III) in ruby.
577. TANABE, Y. and KAMIMURA, H. *J. Phys. Soc. Japan* **13** (1958) 394. Theory for intermediate coupling in octahedral complexes.
577ª. TANABE, Y. *Progr. Theor. Phys. Suppl.* No. 14 (1960) 17. M.O. theory and intermediate coupling in octahedral complexes.
578. TANAKA, N. and TAKAMURA, T. *J. Inorg. Nucl. Chem.* **9** (1959) 15. Cu(II) thiocyanates.
579. TANNER, K. N. and DUNCAN, A. B. F. *J. Am. Chem. Soc.* **73** (1951) 1164. Mo(VI), W(VI) hexafluorides.
580. TAUBE, H. *Chem. Rev.* **50** (1952) 69. Reaction kinetics.
581. TAUBE, H. and MYERS, H, *J. Am. Chem. Soc.* **76** (1954) 2103. Co(III), Cr(II) reaction mechanisms with halogen bridges.
582. TELEP, G. and BOLTZ, D. F. *Anal. Chem.* **24** (1952) 945. Co(III) carbonate complex.
582ª. THOMAS, D. G. *J. Phys. Chem. Solids* **15** (1960) 86. So-called exciton spectrum of Zn(II) oxide.

583. TINKHAM, M. *Proc. Roy. Soc. London* **A236** (1956) 535 and 549. Paramagnetic resonance and hyperfine structure of Mn(II), Fe(II), Co(II) in zinc fluoride.

584. TOMKINSON, J. C. and WILLIAMS, R. J. P. *J. Chem. Soc.* (1958) 1153. Fe(II) with organic ligands.

585. TREES, R. E. *Phys. Rev.* **83** (1951) 756; *Ibid.* **84** (1951) 1089. Correction to the Slater theory of interelectronic repulsion energies.

586. TREES, R. E. *Phys. Rev.* **112** (1958) 165. Parameters for energy levels of gaseous Re.

586ᵃ. TREES, R. E. *J. Opt. Soc. Am.* **48** (1958) 293. Linear theory in atomic spectroscopy.

586ᵇ. TREES, R. E. *Physica* **26** (1960) 353. Gaseous Th atoms.

586ᶜ. TREES, R. E. and JØRGENSEN, C. K. *Phys. Rev.* **123** (1961) 1278. Correlation effects in gaseous 3d-ions.

587. TSUCHIDA, R. *Bull. Chem. Soc. Japan* **13** (1938) 388, 436 and 471. Co(III) amine and anion complexes, the spectrochemical series.

588. TSUCHIDA, R. and KOBAYASHI, M. *Bull. Chem. Soc. Japan* **13** (1938) 476. Co(III) amine and anion complexes.

589. TSUCHIDA, R. and KUROYA, H. *Bull. Chem. Soc. Japan* **15** (1940) 427. Co(III) oxalate and amine complexes.

589ᵃ. TSUCHIDA, R. and YAMADA, S. *Bull. Chem. Soc. Japan* **31** (1958) 813. Pt(II) Magnus' green salt.

590. TSUJIKAWA, I. and SUGANO, S. *J. Phys. Soc. Japan* **13** (1958) 220. Cr(III) in ruby.

590ᵃ. VAN HOUTEN, S. *J. Chem. Phys. Solids* **17** (1960) 7. Semiconductivity in Li(I)Ni(II, III) oxides.

591. VAN VLECK, J. H. *Phys. Rev.* **41** (1932) 208. Ligand field theory applied to magnetochemistry.

592. VAN VLECK, J. H. *J. Chem. Phys.* **3** (1935) 803 and 807. Comparison of the effects of covalent and electrostatic bonding.

593. VAN VLECK, J. H. *J. Phys. Chem.* **41** (1937) 67. Band intensities of lanthanide ions.

594. VAN VLECK, J. H. *J. Chem. Phys.* **7** (1939) 61, 72; *Ibid.* **8** (1940) 787. Jahn–Teller effect.

595. VAREILLE, L. *Bull. Soc. Chim. France* (1955) 1493 and 1496. Fe(III) salicylates and other phenolate complexes.

596. VLCEK, A. A. and BERAN, P. *Collec. Czech. Chem. Comm.* **21** (1956) 1640. Au(III) chlorides.

597. VLCEK, A. A. *Discuss. Faraday Soc.* **26** (1958) 164. Polarography and energy levels of Co(III), Rh(III).

598. VOLBERT, F. *Z. physik. Chem.* **A149** (1930) 382. Ag(I) aquo and ammonia complexes.

599. VON DER LAGE, F. C. and BETHE, H. *Phys. Rev.* **71** (1947) 612. Cubic harmonics and angular functions in octahedral symmetry.

599ᵃ. VON HIPPEL, A. *Z. Physik.* **101** (1936) 680. Electron transfer model of alkali metal halides.

600. WAGGENER, W. C. *J. Phys. Chem.* **62** (1958) 382. Np(III), Np(IV), Np(V), Np(VI).

601. WAGGENER, W. C., MATTERN, J. A. and CARTLEDGE, G. H. *J. Am. Chem. Soc.* **81** (1959) 2958. Addition of Ag(I) and Hg(II) to Cr(III) and Co(III) NCS⁻.

602. WAIND, G. M. and MARTIN, B. J. *Inorg. Nucl. Chem.* 8 (1958) 551. Co(I), Rh(I) dipyridyl complexes.

602ᵃ. WALSH, A. D. *J. Chem. Soc.* **1953** 2260, 2266, 2288, 2296, 2301, 2306, 2318, 2321, 2325 and 2330. Molecular spectra.

603. WATANABE, H. *Progress Theor. Phys.* **18** (1957) 405. Ligand field theory of Mn(II), Fe(III) (⁴S indicated does not exist).

604. WATANABE, K. *J. Chem. Phys.* **26** (1957) 542. Ionization energy of gaseous molecules.

605. WATSON, R. E. *Solid-State and Molecular Theory Group.* M.I.T. Quarterly Progress Reports, January 1958, January 1959 and April 1959. Comparison of orbital energy and Fᵏ integrals of Hartree–Fock radial functions with empirical values.

605ᵃ. WATSON, R. E. *Iron-Series Hartree–Fock Calculations.* Technical Report No. 12 from Solid-State and Molecular Theory Group, M.I.T. 1959.

605ᵇ. WATSON R. E. *Phys. Rev.* **118** (1960) 1036; *Ibid.* **119** (1960) 1934.

605ᶜ. WATSON, R. E. and FREEMAN, A. J. *Phys. Rev.* **120** (1960) 1134. Spin-polarization of Ni(II).

605ᵈ. WATTERS, J. I. and DEWITT, R. *J. Am. Chem. Soc.* **82** (1960) 1333. Ni(II) amine oxalates.

605ᵉ. WEAKLIEM, H. A. and McCLURE, D. S. Private communication. Mn(II), Co(II), Ni(II) in ZnS.

606. WEHNER, P. and HINDMAN, J. C. *J. Am. Chem. Soc.* **72** (1950) 3911. Ru(IV), Ru(VIII).

607. WEHNER, P. and HINDMAN, J. C. *J. Phys. Chem.* **56** (1952) 10. Ru(IV) chloro complexes.

608. WELLS, A. F. *Structural Inorganic Chemistry* (2nd Ed.). Oxford University Press, 1950.

609. WEYL, W. A. *J. Phys. Chem.* **55** (1951) 507 and 512. Fe(II, III) colours. Co(II), Ni(II) sulphides and in ZnS.

609ᵃ. WEYL, W. A. *Coloured Glasses.* Dawson's of Pall Mall, London, 1959.

609ᵇ. WHEALY, R. D. and COLGATE, S. O. *Analyt. Chem.* **28** (1956) 1897. Co(III), Ni(II) den complexes.

610. WHEELER, T. E., PERROS, T. P. and NAESER, C. R. *J. Am. Chem. Soc.* **77** (1955) 3488. Pt(IV) fluorides.

610ᵃ. WHITE, E. A. D. *Quart. Rev. London* **15** (1961) 1. Synthetic gemstones.

611. WIERINGEN, J. S. VAN *Discuss. Faraday Soc.* **19** (1955) 118. Mn(II) hyperfine structure of paramagnetic resonance.

611ᵃ. WILKINSON, G. and COTTON, F. A. *Progress Inorg. Chem.* **1** (1959) 1. Cyclopentadienide complexes.

611ᵇ. WILKINSON, G. *Proc. Chem. Soc.* **1961** 72. Rh(III) amine hydrides.

611ᶜ. WILKINSON, P. G. *J. Mol. Spectr.* **6** (1961) 1. Vacuo ultraviolet spectroscopy. and Rydberg bands.

612. WILLIAMS, F. E. *J. Chem. Phys.* **19** (1951) 457. Tl(I) in alkali halide crystals.

613. WILLIAMS, F. E., SEGALL, B. and JOHNSON, P. D. *Phys. Rev.* **108** (1957) 46. In(I), Tl(I).

614. WILLIAMS, F. E. and JOHNSON, P. D. *Phys. Rev.* **113** (1959) 97. Tl(I).

615. WILLIAMS, R. J. P. *J. Chem. Soc.* (1955) 137. Fe(II), Fe(III) phenanthroline, hydroxyquinolinate complexes etc., Cu(I).

616. WILLIAMS, R. J. P. *J. Chem. Soc.* (1956) 8. Criticism of electrostatic model.

617. WILLIAMS, R. J. P. *Chem. Rev.* **56** (1956) 299. Transitions in porphyrines and other organic ligands.

618. WILLIAMS, R. J. P. *Discuss. Faraday Soc.* **26** (1958) 123. Ligand field stabilization, criticism of electrostatic model.

618a. WILLIAMS, R. J. P. *Ann. Rep. Chem. 1959 London* **56** (1960) 87. Review of transition group spectra.

619. WOLDBYE, F. *Acta Chem.* **12** (1958) 1079. Cr(III) amine, hydroxo, and aquo complexes.

620. WOLFSBERG, M. and HELMHOLZ, L. *J. Chem. Phys.* **20** (1952) 837. M.O. model for tetroxo ions, Cr(VI), Mn(VII).

620a. WYBOURNE, B. G. *J. Chem. Phys.* **32** (1960) 639; *Ibid.* **34** (1961) 279. Nd(III) Er(III).

621. YAMADA, S. *J. Am. Chem. Soc.* **73** (1951) 1182 and 1579. Pd(II) and Pt(II) compounds, co-operative effects in the Magnus salt.

622. YAMADA, S. and TSUCHIDA, R. *J. Am. Chem. Soc.* **75** (1953) 6351. Pt(II) halide and amine complexes.

623. YAMADA, S., SHIMURA, Y. and TSUCHIDA, R. *Bull. Chem. Soc. Japan* **26** (1953) 72. Co(III, IV) peroxo complexes.

624. YAMADA, S. and TSUCHIDA, R. *Bull. Chem. Soc. Japan* **26** (1953) 15. Co(III), hyperchromic series of band intensities.

625. YAMADA, S. and TSUCHIDA, R. *Bull. Chem. Soc. Japan* **29** (1956) 289 and 894; *Ann. Rep. Scient. Works Fac. Sci. Osaka Univ.* **4** (1956) 79. Dichroism of crystals of Cu(II), Pd(II), Pt(II), Pt(IV) complexes. Co-operative effects.

626. YAMADA, S., NAKAHARA, A., SHIMURA, Y. and TSUCHIDA, R. *Bull. Chem. Soc. Japan* **28** (1955) 222. Co(III), dichroism.

626a. YAMADA, S. and TSUCHIDA, R. *Bull. Chem. Soc. Japan* **33** (1960) 98. Cr(III) Co(III) amines, dichroism.

626b. YAMADA, S., YAMASAKI, H., NISHIKAWA, H. and TSUCHIDA, R. *Bull. Chem. Soc. Japan* **33** (1960) 481. Cr(O), Cr(I) benzene and carbonyl complexes.

626c. YAMADA, S., NISHIKAWA, H. and TSUCHIDA, R. *Bull. Chem. Soc. Japan* **33** (1960) 930. Co(III) ammonia NO complex, dichroism.

626d. YAMASHITA, J. and KUROSAWA, T. *J. Phys. Chem. Solids* **5** (1958) 34. Conductivity of NiO and LaMnO₃.

627. YAMATERA, H. *Naturwiss.* **44** (1957) 632. Electron transfer spectra of Co(III) ammonia-halide complexes.

628. YAMATERA, H. *Bull. Chem. Soc. Japan* **31** (1958) 95. Ligand field theory, Co(III) amine and halide complexes.

629. YAMATERA, H., KONDO, Y., OKUNO, H. and UEMURA, T. *Conference on Co-ordination Compounds, Agra, India, 1959.* M.O. theory of Co(III) complexes.

629a. YAMATERA, H. *J. Inorg. Nucl. Chem.* **15** (1960) 50. Co(III) halide pentammine electron transfer bands, intermediate coupling.

630. YONEDA, H. *Bull. Chem. Soc. Japan* **30** (1957) 130. Ni(II), Pt(II) amine complexes.

631. YONEDA, H. *Bull. Chem. Soc. Japan* **30** (1957) 924. Co(III) hydroxylamine complexes.

631a. YONEDA, H. and KIDA, S. *J. Am. Chem. Soc.* **82** (1960) 2139. Co(III) ethanolamine complexes.

631b. YOUNG, J. P. and WHITE, J. C. *Analyt. Chem.* **32** (1960) 799. Cr(III), Co(II), Ni(II), Pr(III), U(IV) fluoride melts.

632. YUSTER, P. H. and DELBECQ, C. J. *J. Chem. Phys.* **21** (1953) 892. Tl(I) in potassium iodide.

632a. ZAHNER, J. C. and DRICKAMER, H. G. Private communication. Pressure effects on Mn(II), Co(II), Ni(II).

633. ZASLOW, B. and RUNDLE, R. E. *J. Phys. Chem.* **61** (1957) 490. Fe(III) tetrachloro complex, Wolfsberg–Helmholz M.O. model.

634. ZDANOV, G. S. and ZVONKOVA, Z. V. *Communications au* XIII *Congrès International de Chimie pure et appliquée, Stockholm*, 1953, p. 175. (Printed in Moscow).

635. ZHVANKO, J. N., MORGENSTERN, Z. L. and SHAMOSKY, L. M. *Optika y Spektroskopia.* **2** (1957) 821. In(I).

636. ZIMMERMAN, G. and STRONG, F. C. *J. Am. Chem. Soc.* **79** (1957) 2063. Chlorine in chloride solutions.

637. ZOLLWEG, R. J. *Phys. Rev.* **111** (1958) 113. Sr(II), Ba(II) oxides, sulphides, selenides, and tellurides.

637ª. ZWICKEL, A. M. and TAUBE, H. *Discuss. Faraday Soc.* **29** (1960) 42. Cr(II) dipyridyl complex.

INDEX